Urban Metabolism and Climate Change

Rahul Bhadouria · Sachchidanand Tripathi ·
Pardeep Singh · P. K. Joshi · Rishikesh Singh
Editors

Urban Metabolism and Climate Change

Perspective for Sustainable Cities

Springer

Editors
Rahul Bhadouria
Department of Environmental Studies
Delhi College of Arts and Commerce
University of Delhi
New Delhi, India

Sachchidanand Tripathi
Department of Botany
Deen Dayal Upadhyaya College
University of Delhi
New Delhi, India

Pardeep Singh
Department of Environmental Studies
PGDAV College
University of Delhi
New Delhi, India

P. K. Joshi
School of Environmental Sciences
Jawaharlal Nehru University
New Delhi, India

Rishikesh Singh
Department of Botany
Panjab University
Chandigarh, India

ISBN 978-3-031-29421-1 ISBN 978-3-031-29422-8 (eBook)
https://doi.org/10.1007/978-3-031-29422-8

This Springer imprint is published by the registered company Springer Nature Switzerland AG
The registered company address is: Gewerbestrasse 11, 6330 Cham, Switzerland

Foreword

Since cities are the first and foremost living spaces for people, enabling healthy living conditions is the top priority for the organization of cities. This means assuming very different living conditions for large parts of the urban population worldwide and improving them through planning and targeted measures. Urban elites around the world have mostly taken care of their needs optimally by spatially organizing their urban living conditions, without worrying about the rest of the urban population. This is not different, e.g., in America than in Africa.

Cities are striving for sustainable development using strategies and models that are often referred to as eco-city or smart city. Achieving these goals is only possible in different ways for different cities depending on their current situation, the actual challenges, and their development efficiently. This is all happening under the already actual stress test of climate change, which cities are facing first and foremost. The urban metabolism is challenged by this in a special way.

Cities are centers of cultural, social, and economic progress. Today, they generate about 90% of the global BIP. An ecologically oriented urban development currently faces three challenges:

1. Securing and promoting the quality of the environment and life for the growing urban population.
2. Reducing the consumption of limited natural resources (resource efficiency).
3. Promoting the ability to adapt to global climate change, but also to other change processes.

Environmental problems such as the supply of clean drinking water or the supply of food are certainly urgent in cities in the group of rapidly growing cities in countries with little capacity to act. In the cities that are hardly growing anymore and in states that can take action to compensate, such problems have been solved or at least significantly reduced, while questions of quality of life and healthy lifestyles are becoming increasingly important. Reducing stress and promoting physical activity in attractive open spaces and the experience of nature in cities are becoming important goals of urban development. In many cities, a major reduction in the need for natural resources and greenhouse gas emissions is an additional goal.

Urban nature in the form of green and blue infrastructure is, like technical infrastructure, infrastructure for the functioning of a livable city. This is being recognized more and more, and the appreciation for urban nature is growing with it. International projects for the development or restoration of different types of urban nature can be found everywhere.

Their driving questions are:

1. How can ecosystem services be determined and quantified more precisely and locally?
2. How can (new) urban nature bring benefits in urban areas with environmental and social problems?
3. How to determine optimally performing quality, quantity, and form for urban nature endowments in different urban structures?
4. How can the direct and indirect effects of the expected global climate change on concrete urban nature connections be determined with regard to urban quality of life?
5. How resistant is designed for urban nature (especially the tree population) to the effects of global climate change and how can this resilience be improved?
6. How can the supply (quantity and quality of ecosystem services) and the management of urban waters be improved by linking them to urban nature?

The expected global climate changes can already be felt in cities, which are "real laboratories" of climate change. On average, cities have a warmer and drier climate than the surrounding area. They are affected by more extreme climate events. Summer temperature extremes are the most stressful part of the urban climate for city dwellers. These will continue to take place in the future. Here, climate moderation to lower the temperature is a desirable goal to avert health hazards from particularly sensitive, vulnerable sections of the population. Cities can take medium- to long-term structural measures to avoid or reduce overheating effects and use urban nature to moderate or optimize the existing urban nature for this purpose.

Within the framework of urban development, technology is usually used to solve problems and reduce risks. Technology as a problem solver is familiar, predictable, well-established, and recognized outside the community of technicians. This is often the reason for preferring technical solutions over environmental engineering solutions. Environmental engineering, which integrates the debate about ecosystem services, has long offered the option of working with nature to solve problems. Solutions are offered outside of the technical domain or fostering the use of nature besides the use of technology. Financial considerations and a broader perspective of urban planners are indispensable/should be considered to make the use of "Nature-Based Solutions" (NbS) possible in cities using natural processes to support problem solutions.

The ideal as a concept on the way to the Green City needs to be adapted to every context. Green Cities should be the result of a transformation process in small steps of existent cities. This conversion process with many different requirements takes place in all areas. It depends on whether and what role urban nature can play.

Only a few societies are currently able to build new "eco-cities", above all China. The eco-city as a Green City can also be rebuilt. However, this remains the exception. In addition, even if several hundred cities worldwide, especially in China, already have this title, the actual realization of their ecological quality often remains incomplete.

The present book *Urban Metabolism and Climate Change—Perspective for Sustainable Cities* takes up exactly this current topic. It is becoming clear that precise analyses with large amounts of data are required for smart and ecological development targets of cities. A separate book section is devoted to this. Urban planning as a means of urban politics still appears to be the most suitable means of design, even if the efficiency of planning varies from region to region. Selected examples, representative of cities with traditional and efficient planning and those with less effective planning, are presented as examples.

It can be expected that this book will make a significant contribution to a better understanding of urban metabolism under the challenges of climate change. It is warmly recommended for reading by professionals and students.

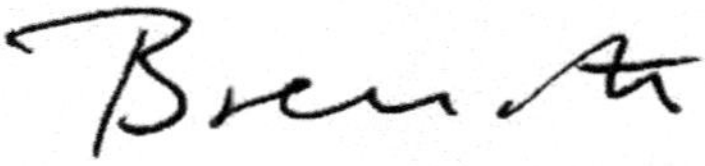

Jürgen Breuste
SURE (Society for Urban Ecology)
President
Salzburg, Austria

Preface

Urbanization is accelerating flow of people, goods, and information across the world. This has a significant impact on flow of energy and materials in and around the cities. In the era of environmental degradation, resource scarcity, environmental pollution, and population growth, the sustainability of a city is of utmost interest and importance. With much wider interests and needs of Sustainable Development Goals, research in urban ecosystem sustainability has taken the central stage. An efficient analysis technique for studying urban ecosystems is urban metabolism (UM). Here, metabolism refers to physiological mechanisms that enable living things to continuously exchange matter and energy with their surroundings in order to function, grow, and reproduce. An urban settlement or a city can be defined as a "superorganism" that has distinct structure and function, with coexistence of biotic and abiotic elements, their interactions, and interrelations processes between society and the natural environment. Hence, UM is a useful concept for assessing and understanding the movements of material and energy within an urban ecosystem and might be discharging to other systems. This would provide a comprehensive insight to understand the prospective sustainable framework, the severity of problems within the same from local to global level, and making policies toward their sustainability.

With half of the world's population living in cities, knowing how cities run and keeping them healthy is crucial to their long-term sustainability. We have witnessed an increase in UM research in the recent years. Contemporary UM research is strongly influenced by biophysical sciences, system theory and thermodynamics, and political economy, while human activities are acknowledged to be essential component of UM analysis. In order to adapt UM to address future concerns, we need a deeper understanding of the connections between mass and energy flows (production and consumption) that form and sustain one another in society. It is crucial to comprehend the temporal and spatial characteristics of services flow and their linkages with material and energy flows in an urban ecosystem. A holistic approach addresses the need for urban sustainability transitions to urban metabolism. Further, cities are responsible for over 70% of global greenhouse gas (GHG) emissions. In a few decades, cities will house the great majority of world's inhabitants. Thus, cities can play a critical role in reducing CO_2 emissions and combating climate change, which is one of the most pressing issues confronting our modern way of life. Cities and urban systems are intricate, self-replicating organisms that require a range of material and energy inputs from outside their boundaries in order to sustain their own metabolism. In the recent years, there has been a growing interest in cities and their energy and material usage, fueled in part by concerns about climate change and in part by social issues resulting from massive migratory movements from rural to urban areas. Patterns of urban growth have an impact on material and energy fluxes within urban settlements, as well as exchanges with the surrounding environment.

Urban metabolism was previously thought of as a practical strategy required for the desired transition to sustainable urban development. But more recently, the growth of information and communication technology and, specifically the development of smart cities, has raised the idea of smart urban metabolism (SUM). SUM is regarded as a technology-driven evolution of the urban metabolism framework with the potential to get beyond some of its current drawbacks. The modeling framework has the potential to be utilized as a planning tool. Land use, transportation, soil-vegetation-atmosphere transfer, and meteorological models are all used to represent urban metabolism. Understanding self-sufficiency, efficiency, and resilience has been made possible by studies of resource stocks and flow exchanges in cities. These studies also offer a perspective for the study of urban systems. It is also possible to think of UM as a branch of an integrated earth system modeling approach with a spatial focus.

The use of a standardized interdisciplinary urban metabolism approach can be used to link the high-level policies of the new urban agenda, the SDGs, and the Paris Agreement across government scales (local, regional, national, and international). Linking urban metabolism to policy could be the missing piece in measuring and changing urban sustainability performance, but it will require development of a multidisciplinary urban metabolism practice. The present book encompasses various aspects of urban metabolism along with modeling and urban metabolism indicators in changing climate scenarios. The book contains a total of 17 chapters:

Chapter 1, by Rahul Bhadouria et al. from India, provides a brief introduction of the topics discussed in this book. Taking a lead from the systematic review from 1990 to 2023, the chapter provides research trend analysis on urban metabolism at the backdrop of changing climate. It discusses the sustainability in urban metabolism, sustainable urban planning, ecosystem services, and disaster resilience so as to provide an interdisciplinary understanding of urban metabolism. The chapter identifies an urgent need to develop new methodological approaches for real time and reliable evaluation of urban metabolism.

Chapter 2, by Riya Raina et al. from India, explores urban metabolism and sustainability and their interlinkages, which is a useful approach for the smart city development. The chapter comments upon the origin and evolution of urban metabolism, its relationship with sustainability of cities, and limitations of urban metabolism concerning urban sustainability.

Chapter 3, by Ariyaningsih et al. from Japan, Taiwan, and Indonesia, discusses the integration of resilience and urban metabolism framework while presenting some case studies that illustrate how such integration can be put into practice and its implications. The chapter emphasizes that the development of theoretical and practical approaches to urban metabolism requires interdisciplinary collaboration.

Chapter 4, by N. S. Nalini and Neesha Dutt from India, presents a case study of urban metabolism in an Indian city—Bengaluru. The chapters suggest that apart from economic processes, urban metabolisms include social, spatial, and environmental processes which should be considered together while developing blueprints for urban expansion.

Chapter 5, by Shivangi Singh Parmar et al. from India, focuses on the understanding of urban metabolism in the context of peri-urban development or urban sprawl with multiple case studies to develop recommendations for the sustainable development of an urban ecosystem.

Ayesha Agha Shah et al. from Bahrain, Italy, the United Arab Emirates, Sweden, and Saudi Arabia in Chap. 6 highlighted ecological aspect of a built environment concerning the sustainable supply of materials and energy required for a desired urban metabolism. The research also deals with a case for conservation and adaptive reuse to be employed as an important indicator for urban metabolism through the maintenance and management of historical urban built environments.

Chapter 7, by Mangalasseril Mohammad Anees and Bhartendu Pandey from the USA, presents a systematic understanding of how urban metabolism aligns with ecosystem concepts and how social dimensions, and associated multi-dimensional and multi-scalar inequalities, add novel characteristics, tradeoffs, and synergies. They provided an analytical framework that gives an outlook for sustainability goals relevant to urban areas including developing ecologically informed solutions and improving resource efficiency in urban areas.

In Chap. 8, Mushtaq Ahmad Dar et al. from India investigate the academic scope of urban metabolism, which includes its relationship with city sustainability and urban planning. They suggest for appropriate urban planning, particularly in developing economies, based on the principles of urban metabolism that can be a sustainable approach for reducing the carbon footprints of cities and mitigating climate change in the coming decades.

Chapter 9, on urban metabolism in the circular bio-economy of tomorrow by G. Venkatesh from Sweden, recommends that the social and economic dimensions of sustainability and thereby the related SDGs must not be overly compromised.

Chapter 10, by Bill Butterworth from the UK, provides a detailed overview of urban wastes and global warming. The chapter highlights the possibilities of reducing irrigation needs and reclaiming deserts along with waste management in urban settlements for financial sustainability.

Chapter 11, on "Transitioning Urban Agriculture to a Circular Metabolism at a Neighbourhood Level" authored by Sharmila Jagadisan and Joy Sen, explores integrated agriculture concepts. This chapter tries to provide an explanation on design and architectural changes that might enhance metabolic functions and restore urban sustainability through urban gardening.

Chapter 12, by G. Venkatesh from Sweden, proposes waste management (urban and otherwise) as both enablers and enabled in the context of SDGs. Readers will learn about the various facets of sustainable development and how effective waste management can create value in a circular economy or bio-economy.

Chapter 13 on emerging approaches for sustainable urban metabolism by G. Gupta et al. provides a detailed view of novel technologies for sustainable urban metabolism.

Chapter 14 by Ombir Singh emphasizes the selection of suitable plant species for plantation in urban areas. This chapter focuses on the species selection in urban forestry—toward urban metabolism.

Chapter 15, by Sunil Bhaskaran et al. from the USA, provides a detailed geospatial analysis for urban metabolism and climate change. It emphasizes that city planners and decision-makers can create new sustainable city design tactics to lessen the negative effects of GHGs and maintain urban metabolism.

Chapter 16 by Ruchira Ghosh and Dipankar Sengupta from the UK examines the application of big data and machine learning as an efficient technique to channelize and manage heterogeneous multi-dimensional datasets, adoption of practices, constructing self-learning machine learning models, and gaining fresh insights via predictive analytics in "smart urban metabolism".

Chapter 17 by Gladys Nkrumah et al. from Ghana provided a detailed overview of policies in Ghana that are related to urban metabolism.

Thus, the book provides a comprehensive understating of the contemporary practical and theoretical concepts along with the emerging issues of urban metabolism which will help readers to get a better insight and problem-solving approach. It highlights some of the major challenges faced by cities in terms of livelihood, agriculture, and other anthropogenic activities. This book will be equally beneficial for undergraduate, postgraduate, research scholars, teachers, environmental scientists, ecologists, agriculturists, urban policymakers, and early career researchers, especially those working in the areas of urban ecology and urban metabolism, non-governmental organizations (NGOs), and government institutions working on sustainable city, urban planning, and urban policymaking.

New Delhi, India — Rahul Bhadouria
New Delhi, India — Sachchidanand Tripathi
New Delhi, India — Pardeep Singh
New Delhi, India — P. K. Joshi
Chandigarh, India — Rishikesh Singh

Contents

Part I
Urban Metabolism and Climate Change: An Introduction

Chapter 1
Urban Metabolism and Global Climate Change: An Overview

Rahul Bhadouria, Sachchidanand Tripathi, Pardeep Singh, P. K. Joshi, and Rishikesh Singh

Abstract Urban centres are increasingly challenged by population increase and the resultant environmental concerns including the urban sprawl and climate change. Moreover, different patterns of urbanization contribute to the changing climate via differences in their urban metabolism represented by energy and matter. Urban metabolic studies in terms of energy and material inflows, outflows, and stocks can be associated with traditional evaluation techniques to help assess the magnitude and potential effects of variety of environmental challenges the world is facing today. Further, urban centres are critical real-time observatories that indicate the impact of anthropogenic activities on global biogeochemical cycles. For example, urban processes have significant and lasting impacts on the global carbon budget. It has also been observed that the technology and infrastructure advancements have fuelled increase in urban inputs and outputs of material and energy. Therefore, more sustainable approaches need to be adopted in changing scenarios for urban planning, particularly for sustainable resource utilization and better waste management practices. Taking a lead from the systematic review from 1990 to 2023, the chapter provides research trend analysis on urban metabolism in the backdrop of changing climate. It discusses the sustainability in urban metabolism, sustainable urban planning, ecosystem services, and disaster resilience so as to provide an interdisciplinary understanding of urban metabolism. The chapter identifies an urgent need to develop new methodological approaches for real-time and reliable evaluation of urban metabolism.

R. Bhadouria
Delhi College of Arts and Commerce, University of Delhi, Delhi, India

S. Tripathi (✉)
Deen Dayal Upadhyaya College, University of Delhi, Delhi, India
e-mail: drsachchidanandtripathi@ddu.du.ac.in

P. Singh
PGDAV College, University of Delhi, Delhi, India

P. K. Joshi
Jawaharlal Nehru University, Delhi, India

R. Singh
Panjab University, Chandigarh 160014, India

R. Bhadouria et al. (eds.), *Urban Metabolism and Climate Change*,
https://doi.org/10.1007/978-3-031-29422-8_1

Keywords Circular bio-economy · Climate change · Energy flow · Material flow · Sustainable development · Urban metabolism · Urbanization

1.1 Introduction

The term "urban metabolism" refers to "exchange processes that produce urban environment" (Kennedy et al. 2011). Urban metabolism indicates the processes by which cities convert raw material, energy, and freshwater into "built environment", "human biomass", and "waste" (Conke and Ferreira 2015). Cities are increasingly challenged by climate and other environmental concerns pertaining to ongoing rapid urbanization and population increase. Urban metabolic studies in terms of energy and material inflows, outflows, and stocks can be associated with traditional evaluation techniques to help assess the magnitude and potential effects of these environmental challenges. Further, urban settlements are critical real-time observatories that indicate the impact of anthropogenic activities on global biogeochemical cycles (Seto et al. 2012). Urban processes have significant and lasting impacts on the global carbon budget. It has been reported that cities contribute to approximately 70% of global greenhouse gas (GHG) emissions (Hendriksen et al. 2011). Hence, the estimation of the total C consumed and sequestered in a city metabolism may indicate its role in stabilizing the ongoing global climate change (Chen et al. 2020). It is anticipated that by 2050, up to 66% of global population would live in urban areas (United Nations 2015). Extensive urbanization with a rapidly increasing human population may lead to a change in the urban environment especially land use patterns, energy budget, overexploitation and depletion of natural resources, degradation of water, air and soil quality, and waste generation including wastewater and GHGs, causing more imbalances in the urban environment (Conke and Ferreira 2015). Therefore, more sustainable approaches are needed to be adopted in changing scenarios for the urban planning, particularly in developing smart cities with sustainable resource utilization and better waste management practices.

It has also been observed that technology and infrastructure advancements have fuelled increase in urban inputs and outputs of material and energy (Kennedy et al. 2007, 2014). Pressure is being placed on the land, housing stock, infrastructure, and environment of the cities due to exceptional urban growth and housing needs (Tainter 2000; Ford et al. 2019; Clark et al. 2022). Under such a scenario, sustainable urban development is a crucial challenge in a world of cities and may be the most important present and future environmental concerns (Kacyira 2012). It is therefore critical to know how urban metabolic systems work towards meeting the sustainability measures of a city (Kacyira 2012; Conke and Ferreira 2015).

This chapter provides a brief insight on different aspects of urban metabolism and sustainable cities against the backdrop of changing climate. The beginning of the chapter outlines different analytical approaches for the metabolism of cities, which is followed by climate change and sustainability, sustainable urban planning, ecosystem services of urban systems and their interlinkages with the urban metabolism, circular economy and its potential role in urban metabolism, digitalization approaches for smart urban metabolism, policies governing urban metabolism, and interdisciplinary understanding of urban metabolism. An up-to-date literature survey was performed using bibliometric analysis tools for the keywords "urban metabolism and climate change". This chapter is a concise account of the information related to different dimensions of urban metabolism and sustainable cities which are described and elaborated later in different chapters of the book.

1.2 Changing Climate and Urban Metabolism: Research Trend Analysis

A literature survey was performed on the first week of February 2023 using a set of keywords, viz. ("urban metabolism" and "climate change") in the "Web of Science Core Collection" database. The search query provided a total of 95 published documents between 1990 and 2023. Out of total 95 documents, 83 articles, 11 review articles, 3 proceeding papers, and 1 book chapter were published in 49 sources. The documents were contributed by 355 authors, with 356 keywords, and received 34.15 average citations per document. Though the literature search was performed from 1990 onwards, only one document was found in 1997 and then continuous publication on the topic was observed from 2010 (2) onwards, and in 2023, already two documents have been published. A maximum of 18 documents have been published in a single year in 2020. The annual production rate observed was 2.7% (Fig. 1.1). For clearly visualizing the research trend on this topic, bibliometric analysis was performed with the R program (ver. R 4.2.1, R Core Team 2022) using *bibliometrix* package (Aria and Cuccurullo 2017) based on the keywords plus database of "Web of Science Core Collection" (Singh et al. 2023). On the basis of the bibliometric analysis, a list of the top 10 countries, 10 leading affiliations, 10 most relevant sources/journals, and 20 leading authors publishing and/or carrying out research on the topic "urban metabolism and climate change" is given in Tables 1.1 and 1.2. Among leading countries (Table 1.1), 50 documents are published from China, followed by USA (29), Australia (26), the Netherlands (19), and UK (19). Among leading institutions/affiliations, the maximum number of documents (8) has been originated from the University of Queensland, 6 from Beijing Normal University, and 4 each from the University of Denmark, the University of Lisbon, and the University of Natural Resources and Life Sciences Vienna (Table 1.1). Elliot T. (4), Kenway S. J. (4), Zhang Y. (4), Kennedy C. (3), and Rugani B. (3) were identified as leading researchers publishing research on urban metabolism and climate change (Table 1.2). Results for country-wise, affiliation-wise, and author-wise analyses showed the predominance of China and European countries in urban metabolism research in the light of changing climate. Among different sources/journals publishing and promoting research on the topic "urban metabolism and climate change", Journal of

Cleaner Production (9), Resources Conservation and Recycling (8), Journal of Industrial Ecology (7), Science of the Total Environment (5), and Landscape and Urban Planning (4) were the leading ones (Table 1.2). The results further showed a considerably higher proportion of journals published by Elsevier with the core theme of sustainable development, and resource management was mainly publishing research articles on the book's theme.

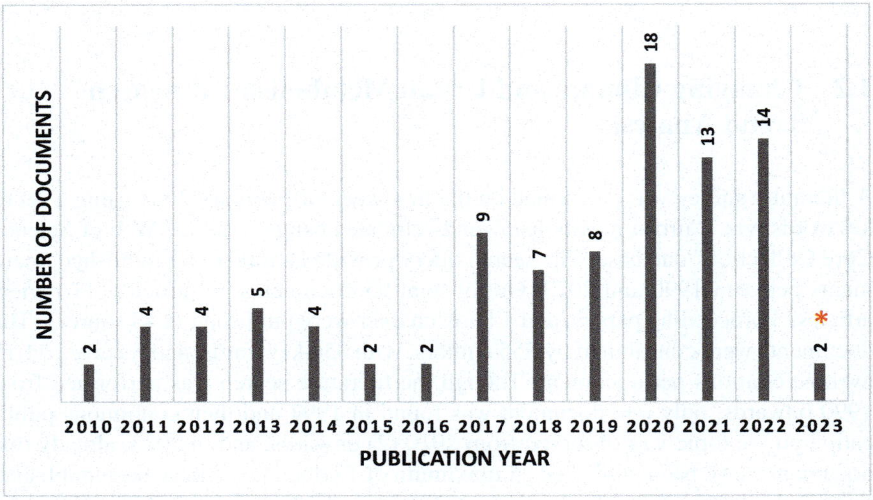

Fig. 1.1 Year-wise publication scenario on the topic "urban metabolism and climate change". *Source* Web of Science Core Collection 2023. *Represents incomplete dataset for the year 2023

Table 1.1 List of top 20 counties and affiliations/universities/institutions supporting/publishing research on the topic "urban metabolism and climate change"

Country/affiliation	No. of articles	Country/affiliation	No. of articles
Top 20 countries			
China	50	Norway	10
The USA	29	Denmark	6
Australia	26	France	6
Netherlands	19	Portugal	6
The UK	19	Sweden	5
Canada	15	Japan	4
Italy	15	Singapore	4
Spain	13	India	3
Austria	11	Luxembourg	3

(continued)

Table 1.1 (continued)

Country/affiliation	No. of articles	Country/affiliation	No. of articles
Germany	11	Peru	3
Top 20 affiliations/universities/institutions			
The University of Queensland, Australia	8	Norwegian University of Life Sciences: NMBU	3
Beijing Normal University, China	6	NTNU: Norwegian University of Science and Technology	3
Technical University of Denmark, Denmark	4	Qingdao University of Science and Technology, China	3
University of Lisbon, Portugal	4	Singapore University of Technology and Design	3
University of Natural Resources and Life Sciences, Vienna, Austria	4	Autonomous University of Barcelona, Spain	3
Griffith University, Australia	3	The University of British Columbia, Canada	3
Humboldt University of Berlin, Germany	3	University of California, Los Angeles, the USA	3
Monash University, Australia	3	University of Toronto, Canada	3
The University of Newcastle, Australia	3	Wageningen University and Research—WUR, the Netherlands	3
North China Electric Power University, China	3	Chalmers University of Technology, Sweden	2

Source Web of Science Core Collection 2023

Table 1.2 List of top 20 authors and sources/journals publishing research on the topic "urban metabolism and climate change"

Authors	No. of articles	Country/affiliation	No. of articles
Top 20 authors			
Elliot T.	4	Dalla Fontana M.	2
Kenway S. J.	4	Facchini A.	2
Zhang Y.	4	Feng K. S.	2
Kennedy C.	3	Goldstein B.	2
Rugani B.	3	Hayat T.	2
Almenar J. B.	2	Holtslag A. A. M.	2
Alsaedi A.	2	Islam K. M. N.	2

(continued)

Table 1.2 (continued)

Authors	No. of articles	Country/affiliation	No. of articles
Brattebo H.	2	Jiang D. L.	2
Chen B.	2	Krausmann F.	2
Choy D. L.	2	Li Y. M.	2
Top 20 sources/journals			
Journal of Cleaner Production	9	Applied Energy	2
Resources Conservation and Recycling	8	Bioresource Technology	2
Journal of Industrial Ecology	7	Cities	2
Science of the Total Environment	5	Renewable and Sustainable Energy Reviews	2
Landscape and Urban Planning	4	Water Research	2
Sustainability	4	Water Science and Technology	2
Sustainable Cities and Society	4	Annual Review of Environment and Resources, Vol 39	1
Ecological Modelling	3	Anthropocene	1
Energy Policy	3	Applied Ecology and Environmental Research	1
Journal of Water and Climate Change	3	Applied Geography	1

Source Web of Science Core Collection 2023

For clear visualization of research themes and patterns on the topic "urban metabolism and climate change" in the last few years, a word cloud diagram, keyword co-occurrence network analysis, and thematic evolution map were formulated with the help of "*bibliometrix*" package in the R program. A detailed description of the above-mentioned features is given in Verma et al. (2021) and Singh et al. (2022). Among the most frequent keywords excluding climate change (25) and urban metabolism (24), the results presented in Fig. 1.2 revealed energy (19), cities (18), sustainability (14), consumption (13), GHG emissions (10), systems (9), water (9), China (8), and metabolism (8) as the top 10 keywords having a higher frequency of occurrence in the literature. Keyword co-occurrence network analysis (Fig. 1.3) results revealed the formation of two separate but interlinked clusters identified as urban metabolism clusters (red) associated with the city, consumption, GHG emissions, system, material flow analysis, life cycle assessment, footprint, CO_2 emissions, model, etc. as one cluster whereas climate change (blue) cluster as the second one which was mainly comprised of cities, sustainability, energy, water, management, systems, metabolism, governance, ecosystem services, mass balance, flow analysis,

Fig. 1.2 Word cloud diagram based on the keyword plus database on the topic "urban metabolism and climate change". Size of the word depicts the frequency of occurrence in the literature. *Source* Web of Science Core Collection 2023

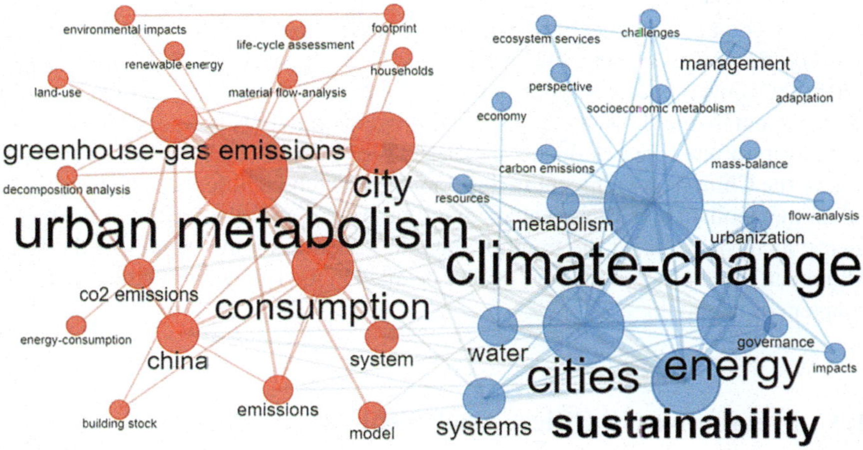

Fig. 1.3 Keywords co-occurrence network plot based on the keyword plus database on the topic "urban metabolism and climate change". Size of bubble represent the frequency of occurrence in the literature. *Source* Web of Science Core Collection 2023

Fig. 1.4 Thematic evolution map based on the keyword plus database on the topic "urban metabolism and climate change". *Source* Web of Science Core Collection 2023

etc. The first red cluster revealed the role of within city performance, whereas the blue cluster provides a broad understanding of the impacts and measures for mitigating climate change through proper resource utilization, management, and governance approaches. Further, the thematic evolution map for the topic "urban metabolism and climate change" for two different time periods, i.e. 1997 to 2015 and 2016 to 2023 revealed two different approaches where the first one is linked with basic approaches for identification and management, whereas the second one is showing more implicative approaches for climate change mitigation. The convergence of different themes of 1997–2015 such as cities, systems, and consumption to a holistic theme of urban metabolism; metabolism to governance; and consumption to strategies themes of the 2016–2023 time period (Fig. 1.4) is observed. Moreover, the emergence of different new themes cum analytical approaches for understanding urban metabolism, i.e. material flow analysis and life cycle analysis is observed for 2016–2023 period where the convergence of carbon emissions is observed for 1997–2015 period. The management theme showed consistency throughout the study period. China emerged in majority of the research during the 2016–2023 period which further revealed the role of the Chinese government in developing research facilities and approaches for urban metabolism and climate change mitigation. Overall, the results of the bibliometric analysis revealed the trend of ongoing research on the themes of the book. Moreover, these observations are further detailed in the following sections and other chapters of the book.

1.3 Metabolism in Cities: Analytical Approaches

Analysing Urban Metabolism

The studies in urban metabolism help in quantifying material and energy flow in the urban ecosystem, thereby contributing to the planning of a sustainable city in future (Kennedy et al. 2007). Also, it indicates that how the sustainability of a city is assessed and reflected by its design and planning (Li and Kwan 2018). The conceptual frameworks adopted to describe physical and social processes in urban centres employ urban climate, governance, and human health-related information in studies and modelling. The urban metabolism analysis is conventionally based on two major approaches—material and energy flow.

1.3.1 Material Flow-Based Approaches

There are three types of material-based analysis: "material flow analysis (MFA)", "life cycle assessment (LCA)", and "ecological footprint assessment (EFA)". The basic principle behind all three methods is that "matter can neither be created nor destroyed but merely transformed". Hence, the input of material and resources in an urban ecosystem should be equal to the output in the form of waste generated, emissions of gases, and produces (Kennedy et al. 2007; Li and Kwan 2018).

Material flow analysis (MFA) has a wider application in urban metabolism studies (Zhang 2013; Thomson and Newman 2018; Bahers et al. 2019). According to Kennedy et al. (2007), the MFA includes three major components: defining the system, quantifying material stock and flow, and analysing the result generated. Therefore, MFA provides a quantitative assessment method for limited urban resources needed to maintain a sustainable urban metabolism. In other words, MFA provides a tool to analyse the flux of resources and their transformation in an urban centre (Thomson and Newman 2018).

The LCA gives a detailed account of a production process, and evaluation of supply chain impacts of resource conversion and its applicability (Chau et al. 2015). According to Chau et al. (2015), various stages of LCA are *goal and scope definition, inventory analysis, impact assessment, interpretation, and applications*. All the stages are interlinked and impact each other. Furthermore, the environmental impact of urban metabolism and urban products can be assessed by applying LCA (Solli et al. 2009; Heijungs et al. 2010). LCA can be applied in two ways, first the process-based LCA which involves input and output processes, sub-processes involved via the supply chain, and assessment of principal components included in the whole process (Li and Kwan 2018). However, the economic input–output LCA (EIO-LCA) emphasizes more on "*calculating, evaluating and predicting the urban inputs and outputs associated with the economic activities in various sectors of the economy*" (Li and Kwan 2018). Often, both the methods are applied in combination to conduct any urban metabolism study, which further helps in the comprehensiveness of the data required for urban modelling of the economic supply chain.

Ecological footprint assessment (EFA) assesses the extent of the land cover of a country, or a city needs to attain its input and output metabolic requirements. According to the urban ecological footprint, there is always a requirement of an optimum land area of biologically productive land to meet the demand for natural resources for urban consumption and the removal of waste generated therein.

According to Kennedy et al. (2007), the actual area required to sustain an urban settlement is twice in magnitude than the relevant urban area, particularly in terms of the consumption of natural resources and to manage the waste generated. Moore et al. (2013) in their study integrated EFA with urban metabolism to assess the actual land required to manage the input and output fluxes of an urban system. The study also compared metabolic load of a city and its overall biophysical carrying capacity, thereby trying to comprehend the sustainability of urban metabolism through EFA.

1.3.2 Energy Flow-Based Approaches

Emergy indices have been designed to evaluate urban metabolism of cities across the globe; however, their measurement depends upon the status of the ecosystem processes and its health and availability of solar energy (Huang and Chen 2009; Carréon and Worrell 2018; Li and Kwan 2018). The ability of nature and humans in providing resources and services is measured in terms of solar in joules. Emergy indices can be commonly utilized to measure socio-economic and environmental values of an urban system and therefore "*seeks to provide a common value basis to study the material and energy flows in urban metabolism*" (Li and Kwan 2018).

The above-mentioned conventional methods to analyse urban metabolism, though are descriptive and largely being in use, may face challenges in the spatio-temporal analysis of urban metabolism of cities in developing countries that are facing great environmental and demographic challenges in the contemporary scenario of changing climate. Similarly, in developed countries which are currently facing a decline in population growth, the conventional analysis methods may have certain limitations in the actual evaluation of urban metabolism. Hence, novel and upgraded methodological approaches are warranted for real-time and reliable evaluation of urban metabolism. For instance, Rosado and his team (Niza et al. 2009; Rosado et al. 2014) provided Urban Metabolism Analyst (UMAn) model to tackle the shortcomings of MFA method. Several new approaches like Integrated Urban Metabolism Analysis Tool (IUMAT) (Mostafavi et al. 2014) and 3D geo-visualization in urban metabolism studies tried to cater the current methodological need; however, they may have their limitation in the application.

1.4 Climate Change and Sustainability in Urban Metabolism

Along with the increase in the size of the city and the high rate of urbanization, unsustainable urban development can cause severe ecological and environmental issues because cities are essential points of interaction between natural and socio-economic setups and places where human activity is concentrated (Kapoor et al. 2020; Singh et al. 2022). In addition, urban areas are becoming extremely vulnerable to risks caused by a wide range of sources, including climate change and urban disasters (Satterthwaite et al. 2009). Furthermore, it is not clear how the urban system relates to one another across different scales within a city or how cities compare to one another in terms of their capacity to survive, recover from, and adapt to various sources of stress (Carter et al. 2015; Verma et al. 2020). In the meantime, the concept of resilience has been around for a long time, and it has seen great benefits in the implementation in urban planning and disaster preparedness (Bush and Doyon 2019). Further, in the recent years, urban metabolism has been extensively adopted as a valuable framework for assessing urban system energy efficiency, material recycling, waste management, and infrastructure (Pincetl et al. 2012; González et al. 2013). The amalgamation of principle of urban resilience and urban metabolism can be of tremendous value to the development of strategies for sustainability (Broto et al. 2012; Dijst et al. 2018). Further, the development of theoretical and practical approaches to urban metabolism requires interdisciplinary collaboration. However, there is a downside to this growth pattern where rapid urbanization coincides with deteriorating livelihood, air pollution, climate change, and many man-made and natural disasters (Revi et al. 2014). Therefore, while promoting such regional and urban development, it is important to manage the impending risk to our ecological environment stemming from rapid urbanization (Kacyira 2012).

Cities are the biggest consumer of global resources (Glaeser et al. 2001; Westphal et al. 2017). Unsustainable consumption of global resources often leads to environmental degradation and ecological imbalance (UNEP 2011; Kandil et al. 2020; Bhadouria et al. 2022). Therefore, to alleviate environmental pressures, sustainable usage of natural resources is imperative to support city development (Freedman 2018; Liu et al. 2022). This intent leads to emergence of concepts like, "safe", "resilient", and "sustainable" metropolises and inclusion of strategies pushing to build such sustainable cities to reduce over-extraction of and address the consequent degradation of environmental resources. Two Sustainable Development Goals (SDGs), SDG-11 (to make urban systems inclusive, safe, resilient, and sustainable) and SDG-12 (to ensure sustainable consumption and production patterns), explicitly place emphasis on this direction (Musango et al. 2020).

Following through on these, city planning objectives serve as a powerful reminder of the significance of environmentally responsible development and disaster preparedness (Lucertini and Musco 2020). Urban metabolism has, thus, been extensively adopted as a useful method for evaluating an urban system's infrastructure, material recycling, waste management, energy efficiency, and contribution to the GHG emission having direct connotation for changing climate. The concept of urban metabolism positively impacts strategizing sustainability and ensuring urban resilience, thus correcting the feedback process within the metabolism of the urban system. This may have direct implications on urban sustainability and well-being, under the current scenario of rapid urbanization across the globe (Van Kamp et al. 2003; McPhearson et al. 2016a, b). Further, it is critical to comprehend how urban communities self-organize and make decisions to manage better resources and understanding the life cycle of resources and how to best repurpose them to benefit the local community (Ulgiati and Zucaro 2019).

The boom in the transportation sector has on the one hand made life more efficient in a household, while the increase in fossil fuel consumption and air pollution on the other (McMichael 2000). Furthermore, the urban design on the mode of transportation may have space demands, air quality impacts, and health effects in the coming days (Cervero 2001). The urban metabolism analysis may be applied efficiently to understand the role of urban GHG emissions in global warming (Conke and Ferreira 2015; Chen et al. 2020). Further, enhancing awareness towards developing urban resilience in the perspective of climate change adaptation must be considered in future policy planning (Demuzere et al. 2014).

1.5 Urban Metabolism and Disaster Resilience

Under the current scenario of changing climate, rampant urbanization, and increasing population, cities across the globe are facing two major challenges to be met: resilience and sustainability in their ecosystem. The occurrence of extreme natural events has increased in the last decade which makes the concept of resilience and sustainability more significant. Therefore, it seems that only a resilient urban system can be sustainable in future (Verma et al. 2020). Further, extreme natural events have many negative consequences pertaining to environmental and socio-economic perspectives which make resilience and sustainability more relevant. The erratic disaster events related to climate change pose great uncertainty and challenge towards risk understanding and management (Singh et al. 2022). In the recent years, it is comparatively difficult to forecast the nature and frequency of hydrometeorological disasters; hence, disaster management authorities and planners find uncertainties to assess and managing these disasters in cities across the world. Unmanaged and ill-planned urbanization, migratory influxes, and related unsustainable exploitation of natural resources particularly in developing nations have worsened the situation (Verma et al. 2020). Furthermore, extreme levels of pollutants in almost all the components of the biosphere have disturbed biogeochemical cycles in urban ecosystems and invited various water-borne and vector-borne diseases and epidemics in the recent years. Under the above-mentioned scenario, urban metabolism can be an important utensil to provide parameters which can be used to measure and analyse the role of urban centres in creating environmental concerns (Kennedy et al. 2011). Furthermore, urban metabolism aids in the formulation of efficacious policies in sustainable urban planning which eventually assist in the development of a resilient urban ecosystem. Moreover, the parameters generated through urban metabolism provide satisfactory criteria of sustainability indicators in terms of scientifically valid data which are comprehendible, non-ambiguous, and comparative over time (Kennedy et al. 2011; Pincetl et al. 2012; Yang et al. 2014). The data generated, therefore, can be utilized by policy planners to develop sustainable plans that may assist constructively in resource conservation, management of waste, and GHG emissions (Álvarez and Julián 2014).

1.6 Urban Metabolism and Sustainable Urban Planning

There are two widely accepted schools of thoughts, the first propagated by Odum (1975), which involves urban metabolism based on of energy equivalents, and the second, which is based on material flow in terms of mass fluxes, can be utilized to develop sustainability indicators for a city and, hence, the measures of overall energy consumption pattern, water flow, material dynamics, and waste output from city's metabolism. This assessment can also be applicable to the quantification of GHG emission from urban centres. The use of urban metabolism studies in urban planning and urban designing is a comparatively juvenile area, and researches are going on across the globe to understand this concept (Kennedy et al. 2011).

Contemporary studies have been emphasizing on implementation of urban metabolism in developing sustainable urban designs (Kennedy et al. 2011). The data obtained from the urban metabolism studies have been utilized to mathematical models having its significance for policy analysis and development. The group focused on MFA has developed many such mathematical models. For instance, modelling platforms like SIMBOX (Baccini and Bader 1996) and STAN (Cencic and Rechberger 2008) include sub-processes like stocks and material flow and outputs. These models may also be applicable to forecast future alterations in urban metabolism arising out of technological and policy innovations. Further, a sustainable urban design must strive to reduce GHG emissions, and urban metabolism matrices being useful in the quantification of GHG emissions may have implications for sustainable urban design. Moreover, studies have been conducted to track material and energy flow to mitigate environmental impacts in an urban design. In the master plans for Dongtan and the Thames Gateway, for instance, Arup's Integrated Resource Modelling (IRM), which is basically a model of urban metabolism, has been used (Kennedy et al. 2011).

In the multidisciplinary researches of cities, urban metabolism is an interesting topic of study. It is suggested that different metabolic components be evaluated at various scales and that urban metabolism be monitored as part of planning practice. In order to identify research gaps and needs in the field of urban metabolism, the constraints of the same must be carefully examined in the light of urban sustainability (Cui 2018; Leal Filho et al. 2020).

1.7 Ecosystem Services and Urban Metabolism

In an urban system, ecosystem services are critical in resource cycling and GHG emission abatement and, therefore, provide a range of techniques and strategies for the optimization of urban metabolism (Perrotti 2020; Cárdenas-Mamani and Perrotti 2022). However, it has been observed that this range of techniques based on the ecosystem services concept is by and large underexploited in urban metabolism studies, even after the considerable progress in research analysing and classifying the ecosystem services (Cárdenas-Mamani and Perrotti 2022). Urban policy decisions have been benefited from the use of the notion of ecosystem services, such as the adoption of nature-based solutions (Almenar et al. 2021). Further, it suggests that integration of the concept of ecosystem services with urban metabolism research may help in mitigating the methodological lagging and strengthen the modelling of the urban system (Newell et al. 2019; Winker et al. 2019; Cárdenas-Mamani and Perrotti 2022).

1.8 Urban Metabolism and Circular Bio-economy

A bio-economy is focused on generating biological resources renewable in nature on land and water, and a value addition to be given to transformation of these resources and waste produced, in terms of bio-based products and bioenergy (Venkatesh 2022). While resource conservation is the driving force behind a circular economy and renewable is the essential term in a bio-economy, a circular bio-economy combines the "best of these two worlds" (D'Amato et al. 2017; Venkatesh 2022). Urban metabolism in a circular bio-economy will undoubtedly turn into something quite powerful for addressing a variety of difficulties, including tackling climate change and its effects and achieving numerous SDGs in the process. Needless to state, as all know, challenges lurk where opportunities abound to replace the take-make-use-dispose paradigm of a linear economy with a grow-make-use-share-partake-restore paradigm of a circular bio-economy. These approaches pave the way for sustainable as well as smart urban metabolism.

1.9 Smart Urban Metabolism (SUM)

Smart urban metabolism (SUM) is a contemporary conception of urban metabolism which includes modern-day technologies dealing with the complex challenges of growing smart cities (Allam 2020a, b, c). Traditionally, urban metabolism deals with the influx-efflux of energy and the flow of materials through urban space. However, with the growing needs of smart cities, these flow patterns are transiting as a complex network and are subject to interdisciplinary understanding. Furthermore, data availability is a major challenge faced by city planners due to the lack of data inventories and appropriate data management solutions to handle massive datasets, arising from these complex flow patterns. This is ensuing to the inefficient adaptation of urban metabolism approaches, especially in developing economies. Thus, the situation remains grave when it comes to resource management of a smart city, and how urban areas may additionally deal with intricate issues like climate change when they are striving to understand their own material and energy cycling. SUM can potentially be an effective approach for identifying complex issues in the flow patterns of energy and material in an urban space (Allam 2020a, b, c).

1.10 Digitalization: An Approach to Circular and Smart Urban Metabolism

The adverse consequences of the linear and traditional urban approach have now been realized by policymakers and city planners (European Commission 2020a, b; D'Amico et al. 2022). Therefore, digitization of circularity has been recommended to take urban metabolism towards automation and sustainability (Ingrao et al. 2018; D'Amico et al. 2022). Furthermore, United Nations has recognized that digitalization may be a potent tool to drive the circularity in urban metabolism and will be instrumental in achieving SDGs by 2030 (United Nations 2015; De Pascale et al. 2021). Further, circularity in urban metabolism strives to achieve a holistic connection among the natural, built, and digital environments (Mikalef et al. 2020; D'Amico et al. 2022).

1.11 Governing the Urban Metabolism: Policy Approaches

The promotion of responsible and balanced use of resources is the focus of an increasing number of policies being enacted at the international level. International political instruments, such as the SDGs and the New Urban Agenda, direct cities to focus towards an environmentally sustainable urban ecosystem. This mandate has provided the impetus to research, particularly in the area of urban climate governance (Heidrich et al. 2016; Bansard et al. 2017; Reckien et al. 2018; Heikkinen et al. 2020). Urban sustainability and resilience can be aided by the centrality of urban metabolism and its ability to influence city policies and designs directly (Bristow and Kennedy 2013; Bristow and Mohareb 2020). It has been realized that most of the challenges in governing environmental flows are due to gaps in accountability and loopholes in strategic planning and its implementation. The urban metabolism approaches integrated to environmental governance may strengthen accountability in terms of responsibility, transparency, assessment and participation, and augment urban policy planning and implementation.

1.12 Interdisciplinary Understanding of Urban Metabolism

Urban metabolism indicates the exchange processes that give rise to an urban environment and gave impetus to find out ways to make cities more sustainable. However, it has invited criticism for some specific types of social and economic systems that have prioritized or marginalized some peculiar type of flow patterns with urban systems. The current scenario of changing climate and increased urbanization with an increase in urban population has compelled to develop an alternative understanding of the functioning of an urban system. For instance, analysis of urban ecology and material flow (MFA) may be instrumental in the formulation of social and environmental policies (Dijst et al. 2018). Furthermore, analysis focused on the metabolism of the urban economy suggests the role of environmental and social resources in running the growth of the urban economy, along with their spatial manifestation pattern.

1.13 Conclusion and Future Perspectives

Currently, more than half of the global population is residing in cities. The majority of cities in the globe are seeing significant growth, particularly those that are located near large urban centres. Therefore, it is crucial that efforts are made to make cities and other human settlements safer and more resilient. There is also a perceived need to find, test, and adopt policies that could improve the physical environment and the harmony between urban residents and ecosystems. To create more sustainable cities and human settlements, the SDGs, particularly SDG-11, provide novel opportunities and thrust. With aspects such as life cycle analysis, input–output analysis, and ecological network analysis, studies have been focused on understanding interrelationship between the components of the urban system and direct and indirect resource consumption and assess how the human system affects the environment. It will be important to strengthen the theoretical underpinnings of this field of study, standardize and unify the methodology to enable comparisons between studies, examine and combine the outcomes of analyses at various sizes, and promote collaboration across a wide range of disciplines. A systems engineering approach may be adopted to standardize and bring conformity in the methods for categorizing and quantifying data to develop databases, to strengthen research, analysis, and evaluation, hence providing multiscale data needed for such analysis databases thereby augmenting our understanding towards scale effect and principles of urban metabolism, that will be instrumental in development of sustainable policies and designs for urban systems. Discussions of this nature will be more fruitful if experts from fields not typically represented in resource management decision-making can participate. For instance, it would be fascinating to explore the literature in future research to find out how psychologists and sociologists try to address the human aspects of urban issues. Then, we may apply these ideas to the current technological and ecological paradigms. This may offer more consistent, unbiased, and useful direction for political systems, economic growth, ecological restoration in cities, and ecological consciousness.

Acknowledgments RS gratefully acknowledges the financial support from DST-SERB, New Delhi, India as National Post-Doctoral Fellowship Scheme (Grant number: PDF/2020/001607).

References

Allam Z (2020) Urban governance and smart city planning: lessons from Singapore. Emerald Group Publishing

Almenar JB, Elliot T, Rugani B, Philippe B, Gutierrez TN, Sonnemann G, Geneletti D (2021) Nexus between nature-based solutions, ecosystem services and urban challenges. Land Use Policy 100:104898

Álvarez D, Julián C (2014) Metabolismo urbano: herramienta para la sustentabilidad de las ciudades. Interdisciplina 2(2):51–70

Aria M, Cuccurullo C (2017) Bibliometrix: an R-tool for comprehensive science mapping analysis. J Informet 11:959–975

Baccini P, Bader HP (1996) Regionaler stoffhaushalt: erfassung, bewertung und steuerung. Spektrum Akademischer Verlag, Heidelberg, pp 1–420

Bahers JB, Barles S, Durand M (2019) Urban metabolism of intermediate cities: the material flow analysis, Hinterlands and the Logistics-Hub function of Rennes and Le Mans (France). J Ind Ecol 23(3):686–698

Bansard JS, Pattberg PH, Widerberg O (2017) Cities to the rescue? Assessing the performance of transnational municipal networks in global climate governance. Int Environ Agreements Politics Law Econ 17:229–246

Bhadouria R, Upadhyay S, Tripathi S, Singh P (eds) (2022) Urban ecology and global climate change. Wiley, UK

Bristow DN, Kennedy CA (2013) Urban metabolism and the energy stored in cities: implications for resilience. J Ind Ecol 17(5):656–667

Bristow DN, Mohareb EA (2020) From the urban metabolism to the urban immune system. J Ind Ecol 24(2):300–312

Broto VC, Allen A, Rapoport E (2012) Interdisciplinary perspectives on urban metabolism. J Ind Ecol 16(6):851–861

Bush J, Doyon A (2019) Building urban resilience with nature-based solutions: how can urban planning contribute? Cities 95:102483

Cárdenas-Mamani Ú, Perrotti D (2022) Understanding the contribution of ecosystem services to urban metabolism assessments: an integrated framework. Ecol Indic 136:108593

Carréon JR, Worrell E (2018) Urban energy systems within the transition to sustainable development. A research agenda for urban metabolism. Resour Conserv Recycl 132:258–266

Carter JG, Cavan G, Connelly A, Guy S, Handley J, Kazmierczak A (2015) Climate change and the city: building capacity for urban adaptation. Prog Plan 95:1–66

Cencic O, Rechberger H (2008) Material flow analysis with software STAN. In: EnviroInfo, pp 440–447

Cervero R (2001) Integration of urban transport and urban planning. The challenge of urban government: policies and practices, pp 407–427

Chau CK, Leung TM, Ng WY (2015) A review on life cycle assessment, life cycle energy assessment and life cycle carbon emissions assessment on buildings. Appl Energy 143:395–413

Chen F, Chen Y (2022) Urban metabolism and spontaneous architectural growth: a sustainable strategy featuring participatory co-construction by multiple stakeholders. Buildings 12(3):352

Chen IC, Wang YH, Lin W, Ma HW (2020) Assessing the risk of the food-energy-water nexus of urban metabolism: a case study of Kinmen Island, Taiwan. Ecol Indic 110:105861
Clark LP, Tabory S, Tong K, Servadio JL, Kappler K, Xu CK et al (2022) A data framework for assessing social inequality and equity in multi-sector social, ecological, infrastructural urban systems: focus on fine-spatial scales. J Ind Ecol 26(1):145–163
Conke LS, Ferreira TL (2015) Urban metabolism: Measuring the city's contribution to sustainable development. Environ Pollut 202:146–152
Cui X (2018) How can cities support sustainability: a bibliometric analysis of urban metabolism. Ecol Indic 93:704–717
D'Amato D, Droste N, Allen B, Kettunen M, Lähtinen K, Korhonen J, Leskinen P, Matthies BD, Toppinen A (2017) Green, circular, bio economy: a comparative analysis of sustainability avenues. J Clean Prod 168:716–734
D'Amico G, Arbolino R, Shi L, Yigitcanlar T, Ioppolo G (2022) Digitalisation driven urban metabolism circularity: a review and analysis of circular city initiatives. Land Use Policy 112:105819
Demuzere M, Orru K, Heidrich O, Olazabal E, Geneletti D, Orru H, Bhave AG, Mittal N, Feliú E, Faehnle M (2014) Mitigating and adapting to climate change: multi-functional and multi-scale assessment of green urban infrastructure. J Environ Manage 146:107–115
De Pascale A, Arbolino R, Szopik-Depczy´nska K, Limosani M, Ioppolo G (2021) A systematic review for measuring circular economy: the 61 indicators. J Clean Prod 281, 124942
Dijst M, Worrell E, Böcker L, Brunner P, Davoudi S, Geertman S, Harmsen R, Helbich M, Holtslag AA, Kwan MP, Lenz B (2018) Exploring urban metabolism—towards an interdisciplinary perspective. Resour Conserv Recycl 132:190–203
European Commission (2020a) Circular economy action plan. Brussels
European Commission (2020b) Roadmap circular resource efficiency management plan. https://ec.europa.eu/futurium/en/system/files/ged/roadmap_circular_resource_efficiency_management_plan_v6.pdf
Ford A, Barr S, Dawson R, Virgo J, Batty M, Hall J (2019) A multi-scale urban integrated assessment framework for climate change studies: a flooding application. Comput Environ Urban Syst 75:229–243
Freedman B (2018) Chapter 12: resources and sustainable development. Environ Sci Can Perspect 1–37
Glaeser EL, Kolko J, Saiz A (2001) Consumer city. J Econ Geogr 1(1):27–50
González A, Donnelly A, Jones M, Chrysoulakis N, Lopes M (2013) A decision-support system for sustainable urban metabolism in Europe. Environ Impact Assess Rev 38:109–119
Heidrich O, Reckien D, Olazabal M, Foley A, Salvia M, de Gregorio Hurtado S et al (2016) National climate policies across Europe and their impacts on cities strategies. J Environ Manage 168:36–45
Heijungs R, Huppes G, Guinée JB (2010) Life cycle assessment and sustainability analysis of products, materials and technologies. Toward a scientific framework for sustainability life cycle analysis. Polym Degrad Stab 95(3):422–428
Heikkinen M, Karimo A, Klein J, Juhola S, Ylä-Anttila T (2020) Transnational municipal networks and climate change adaptation: a study of 377 cities. J Clean Prod 257:120474
Hendriksen B, de Boer Y, Copius Peereboom E, Jansen S, Ballantine J (2011) CDP cities 2011: global report on C40 cities. Carbon Discl Project (CDP)
Huang SL, Chen CW (2009) Urbanization and socioeconomic metabolism in Taipei: an emergy synthesis. J Ind Ecol 13(1):75–93
Ingrao C, Messineo A, Beltramo R, Yigitcanlar T, Ioppolo G (2018) How can life cycle thinking support sustainability of buildings? Investigating life cycle assessment applications for energy efficiency and environmental performance. J Clean Prod 201:556–569
Kacyira AK (2012) Addressing the sustainable urbanization challenge. UN Chron 49(2):58–60

Kandil RA, Sarhan A, AbdelGalil RE (2020) Analysis of ecological balance issue for the built-up land and cropland footprints in Alexandria City, Egypt during this time-series (2005–2019). Planning 15(6):911–920
Kapoor V, Tripathi S, Devi RS, Srivastava P, Bhadouria R (2020) Ecological economics of an urban settlement: an overview. In: Verma P, Singh P, Singh R, Raghubanshi AS (eds) Urban Ecology: Emerging Patterns and Social-Ecological Systems. Elsevier. pp 91–110
Kennedy C, Cuddihy J, Engel-Yan J (2007) The changing metabolism of cities. J Ind Ecol 11:43–59. https://doi.org/10.1162/jie.2007.1107
Kennedy C, Pincetl S, Bunje P (2011) The study of urban metabolism and its applications to urban planning and design. Environ Pollut 159(8–9):1965–1973
Kennedy C, Stewart ID, Ibrahim N, Facchini A, Mele R (2014) Developing a multi-layered indicator set for urban metabolism studies in megacities. Ecol Indic 47:7–15. https://doi.org/10.1016/j.ecolind.2014.07.039
Leal Filho W, Marisa Azul A, Brandli L, Gökçin Özuyar P, Wall T (eds) (2020) Sustainable cities and communities. Springer International Publishing, Cham
Li H, Kwan MP (2018) Advancing analytical methods for urban metabolism studies. Resour Conserv Recycl 132:239–245
Liu H, Saleem MM, Al-Faryan MAS, Khan I, Zafar MW (2022) Impact of governance and globalization on natural resources volatility: the role of financial development in the Middle East North Africa countries. Resour Policy 78:102881
Loiseau E, Roux P, Junqua G, Maurel P, Bellon-Maurel V (2013) Adapting the LCA framework to environmental assessment in land planning. Int J Life Cycle Assess 18:1533–1548
Lucertini G, Musco F (2020) Circular urban metabolism framework. One Earth 2(2):138–142
McMichael AJ (2000) The urban environment and health in a world of increasing globalization: issues for developing countries. Bull World Health Organ 78:1117–1126
McPhearson T, Haase D, Kabisch N, Gren Å (2016a) Advancing understanding of the complex nature of urban systems. Ecol Indic 70:566–573
McPhearson T, Pickett ST, Grimm NB, Niemelä J, Alberti M, Elmqvist T, Weber C, Haase D, Breuste J, Qureshi S (2016b) Advancing urban ecology toward a science of cities. Bioscience 66(3):198–212
Mikalef P, Boura M, Lekakos G, Krogstie J (2020) The role of information governance in big data analytics driven innovation. Inf Manag 57(7):103361
Moore J, Kissinger M, Rees WE (2013) An urban metabolism and ecological footprint assessment of Metro Vancouver. J Environ Manage 124:51–61
Mostafavi N, Farzinmoghadam M, Hoque S, Weil B (2014) Integrated urban metabolism analysis tool (IUMAT). Urban Policy Res 32(1):53–69
Musango JK, Currie P, Smit S, Kovacic Z (2020) Urban metabolism of the informal city: probing and measuring the 'unmeasurable' to monitor sustainable development goal 11 indicators. Ecol Indic 119:106746
Newell JP, Goldstein B, Foster A (2019) A 40-year review of food–energy–water nexus literature and its application to the urban scale. Environ Res Lett 14(7):073003
Niza S, Rosado L, Ferrao P (2009) Urban metabolism: methodological advances in urban material flow accounting based on the Lisbon case study. J Ind Ecol 13(3):384–405
Odum EP (1975) Diversity as a function of energy flow. In: Unifying concepts in ecology. Springer, Dordrecht, pp 11–14
Perrotti D (2020) Urban metabolism: old challenges, new frontiers, and the research agenda ahead. In: Verma P, Singh P, Singh R, Raghubanshi AS (eds) Urban Ecology: Emerging Patterns and Social-Ecological Systems. Elsevier. pp 17–32
Pincetl S, Bunje P, Holmes T (2012) An expanded urban metabolism method: toward a systems approach for assessing urban energy processes and causes. Landsc Urban Plan 107(3):193–202
Reckien D, Salvia M, Heidrich O, Church JM, Pietrapertosa F, De Gregorio-Hurtado S et al (2018) How are cities planning to respond to climate change? Assessment of local climate plans from 885 cities in the EU-28. J Clean Prod 191:207–219

Revi A, Satterthwaite DE, Aragón-Durand F, Corfee-Morlot J, Kiunsi RBR, Pelling M, Roberts DC, Solecki W (2014) Urban areas. In: Field CB, Barros VR, Dokken DJ, Mach KJ, Mastrandrea MD, Bilir TE, Chatterjee M, Ebi KL, Estrada YO, Genova RC, Girma B, Kissel ES, Levy AN, MacCracken S, Mastrandrea PR, White LL (eds) Climate change 2014: impacts, adaptation, and vulnerability. part A: global and sectoral aspects. Contribution of working group II to the fifth assessment report of the intergovernmental panel on climate change. Cambridge University Press, Cambridge, UK and New York, NY, USA, pp 535–612

Rosado L, Niza S, Ferrão P (2014) A material flow accounting case study of the Lisbon metropolitan area using the urban metabolism analyst model. J Ind Ecol 18(1):84–101

Satterthwaite D, Huq S, Reid H, Pelling M, Lankao PR (2009) Adapting to climate change in urban areas: the possibilities and constraints in low-and middle-income nations. In: Satterthwaite D, Huq S, Reid H, Pelling M, Lankao PR (eds) Vol 1. Adapting Cities to Climate Change. pp. 1–45

Seto KC, Güneralp B, Hutyra LR (2012) Global forecasts of urban expansion to 2030 and direct impacts on biodiversity and carbon pools. Proc Natl Acad Sci 109(40):16083–16088

Singh R, Kumari T, Verma P, Singh BP, Raghubanshi AS (2022) Compatible package-based agriculture systems: an urgent need for agro-ecological balance and climate change adaptation. Soil Ecol Lett 4(3):187–212

Singh VK, Singh R, Rajput VD, Singh VK (2023) Halophytes for the sustainable remediation of heavy metal-contaminated sites: recent developments and future perspectives. Chemosphere 313:137524

Solli C, Reenaas M, Strømman AH, Hertwich EG (2009) Life cycle assessment of wood-based heating in Norway. Int J Life Cycle Assess 14:517–528

Tainter JA (2000) Problem solving: complexity, history. Sustainability. Popul Environ 22:3–41

Thomson G, Newman P (2018) Urban fabrics and urban metabolism–from sustainable to regenerative cities. Resour Conserv Recycl 132:218–229

Ulgiati S, Zucaro A (2019) Challenges in urban metabolism: sustainability and well-being in cities. Front Sustain Cities 1:1

UNEP (2011) International resource panel, United Nations Environment Programme. Sustainable consumption and production branch, 2011. Decoupling natural resource use and environmental impacts from economic growth. UNEP/Earthprint

United Nations (2015) Transforming our world: the 2030 agenda for sustainable development. United Nations General Assembly, New York, NY, USA, pp 1–41

Van Kamp I, Leidelmeijer K, Marsman G, De Hollander A (2003) Urban environmental quality and human well-being: towards a conceptual framework and demarcation of concepts; a literature study. Landsc Urban Plan 65(1–2):5–18

Venkatesh G (2022) Circular bio-economy—paradigm for the future: systematic review of scientific journal publications from 2015 to 2021. Circular Econ Sustain 2(1):231–279

Verma P, Singh P, Singh R, Raghbanshi AS (2020) Urban ecology: emerging patterns and social-ecological systems. Elsevier. https://doi.org/10.1016/C2018-0-05562-0

Verma P, Kumari T, Raghubanshi AS (2021) Energy emissions, consumption and impact of urban households: a review. Renew Sust Energ Rev 147:111210

Web of Science Core Collection (2023). https://www.webofscience.com/wos/woscc/summary/a77e95f4-1f77-40a9-abb5-501fdb184558-6f008c53/relevance/1(overlay:export/exp. Accessed on 08 Feb 2023

Westphal MI, Martin S, Zhou L, Satterthwaite D (2017) Powering cities in the global south: how energy access for all benefits. World Resources Institute, Wahsington DC, Estados Unidos

Winker M, Gehrmann S, Schramm E, Zimmermann M, Rudolph-Cleff A (2019) Greening and cooling the city using novel urban water systems: a European perspective. In: Approaches to water sensitive urban design. Woodhead Publishing, pp 431–454

Yang Z, Zhang Y, Li S, Liu H, Zheng H, Zhang J, Liu G (2014) Characterizing urban metabolic systems with an ecological hierarchy method, Beijing, China. Landsc Urban Plan 121:19–33

Zhang Y (2013) Urban metabolism: a review of research methodologies. Environ Pollut 178:463–473

Zhang Y, Yang Z, Yu X (2015) Urban metabolism: a review of current knowledge and directions for future study. Environ Sci Technol 49(19):11247–11263

Part II
Urban Metabolism and Sustainability: Case Studies

Chapter 2
Interlinkages Between Urban Metabolism and Sustainability: An Overview

Riya Raina, Mushtaq Ahmad Dar, Rishikesh Singh, Daizy R. Batish, and Harminder Pal Singh

Abstract Urban metabolism is a concept in which the city is examined using biological principles that relate to the internal mechanisms by which living creatures maintain a constant interchange of matter and energy with their surroundings to permit their growth and development. In multidisciplinary studies of cities, urban metabolism is one of the potential fields of study. Urban sustainability refers to the urban planning measures implemented to enhance and build urban areas without depleting their resources indefinitely. To understand the in-depth features of urban metabolism and its potential for examining the sustainability of cities, the concept's historical development as well as its theoretical underpinnings are required. In order to address and identify the various research gaps and needs in the metabolism of cities, this chapter is intended to discuss the origin and evolution of urban metabolism, its relationship with the sustainability of cities, and the limitations of urban metabolism in relation to urban sustainability. There exists an interlinkage between the urban metabolism and sustainability within a city with respect to judicious resource utilization, energy flow, waste management, and social and economic development. The assessment of the urban metabolism and sustainability of a city can be a useful approach for smart city development. Moreover, there is a dire need to promote the circular urban metabolism for developing sustainable cities.

Keywords Agenda 21 · Circular urban metabolism (CUM) · Cities-4-People project · Climate change · Greenhouse gases · Linear metabolism · Material flow analysis (MFA) · People-Oriented Transport and Mobility (POTM) · Sustainability indicators

Riya Raina and Mushtaq Ahmad Dar contributed equally.

R. Raina · H. P. Singh (✉)
Department of Environment Studies, Panjab University, Chandigarh 160014, India
e-mail: hpsingh_01@yahoo.com

M. A. Dar · R. Singh · D. R. Batish
Department of Botany, Panjab University, Chandigarh 160014, India

R. Bhadouria et al. (eds.), *Urban Metabolism and Climate Change*,
https://doi.org/10.1007/978-3-031-29422-8_2

2.1 Introduction

The pace of population growth and simultaneous urbanization revealed that by the end of 2050, two-thirds of the world's human population (approximately 6.5 billion) will be inhabiting urban areas (Verma et al. 2020). The decisions made at the municipal, metropolitan, regional/sub-national, and national levels for urban expansion and related activities have a substantial impact on sustainable urbanization. For example, the effects of different processes and functions in cities on social and economic outcomes, human health and welfare, and both global and local natural surroundings will have considerable impacts on planetary boundaries and sustainable development (Webb et al. 2018). It is widely known that cities contribute to climate change and that potential mitigation and adaptation measures are required for attaining sustainability in the cities (Verma and Raghubanshi 2018; Sharifi 2021). However, because cities are complicated and dynamic systems, it is vital to identify flexible options and pathways when making decisions. Urbanization is the catalyst for economic development because it makes better use of resources like surplus labor and land. One of the key components of structural changes that take place when countries start down the path of improvement is the relocation of people (from rural to urban regions) in quest of better employment options as well as other opportunities (Roy and Thangaraj 2022). The workers' movement raises their income and standard of living while still assisting them in joining the "consuming class". This large consumer class is better able to make large purchases because of their increasing purchasing power, which in turn fuels the industrialization and urbanization processes (Joshi 2021). Thus, urbanization has improved people's living standards in addition to contributing to their increased incomes; hence, many developing nations like India may actively encourage urbanization. However, India presently faces a wide range of difficulties as a consequence of its large urban dweller population. Two major challenges include: (i) acceleration of the rate of urbanization; and (ii) implementation of the effective or planned urbanization in order to reap the most benefits (Tripathi 2021). This necessitates a greater involvement of researchers in urban planning, and management methods for the development of a city (Kennedy et al. 2011; McPhearson et al. 2016). Researchers may assist innovation, evolutionary co-design, and adaptive management of our cities by cooperating with the urban stakeholders to generate knowledge, carefully documenting and communicating learning to decision-makers.

It is important to understand how urban environments influence local climates and how climatic variability could have a big impact on urban settings as the population moves from rural to urban areas. Rapid urbanization has led to rising resource needs and pollutant flows throughout the planet. Urban economic networks extract, transform, and transport more than 40 billion tons of materials annually (as of 2010), accounting for an estimated 80% of world energy use. Cities are also accountable for 65% of the world's environmental emissions as a result of this mobilization of both energy and matter (Swilling et al. 2018). Additionally, the intricacy of the metropolitan system is increased by natural catastrophes. This is particularly important in cities that are experiencing fast development characterized by inadequate planning and a lack of basic urban public services (Bahers et al. 2022). Sustainability and resilience are thought to be powerful tools for dealing with hazards and supporting urban planning processes (Pirlone et al. 2020). Sustainability is a "complex and contentious" issue in terms of "interpretation and application", and when the words "sustainable" and "development" are combined, the emphasis is on economic development rather than comprehensive sustainability (Purvis et al. 2019; Toli and Murtagh 2020). Fair, livable, and financially viable are the three pillars of sustainable development (Pissourios 2013). Since its beginnings in the economic and ecological philosophies, sustainability has indeed been extensively included in urban development (Zeng et al. 2022). On the other hand, resilience is the ability of a system to 'bounce back' or revert to a pre-stressed, stable state. The capacity of the urban community to rebound from hazards is another way to define resilience (Sarker et al. 2020). A community's or city's potential to alter in response to numerous internal and external risks is known as resilience. Urban sustainability is concerned with an urban environment's ability to maintain the desired result over time. Given the consequences of climate change, the urban community ought to be able to think resiliently in order to help address environmental problems (Zeng et al. 2022).

Urban metabolism (UM) is a concept that studies how cities interact with the environment, particularly how they use resources and produce pollutant fluxes, as well as the associated sociological, economic, and environmental challenges related to these flows (Kapoor et al. 2020; Bahers et al. 2022). Although the idea of urban metabolism has its roots in the eighteenth century, it has gained considerable importance in the research community over the past few years (Barles 2010). Theodor Schwann first proposed the idea of urban metabolism in the nineteenth century, and Karl Marx's economic theory outlines its historical antecedents (Restrepo and Morales-Pinzón 2018). However, Abe Wolman's work was the first to introduce the idea of urban metabolism (Wolman 1965). In broader aspects, urban metabolism is the study of energy storage, its outputs and inputs into the urban region, as well as the quantification of the removal of wastes and nutrients from an urban area (Brunner and Rechberger 2016). The study of urban metabolism gives the parameters and indicators that enable researchers to assess and measure the environmental impacts caused by urban systems, i.e., cities, and therefore helps to generate certain effective policies that could be helpful in urban planning and development (Pincetl et al. 2012). The idea of urban metabolism has been correlated to the metabolism of living organisms and their interaction with the ecosystem. Similarly, the cities in urban metabolism are considered living organisms that consume resources from their surroundings and expel wastes to their nearby areas (Cui 2018). However, cities themselves constitute a whole ecosystem as they are home to a huge number of organisms (Savard et al. 2000). Therefore, the idea that cities are similar to ecosystems would be more appropriate. Moreover, this comparison of cities with natural ecosystems can be used for developing sustainable urban ecosystems because these are usually energy self-sufficient and can conserve mass and energy through recycling by decomposers (Moscovici et al. 2015). If cities attain such qualities, they would be more supportable and sustainable (Pincetl 2012).

The information and guidance obtained from the studies on urban metabolism have ensured the sustainability of cities. This would optimistically influence the representatives to develop strategies that are helpful in the preservation of resources, and the reduction of waste production and greenhouse gases (GHGs). For instance, the European Union's Seventh Framework Programme (FP7) project BRIDGE (sustainable urban planning and decision support accounting for urban metabolism) has established a decision support system (DSS) for sustainable urban planning (Chrysoulakis et al. 2013). DSS helps to evaluate the potential impacts of urban metabolism components, *viz.*, carbon, energy, and pollutant changes, and can offer a quantifiable assessment of their sustainability performances (González et al. 2013). In this way, it is helpful to formulate various policy recommendations that could further promote the effective usage of resources and augment the ecological quality of the cities. Moreover, studies related to urban metabolism are contemplated as an important requisite for planning projects that target urban sustainability in cities (Perrotti 2020). The present chapter encompasses the theoretical foundations of urban metabolism, its origin and evolution, processes in the urban metabolism concept for urban sustainability, interlinkages between urban metabolism and sustainability, as well as limitations of the concept and challenges for urban sustainability.

2.2 Concept of Urban Metabolism: Origin and Evolution

Researchers have compared urban areas to biological organisms (Wolman 1965). As we know, organisms in the natural environment need energy and resource inputs, and they then utilize this energy in doing work and excreting waste, the same as cities do (Barrera et al. 2018). Urban metabolism can be defined as *"the sum total of the technical and socioeconomic processes that occur in cities, resulting in growth, the production of energy, and the elimination of waste"* (Kennedy et al. 2007; Lucertini and Musco 2020). The concept basically appeared in the late twentieth century as an organized method to understand urban pathways of resource utilization, production of waste, and their influence on the surroundings (Lanau et al. 2021). Some ecologists neglected the comparison made between the metabolism of cities and that of an organism, as they believed that only living organisms can have metabolisms, and declared the urban metabolism concept an inappropriate biological analogy (Pincetl et al. 2012; Webb et al. 2018). However, some have recommended that cities be more similar to ecosystems (Ko and Chiu 2020). Yet urban metabolism is an integrated and multidisciplinary platform that studies the matter and energy flows in urban areas (cities) as complicated systems as different societal, economic, and environmental factors shape them. Therefore, urban metabolism gives us a figurative outline to observe the exchanges of human and natural systems (EsmaeilpourZanjani et al. 2020).

The concept of urban metabolism is well known in the theories of the socio-economic distribution of neo-Marxist (Özdemir 2021). Marx used the term metabolism to define the processes between man and nature in which man regulates, contemplates, and controls the metabolism among him and the environment (Wang 2020). Various other theorists on urban metabolism, *viz.*, Kaika, Heynen, and Sywngedeou, have also approached this concept of urban metabolism from a neo-Marxist perspective (Ko and Chiu 2020; Ulgiati et al. 2021). They have used Marx's approach to understand and analyze the dynamic relationship between human beings and nature. Earlier to the petroleum age, humans used to exert physical and animal labor for their survival (in terms of food and shelter) on the Earth. In this process, they altered the various physical forces in order to supply the metabolism of human activities. Marx explained that man is a natural being itself and is dependent on nature for his material needs. In this process, he alters and transforms nature and eventually changes the Earth's systems, of which 'climate change' is possibly the most vivid example (Pincetl et al. 2012). Further developments in the concept of urban metabolism are outlined in the following sub-section.

2.2.1 Urban Metabolism in the Twentieth Century

The first precise application of this concept was given by Wolman in 1965 in his book "*The Metabolism of Cities*", in which he represented the metabolism of an imaginary city, i.e., the United States (Wolman 1965). The Wolman study was promulgated at a time when uncertainties about the impact of anthropogenic activities on the surroundings were increasing rapidly. The study gained massive importance among environmental activists who were concerned with the growing human population and its effect on the ecosystem's ability to provide resources. Wolman's key innovation was that he presented the city as an ecosystem, and therefore, he focused more on the "ecology of cities" than the "ecology in cities", i.e., he emphasized more on urban areas' (cities) processes and exchange of energy or matter relative to their surroundings (Oliveira and Vaz 2021). He established a model of an imaginary American city with one million people, calculated the actual inputs and outputs of materials and waste from that hypothetical urban system, and therefore, showed urban metabolism as a quantitative entity. He advanced the idea that urban metabolism was no longer controlled by political or geographic limits. He explained that cities have affected the environment on broad as well as local scales by using resources and producing waste and pollution (Paul et al. 2018). He was concerned about the growing population and its effects on natural systems and resources. Published in *Science*, Wolman was able to connect with a large audience by emphasizing the need for freshwater supplies for cities and demonstrating how urban systems (cities) draw their inputs from the environment and leave behind a trail of effects outside of their immediate physical bounds (Kingsland 2019).

2.2.2 Measurement Methods of Urban Metabolism

After the ground-breaking effort by Wolman, urban metabolism progressed into two different approaches: mass balance accounting and Odum's emergy method (Pincetl et al. 2012). The first one discusses industrial ecology and is also associated with engineering fields. Mass balancing is an approach that accounts for materials entering and leaving the system and uses tools such as material flow analysis (MFA) to keep track of the flow of materials as well as energy through the urban system (Cui et al. 2019). It also keeps track of the energy required to convert raw materials into material goods to fulfil the demands and needs of people. A new approach that has largely influenced the urban metabolism concept is the introduction of energy flows by Odum (1983). He quantified the available solar energy that was used directly or indirectly to produce a product, and used that figure to define the metabolism in terms of solar energy equivalents and account for metabolic fluxes. He called this concept *emergy*, which is quantified in terms of solar emergy joules (SEJ). This method highlights standard units for both material and waste flows in any biophysical system. Though theoretically feasible, it would be difficult to characterize all urban processes in common units. The energy-material flux method was therefore accepted as a more typical urban metabolism approach due to the complexity of this methodology and its limited use as a result of the conversion of flows to SEJ metric. It emphasizes calculating an urban system's materials and energy flows, irrespective of units.

Due to the surge in the global impacts of urban areas (cities), they have become the main centers for metabolism studies. A study of urban metabolism includes the measurement and quantification of energy, material, and water flows into the city as well as stocks and flows outside of the urban area. Therefore, we can develop an environmental outline of an urban area as these flows cause degradation of the environment as well as the scarcity of some resources (Wei et al. 2015). Once the metabolic profile of a city is established, it enables us to identify key elements and, thus, helps to formulate effective policies related to some major urban issues, such as waste production due to higher resource demands and scarcity of water supply due to increased water demand. Some other issues are energy and climate change due to the increased emission of GHGs and air pollution instigated by the release of nitrogen, phosphorous, metals, SOx, NOx, and particulate matter (Lv et al. 2020). Changes in landscapes due to the relocation of huge amounts of building materials and impaired environmental quality because of the emissions of the above-mentioned pollutants have an extremely negative influence on the quality of natural ecosystems and human health. For that reason, urban metabolism analysis is necessary to tackle such environmental issues, particularly in areas where fast urbanization is taking place, such as in developing nations (Wei et al. 2015).

The urban metabolism studies include quantification approaches such as MFA, mass balance, and life-cycle assessment (LCA). MFA includes the principle of mass conservation and measures the total materials entering a system, such as material flows inside a system, and the material flows in the form of pollution or waste that are leaving the system (González-García et al. 2021). The materials that enter the urban system are consumed, and biophysical structures such as human bodies, buildings, machines, roads, artefacts, agricultural and livestock products, and export products such as waste are formed. Mass balance is grounded on the belief that substances can neither be created nor destroyed but can only be transformed (Voukkali and Zorpas 2022). Therefore, according to this principle, the total input into a system equals the total output from the system. As explained by Einstein, energy is the transformation of mass, and therefore, standard units such as kilograms, joules, or tons are used to describe the energy flows in the mass balance approach. A cradle-to-grave approach is used in LCA, which includes a regular analysis of the possible environmental impacts of goods or services throughout their life cycle and is recorded (Akizu-Gardoki et al. 2022). By modelling, the energy fluxes entering and leaving the urban system, quantification of urban metabolism in integration with the LCA enhances the ability to compute the environmental impacts of cities (Elliot et al. 2018). LCA provides a hands-on set of methods and tools for measuring the material flows within the urban system. All these methodologies, whether alone or in conjunction, involve the quantification of energy flow, that is, nutrients, raw materials, and food that enter the urban system, as well as the wastes that leave the system, in standard units such as kilograms, tons, and joules. The LCA method has been brought into the ISO 14044 standards and accepted by various LCA practitioners (Talwar and Holden 2022). There are various softwares such as GaBI (PE International, Germany) and SimaPro (PRé Consultants, The Netherlands), that were developed for conducting LCA (Tintelecan et al. 2019). An urban area's physical metabolism may be attained by measuring energy flows. It includes the measurement of energy, materials, nutrients, water, and wastes, and therefore helps researchers to understand various phenomena in cities all over the globe.

2.3 Urban Sustainability

Some of the important events in sustainability research include the 1972 United Nations Conference on the Human Environment, where the international communities met and discussed the challenges related to the environment and its development. Further, the 1987 Brundtland Report gave the definition of sustainable development (Brundtland 1987). The Earth Summit or Rio Declaration that took place in Rio de Janeiro, Brazil (1992), introduced the concept of sustainable human development and posed vital goals for its measurement. During the 1992 Earth Summit, Agenda 21, which calls for developing sustainability indicators, was adopted. Further, in 2002, the implementation of Agenda 21 was reaffirmed at the Johannesburg Earth Summit, which emphasized hands-on methods for executing sustainable development. Sustainable development addresses human needs and increases the quality of life. Concurrently, natural resources must be exploited at a rate that can be sustained by the regenerative ability of the ecosystem. Overall, sustainability is based on the three major pillars of economic, social, and environmental sustainability (Tsalis et al. 2022). The health of an ecosystem is a prerequisite for its sustainable development (Carréon and Worrel 2018). Considering these facts, Sustainable Development Goals (SDGs) with 17 broad goals and 169 interrelated objectives were established on national precedence (Salvia et al. 2019). The SDGs were approved in 2015 and the specific aim of this approach was to attain sustainability in both developing and developed nations for the next 15 years. The SDGs are accompanied by indicators that are supposedly meant to be universal, but basically, they are not relevant to every nation (Vandemoortele 2017). For example, SDG-11 talks about the safe and sustainable development of cities and communities, which also includes human settlement planning and its management as well as affordable housing and sustainable transport. Sustainable development can only be attained by effectively renovating the ways that shape our urban spaces (Sarbu et al. 2021).

The idea of urban metabolism is strongly associated with sustainable development because both emphasize persistent ecosystem facilities to support humans by supplying materials and goods. In other words, sustainable development in an urban area must be viable, equitable, and liveable (Vidal et al. 2020). Over the last few decades, urbanization has turned out to be one of the most imperative subjects that describe the human association with ecosystems (Huang et al. 2015). Cities and the urban population play a vital role in developing urban sustainability. Cities are defined as super organisms that have a spatial structure and dimensions where both living and non-living organisms interact and coexist (Wolfram et al. 2016). This coexistence produces some inter-relational processes, such as involvement in the biogeochemical cycles and the flow of material, energy, and waste within the urban system. Therefore, evaluating the growth of a city towards sustainable or unsustainable urban development has become an important area of research and needs quantification with the help of appropriate sustainability indicators (Verma and Raghubanshi 2018). Due to the increased complexity and difficulty in an energy system's elements such as transportation, delivery systems, utilization of food, supply of freshwater, utilization of goods and services, and waste generation and handling, cities are accountable for about 60% of energy consumption and 70% of anthropogenic GHG emissions (Maranghi et al. 2020). All these exchanges and emissions contribute strongly to climate change. With the increased urbanization, cities would need to develop more sustainable methods, and the increased population would need some novel ways to handle urban living. Thus, some new solutions are required to identify problems such as inefficient resource management, social exclusion, high energy consumption, declining human well-being, overcrowding, and environmental degradation (Perrotti 2020). In this regard, urban sustainability is defined as the urban planning actions and principles that build and improve urban areas without aggressively consuming the resources endlessly (Restrepo and Morales-Pinzón 2018). A sustainable city would, therefore, be one that could minimize the expenditure of fossil energy and waste generation through recycling and recirculation (Broekhoven and Vernay 2018).

2.4 Inter-Linkages Between Urban Metabolism and Sustainability

A city cannot stand alone, and its sustainability largely depends on the quality of surrounding urban systems, agricultural areas, and other countries (Girardet 2017). The urban metabolism can be of two types, *viz.*, linear and circular metabolism. A linear metabolism is defined as a situation where there is no link between the inputs of resources and the outputs of waste. On the other hand, circular metabolism refers to a state where the outputs of the city's metabolism are utilized as inputs in the production system (Girardet 1996). The sustainability of capitals in urban metabolism is grounded on the effectiveness of some processes and the elimination of linear urban metabolism in favour of circular urban metabolism (CUM). Thus, it becomes critical to account for the circuitous utilization of resources and discharge of contaminants outside the city. Keeping track of urban inputs and outputs due to domestic consumption helps us recognize the dependency of a city on other areas, but this consumption-based method is not always helpful, and thus, consumption- and production-based quantifications are needed for the development of suitable urban policies (Maranghi et al. 2020). In a circular metabolism, the cities encompass balance with the ecosystem and therefore face fewer environmental impacts. CUM involves an initiative that incorporates energy and material inputs and outputs used in different functions such as cooking, food consumption, and production to generate other local cycles of energy and material so as to reduce the ecological burden of the existing city areas. A simple illustration of CUM is the utilization of organic waste to yield biogas and compost through anaerobic digestion. However, the lack of suitable data is another major obstacle in developing the metabolic profile for an urban area (Verma et al. 2021). To attain sustainability, cities have to put some serious efforts into collecting data related to urban metabolism. Also, research is required to build up some methodologies for evaluating gaps based on existing data (Cui et al. 2019). Since cities are intricate urban systems, tracking the data of flows, inputs, and outputs of the cities would require some efficient policies and strategies, so in this regard, the integrated models are predominantly helpful for detecting trade-offs and suitable answers to numerous ecological problems beyond "end of pipe" approaches (Wei et al. 2015). The interlinkage between urban metabolism and the sustainability of a city has been represented in Fig. 2.1. For the sustainability of a city, there should be some mechanisms through which the outputs generated after the metabolism of raw materials can be re-utilized, either directly in the city system or indirectly through the input system for re-circulation to the city.

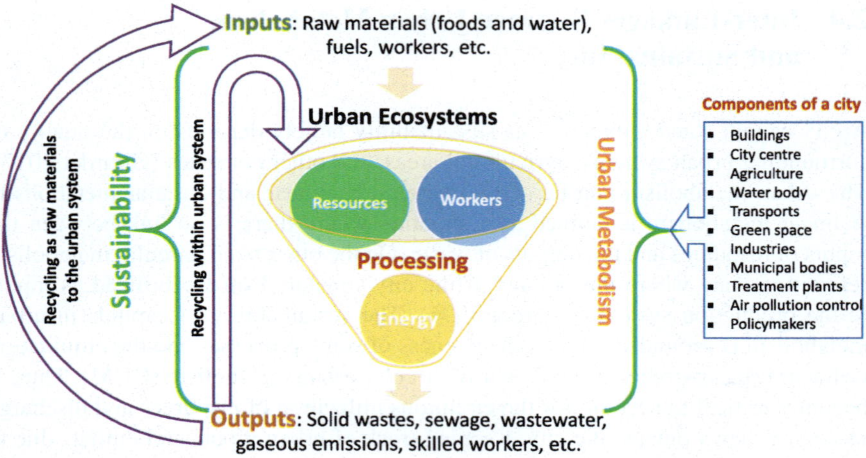

Fig. 2.1 Interlinkages between urban metabolism and the sustainability of a city

2.5 Phases of Urban Metabolism Leading Towards Sustainability

Urban metabolism offers a thorough framework, concepts, methodologies, and policy-making to help investigate the material and energy transitions in cities (Cui 2018). If we desire that a city's development occur without increasing the amount of matter and energy beyond the biosphere's capability, then research related to urban metabolism has a liability to expose how urban areas could achieve sustainability (Cui 2018). The geological transport of matter and energy among different cities led to their interdependence on each other. This, therefore, serves as a threat to upcoming generations. Accordingly, it is tough to recognize whether an urban area is sustainable, particularly for developed cities. Understanding the involvement of an urban area to sustainability is, however, convenient and depends largely on the city's consumption and production patterns, health and welfare levels, and disposal amounts (Bengtsson et al. 2018; Bibri and Krogstie 2020; Perrotti 2020). Determining the city's contribution to sustainability involves the following efforts:

1. Studies of the metabolism of a city help find sustainable development solutions through resource and flow modeling, and network translation (Wang 2023). The patterns in urban metabolism modify with time, and the chief drivers of these alterations in the city's metabolism have been considered from time to time in many studies. These studies assess how urban metabolism affects the environment and surroundings, as well as its effects in the present and the future (Dijst et al. 2018). These studies on urban metabolism are required as the flows of material and energy may vary markedly with time. For instance, material needs, production and consumption patterns, technologies, and underlying economic activities can vary and modify at a temporal scale (Cui 2018). Environmental

impacts on a city with a large number of construction projects compared to a city with a lesser number of construction projects can vary because of the increased demands for landscapes and aesthetics. These studies then help the policymakers to propose solutions for sustainable development and the modeling of energy flows and network translation. The time series methods help the policymakers understand the urban consumption and production trends and, therefore, elucidate sustainable urban planning alternatives from an ecological or environmental perspective (Wu et al. 2018). Furthermore, these trends facilitate the effective investigation of the progression and evolution of urban metabolism, which therefore gives sensible and practical information for studying its role in sustainability.

2. Because anthropogenic activities have an important influence on the sustainability of cities by interfering with the metabolisms of materials, energy, and elements like nitrogen, carbon, and phosphorous, the biogeochemical cycle of these metabolic elements was described, and the related environmental pollution problems were discussed in various studies (Gu et al. 2012). Metabolic elements are more dynamic than the materials used in construction activities, such as steel, cement, gravel, and sand, and therefore, these metabolic elements flow more rapidly within complex urban systems (Huang et al. 2018). For instance, the studies of water metabolism in an urban area are complex and more dynamic as the accessibility of water is largely constrained by wastewater discharge, water table fall, groundwater depletion, and seawater intrusion. Carbon metabolism can also massively affect the urban metabolism of cities as it is very much interrelated with human activities (Xia et al. 2018). About 70–80% of carbon is emitted by an urban area; therefore, accounting of carbon-flow modeling and studies on carbon footprinting essential in attaining sustainability in cities (Verma et al. 2021). Also, cities typically have a high dependence on energy; thus, studies on carbon metabolism help to evaluate the environmental quality and policymaking. In addition, nitrogen and phosphorous flows are also significant for the sustainability of the city's metabolism, as a major chunk of these elements are being released from agricultural run-offs, solid waste landfills, and fossil fuel combustion (Guan et al. 2021). Phosphorous is being released in cities via waste generation, ore extraction, and resource use, which leads to different ecological problems. Air pollution monitoring in the cities, such as organic pollution load, and concentrations of fine and coarse particulate matter, has also been studied in urban metabolism. All these studies and research give us a general view of the impacts relating to air, water, and soil in a combined system.
3. In a city, various urban metabolism mechanisms are linked together and therefore have a complex structure. Simplifying these complex components of urban metabolism components is essential to achieving sustainability. This, therefore, demands a need to describe the inner workings of urban metabolism, which involves the internal mechanisms of urban metabolism as well as flow of matter and energy that are produced and circulated (Fróes and Lasthein 2020). The transportation components in urban metabolism emit a huge amount of GHGs; thus, a sustainable revolution of the substructure in urban areas is essential to

changing the present trend of increasing GHG emissions. The urban substructure on a large scale consists of budget, volume, and timelines. Its intricacy could appear daunting to local people; nevertheless, both citizens and infrastructure are directly and closely related and mutually dependent (Puerari et al. 2018). Urban Transition Labs, Urban Living Labs, and Learning Alliance concepts define the transformation of cities' infrastructure via local stakeholder co-creation processes, and this may lead to some of the key benefits such as encouraging creative and innovative ideas, intensifying social inclusivity, increasing engagement, increasing ownership, etc. Also, co-creating with various stakeholders helps to promote a sturdy flow of resources throughout the urban metabolism (Nevens et al. 2013; Khan et al. 2017). In this aspect, it also supports the typical healthy urban environment for the residents. The formation of healthy cities involves a well-operated technical urban sub-structure, which eventually is the outcome of human activity and design. Involvement of co-creation techniques as well as a co-production approach in urban planning and metabolism can also lead to sustainable cities.

4. The Cities-4-People project is a new method of attaining healthier cities and urban metabolism. It focuses on some new, innovative methods for locally and regionally supported, sustainable, and supportable mobility solutions (Fróes and Lasthein 2020). As reported by the World Health Organization (WHO), people from several disciplines can efficiently work on a way to build sustainable, vital, and healthy cities. This goal can be achieved by transforming urban infrastructure and altering the modes of transportation, thereby enhancing the livability of inhabitants by transforming their ways of transportability from private cars to biking, public transportation, and walking (Noring et al. 2018). This is the exact principle behind the Cities-4-People project, in which stakeholders from five urban areas have been engaged to alter the top-down process approach to transportation and mobility into a more inclusive and dynamic bottom-up approach. These stakeholder groups, more precisely from academia, government, and industry, work together via several interventions to tackle cities' main mobility problems that were recognized earlier through quantitative and qualitative study. The main notion in this mission was People-Oriented Transport and Mobility (POTM), which offered novel ways to get targeted and sustainable answers to the needs of the public (Fróes and Lasthein 2020). POTM included the combination of the latest digital and communal technologies under a comprehensive approach to get answers that had a little environmental footprint, a sharing mindset, and the potential to resolve actual urban mobility problems. Cities-4-People was spread-out in European regions and established by bringing together authorities, citizens, and innovation experts (Liu et al. 2015).
5. The use and reuse of resources for societal welfare and their impacts on a city's health have been mentioned in some key studies. Evaluating source consumption, waste production, and human livability and welfare in different cities allows us to conclude if urban areas have helped policymakers on sustainability (Newman 1999). As a result of this perception, cities have to compute their contributions in terms of reducing metabolic flows while upgrading human livability (Cui

2018). Researchers have designed a model connecting socio-economic activities and underlying environmental circulation processes to the intricate and complex city's landscape pattern (Zhou et al. 2021). This model provides the details of matter and energy inputs, helps to recognize the important infrastructures in urban metabolism, and improves urban metabolic efficiency. During a water shortage in a city, these models promote the concept of regenerative use, which includes rainwater and groundwater collection and recycling wastewater. Studies have confirmed that these techniques are accessible with present methods and can help in habitat water utilization as well as aquifer recharge (Thomson and Newman 2018). Also, elements such as nitrogen, carbon, and phosphorus can be recycled and can supply an adequate growing medium for agriculture. Cities are supposed to be capable of making contributions to sustainability through local solutions that tackle numerous challenges such as resource insufficiency through material use, lesser biodiversity loss, and climate change through reduced energy use (Cui 2018).

2.6 Challenges for Urban Sustainability

The environmental footprint of urban areas expands much ahead of their administrative boundaries by means of the assimilative and productive services of the ecosystem, facilitating the flow of energy and material (Verma and Raghubanshi 2018). Urbanization is a large section of human residents living in cities. As cities grow, anthropogenic activities increase, which, therefore, causes a decrease in natural resources and ultimately causes a threat to urban sustainability (He et al. 2018). Mori and Christodoulou (2012) have stated that cities encompass both economic and social collisions on sustainability as their environmental externalities are not limited to their specific areas. Therefore, this concept becomes very important in understanding the sustainability of cities as they expand outside of their administrative boundaries. Urban sustainability indicators are suitable ways for evaluating urban sustainability because of their huge acceptance and several options in combination with software (Kaur and Garg 2019). There is a broad range of sustainability indicators and indices that are in use; however, these indicators vary with respect to the goals and needs of different cities and regions. Indicators are variables that are used to assess a predetermined phenomenon, whereas an index is comprised of a mixture of different indicators. Indices help in simplifying the intricacy of an indicator; however, indices are more intricate and complex to calculate. For that reason, indices require an assemblage of various indicators to be assessed (Mori and Christodoulou 2012). For example, if we want to evaluate the carbon dioxide production of a particular urban area, it is important to assess the emissions coming from different sources. The sources could be buildings, industries, and transportation, which form different sub-indices. A sub-index may have different sub-indices until a simpler indicator is achieved. In this way, the carbon dioxide emission from a private car can be considered an indicator for the entire carbon dioxide emission;

however, several other indicators may also be important to assess all the emissions in an urban area, which, therefore, makes it tricky to get a valid figure (Corredor-Ochoa et al. 2020).

In order to reduce and simplify the issues during the implementation of indicators, Verma and Raghubanshi (2018) classified the urban sustainability challenges into two broad categories: external and internal challenges, through which sustainable development indicators can be studied. Internal challenges focus more on the problems in the methodologies used for developing indicators. External challenges include problems that resist the execution of these sustainability indicators and are usually characterized by policy lethargy, a lack of sufficient data, and the unwillingness of higher authorities to implement suitable indicators. The steps required during the framework of sustainability indicators are: Preliminary assessment (A), Setting goals (B), Indicator selection (C), Setting baseline (D), Selecting targets (E), External challenges (F), Application (F), Evaluation (G), Reporting findings (G), and Sustaining the indicators (J). Unclear methodologies may produce sustainability indices with unsustainable issues. Also, the development of indicators must involve scientific and technical solutions and should be unaffected by social challenges, which comprise moral and ethical dimensions. Furthermore, the sustainability indicators should focus on simple criteria such as easy calculations, appropriate methodologies, data availability, methodological commensurability, easy reach into policy decisions, and scientifically determined thresholds. Also, owing to the abundance of sustainability indicators, there is a requirement to establish the relevant and important ones. And in order to obtain a basic and relevant indicator system, the obstacles and challenges encountered during the development and application of these indicators need to be resolved.

2.7 Conclusions and Future Directions

The basic assumptions of urban metabolism, its genesis and development, the processes underlying the concept of urban sustainability, the interlinkages between urban metabolism and sustainability, and the limitations of urban sustainability are all covered in this chapter. We also discussed how cities contribute to sustainability, how urbanization has changed people's consumption habits, how pollution from the biogeochemical cycle of materials affects cities, and how to more effectively use the components of urban metabolism in policy-making to advance urban development. The study of urban metabolism is crucial to creating sustainability in the cities. These investigations might also open up fresh insights on how urban areas can grow without contributing to environmental concerns more quickly. Despite the enormous difficulties in reconciling the equally important issues of eco-environmental preservation and urban expansion, cities should always be aware of their inputs to sustainability from the point of view of urban metabolism. Despite an increase in research on urban metabolism over the last two decades, there continue to be numerous unanswered questions, such as:

1. Large cities must first establish a comprehensive and all-encompassing structure that makes the collection and sharing of data possible. If we are to consider sustainability seriously, then the significance of a single city and its worldwide influence is expanding, and there is an increasing need to record the corresponding material flows. As a result, the framework should be necessarily practical and standardized to enable the measurement of metabolism and to enable comparisons between other cities, particularly those in developing nations.
2. Urban metabolism studies should investigate the specifics inside an urban system to assist urban planners. Research must be combined at many scales, chiefly with specific geographic areas, human activities, and economic circumstances, in order to gain insight into the issues that arise throughout urban growth. This depends not just on attempts to close the data gap but also on the integration and use of cutting-edge technologies like geographic information systems, remote sensing, and information networks.
3. To effectively serve all of the stakeholders in a city, urban metabolism research with a more interdisciplinary component may be established. Urban metabolism encompasses a variety of viewpoints on biology, technology, sociology, and economics. By combining these viewpoints, it is feasible to perform additional studies utilizing more comprehensive methodological options. By incorporating policymakers and various local stakeholders in problem-solving, for instance, sociological and psychological approaches should be taken into consideration, which is in line with the initiatives now being made by some cities.

Overall, there is still more work to be done, and these suggestions can aid in coordinating data and methodological decisions in the next studies on sustainable cities.

Acknowledgements MAD is thankful to Council of Scientific and Industrial Research (CSIR), New Delhi, India for providing financial assistance as Senior Research Fellowship. RS gratefully acknowledges the financial support from DST-SERB, New Delhi, India as National Post-Doctoral Fellowship Scheme (Grant number: PDF/2020/001607).

References

Akizu-Gardoki O, de Ulibarri B, Iturrondobeitia M, Minguez R, Lizundia E (2022) Ecodesign coupled with life cycle assessment to reduce the environmental impacts of an industrial enzymatic cleaner. Sustain Prod Consum 29:718–729

Bahers JB, Athanassiadis A, Perrotti D, Kampelmann S (2022) The place of space in urban metabolism research: towards a spatial turn? A review and future agenda. Landsc Urban Plan 221:104376

Barles S (2010) Society, energy and materials: the contribution of urban metabolism studies to sustainable urban development issues. J Environ Plan Manag 53:439–455

Barrera PP, Carreón JR, de Boer HJ (2018) A multi-level framework for metabolism in urban energy systems from an ecological perspective. Resour Conserv Recycl 132:230–238

Bengtsson M, Alfredsson E, Cohen M, Lorek S, Schroeder P (2018) Transforming systems of consumption and production for achieving the sustainable development goals: moving beyond efficiency. Sustain Sci 13:1533–1547
Bibri SE, Krogstie J (2020) Smart eco-city strategies and solutions for sustainability: the cases of Royal Seaport, Stockholm, and Western Harbor, Malmö, Sweden. Urban Sci 4:11
Broekhoven SV, Vernay AL (2018) Integrating functions for a sustainable urban system: a review of multifunctional land use and circular urban metabolism. Sustainability 10:1875
Brundtland G (1987) Report of the world commission on environment and development: our common future. United Nations General Assembly Document A/42/427
Brunner PH, Rechberger H (2016) Handbook of material flow analysis: for environmental, resource, and waste engineers. CRC Press
Carréon JR, Worrell E (2018) Urban energy systems within the transition to sustainable development. A research agenda for urban metabolism. Resour Conserv Recycl 132:258–266
Chrysoulakis N, Lopes M, San José R, Grimmond CSB, Jones MB, Magliulo V. Klostermann JEM, Synnefa A, Mitraka Z, Castro EA, González A, Vogt R, Vesala T, Spano D, Pigeon G, Free-Smith P, Staszewski T, Hodges N, Cartalis C (2013) Sustainable urban metabolism as a link between bio-physical sciences and urban planning: the BRIDGE project. Landsc Urban Plan 112:100–117
Corredor-Ochoa Á, Antuña-Rozado C, Fariña-Tojo J, Rajaniemi J (2020) Challenges in assessing urban sustainability. In: Verma P, Singh P, Singh R, Raghubanshi AS (eds) Urban ecology: emerging patterns and social-ecological systems. Elsevier, pp 355–374
Cui X (2018) How can cities support sustainability: a bibliometric analysis of urban metabolism. Ecol Indic 93:704–717
Cui X, Wang X, Feng Y (2019) Examining urban metabolism: a material flow perspective on cities and their sustainability. J Clean Prod 214:767–781
Dijst M, Worrell E, Böcker L, Brunner P, Davoudi S, Geertman S, Harmsen R, Helbich M, Holtslag AAM, Kwan M, Lenz B, Lyons G, Mokhtarian PL, Newman P, Perrels A, Ribeiro A, Carreón JR, Thomson G, Zeyringer M (2018) Exploring urban metabolism—Towards an interdisciplinary perspective. Resour Conserv Recycl 132:190–203
Elliot T, Rugani B, Almenar JB, Niza S (2018) A proposal to integrate system dynamics and carbon metabolism for urban planning. Procedia CIRP 69:78–82
EsmaeilpourZanjani N, Salem Y, Qadi S, Ham-mouda S (2020) The study of urban metabolism and its role in urban planning. J Urban Manag Energy Sustain 2:81–96
Fróes I, Lasthein MK (2020) Co-creating sustainable urban metabolism towards healthier cities. Urban Transformations 2:1–19
Girardet H (1996) The Gaia Atlas of Cities: new directions for sustainable urban living. UN-HABITAT, London
Girardet H (2017) Regenerative cities. In: Shmelev S (ed) Green economy reader. Springer, Cham, pp 183–204
González A, Donnelly A, Jones M, Chrysoulakis N, Lopes M (2013) A decision-support system for sustainable urban metabolism in Europe. Environ Impact Assess Rev 38:109–119
González-García S, Caamaño MR, Moreira MT, Feijoo G (2021) Environmental profile of the municipality of Madrid through the methodologies of Urban Metabolism and life cycle analysis. Sustain Cities Soc 64:102546
Gu B, Dong X, Peng C, Luo W, Chang J, Ge Y (2012) The long-term impact of urbanization on nitrogen patterns and dynamics in Shanghai, China. Environ Pollut 171:30–37
Guan Y, Bai J, Tian X, Zhi L, Yu Z (2021) Integrating ecological and socio-economic systems by carbon metabolism in a typical wetland city of China. J Clean Prod 279:123342
He BJ, Zhao DX, Zhu J, Darko A, Gou ZH (2018) Promoting and implementing urban sustainability in China: an integration of sustainable initiatives at different urban scales. Habitat Int 82:83–93
Huang L, Wu J, Yan L (2015) Defining and measuring urban sustainability: a review of indicators. Landsc Ecol 30:1175–1193

Huang W, Huang Y, Lin S, Chen Z, Gao B, Cui S (2018) Changing urban cement metabolism under rapid urbanization–A flow and stock perspective. J Clean Prod 173:197–206
Joshi S (2021) Key Determinants of urbanization in India: Evidence from state-level data. In: Kateja A, Jain R (eds) Urban growth and environmental issues in India. Springer, Singapore, pp 55–67
Kapoor V, Tripathi S, Devi RS, Srivastava P, Bhadouria R (2020) Ecological economics of an urban settlement: an overview. In: Verma P, Singh P, Singh R, Raghbanshi AS (eds) Urban ecology: emerging patterns and social-ecological systems. Elsevier, pp 91–110
Kaur H, Garg P (2019) Urban sustainability assessment tools: a review. J Clean Prod 210:146–158
Kennedy C, Cuddihy J, Engel-Yan J (2007) The changing metabolism of cities. J Ind Ecol 11:43–59
Kennedy C, Pincetl S, Bunje P (2011) The study of urban metabolism and its applications to urban planning and design. Environ Pollut 159:1965–1973
Khan Z, Dambruch J, Peters-Anders J, Sackl A, Strasser A, Fröhlich P, Templer S, Soomro K (2017) Developing knowledge-based citizen participation platform to support Smart City decision making: The Smarticipate case study. Information 8:47
Kingsland S (2019) Urban ecological science in America: The long march to cross-disciplinary research. In: Pickett STA et al (eds) Science for the sustainable City. Yale University Press, New Haven, pp. 24–46. https://doi.org/10.12987/9780300249385-005
Ko YY, Chiu YH (2020) Empirical study of urban development evaluation indicators based on the urban metabolism concept. Sustainability 12:7129
Lanau M, Mao R, Liu G (2021) Cities as organisms: urban metabolism of the four main Danish cities. Cities 118:103336
Liu Q, Han Y, Liddawi S (2015) Key factors of public attitude towards sustainable transport policies: a case study in four cities in Sweden. Master Thesis. Blekinge Institute of Technology, Karlskrona, Sweden
Lucertini G, Musco F (2020) Circular urban metabolism framework. One Earth 2:138–142
Lv H, Yang L, Zhou J, Zhang X, Wu W, Li Y, Jiang D (2020) Water resource synergy management in response to climate change in China: from the perspective of urban metabolism. Resour Conserv Recycl 163:105095
Maranghi S, Parisi ML, Facchini A, Rubino A, Kordas O, Basosi R (2020) Integrating urban metabolism and life cycle assessment to analyse urban sustainability. Ecol Indic 112:106074
McPhearson T, Parnell S, Simon D, Gaffney O, Elmqvist T, Bai X, Roberts D, Revi A (2016) Scientists must have a say in the future of cities. Nature 538:165–166
Mori K, Christodoulou A (2012) Review of sustainability indices and indicators: towards a new city sustainability index (CSI). Environ Imp Assess Rev 32:94–106
Moscovici D, Dilworth R, Mead J, Zhao S (2015) Can sustainability plans make sustainable cities? The ecological footprint implications of renewable energy within Philadelphia's Greenworks Plan. Sustain Sci Pract Policy 11:32–43
Nevens F, Frantzeskaki N, Gorissen N, Loorbach D (2013) Urban transition labs: co-creating transformative action for sustainable cities. J Clean Prod 50:111–122
Newman PW (1999) Sustainability and cities: extending the metabolism model. Landsc Urban Plan 44:219–226
Noring L, Passerini G, Marchettini N (2018) Contextualising Mobility Variables. Sustain Dev Planning X. WIT Trans Ecol Environ 217:165–173
Odum HT (1983) Systems ecology: an introduction. Wiley, United States
Oliveira V, Vaz R (2021) Is the concept of urban metabolism useful for planning practice? Eur Plan Stud 29:411–424
Özdemir S (2021) Investigating Complexity of İzmir Region by fractal analysis. Ph.D. Thesis. Middle East Technical University, Ankara, Turkey
Paul R, Kenway S, McIntosh B, Mukheibir P (2018) Urban metabolism of Bangalore city: a water mass balance analysis. J Ind Ecol 22:1413–1424
Perrotti D (2020) Urban metabolism: old challenges, new frontiers, and the research agenda ahead. In: Verma P, Singh P, Singh R, Raghbanshi AS (eds) Urban ecology: emerging patterns and social-ecological systems. Elsevier, pp 17–32

Pincetl S (2012) Nature, urban development and sustainability–what new elements are needed for a more comprehensive understanding? Cities 29:S32–S37

Pincetl S, Bunje P, Holmes T (2012) An expanded urban metabolism method: toward a systems approach for assessing urban energy processes and causes. Landsc Urban Plan 107:193–202

Pirlone F, Spadaro I, Candia S (2020) More resilient cities to face higher risks. The Case of Genoa. Sustainability 12:4825

Pissourios IA (2013) An interdisciplinary study on indicators: a comparative review of quality-of-life, macroeconomic, environmental, welfare and sustainability indicators. Ecol Indic 34:420–427

Puerari E, De Koning JIJC, Von Wirth T, Karré PM, Mulder IJ, Loorbach DA (2018) Co-creation dynamics in urban living labs. Sustainability 10:1893

Purvis B, Mao Y, Robinson D (2019) Three pillars of sustainability: in search of conceptual origins. Sustain Sci 14:681–695

Restrepo JDC, Morales-Pinzón T (2018) Urban metabolism and sustainability: precedents, genesis and research perspectives. Resour Conserv Recycl 131:216–224

Roy NC, Thangaraj V (2022) Dynamic interaction of urban development and rural–urban migration: an application of integrated urban metabolism analysis tool (IUMAT) for sustainable city planning. Digit Policy Regul Gov 24:118–140

Salvia AL, Filho WL, Brandli LL, Griebeler JS (2019) Assessing research trends related to sustainable development goals: local and global issues. J Clean Prod 208:841–849

Sarbu R, Alpopi C, Burlacu S, Diaconu S (2021) Sustainable urban development in the context of globalization and the health crisis caused by the covid-19 pandemic. Globalization Its Socio-Econ Consequences 92:01043

Sarker MNI, Peng Y, Yiran C, Shouse RC (2020) Disaster resilience through big data: way to environmental sustainability. Int J Disaster Risk Reduct 51:101769

Savard JPL, Clergeau P, Mennechez G (2000) Biodiversity concepts and urban ecosystems. Landsc Urban Plan 48:131–142

Sharifi A (2021) Co-benefits and synergies between urban climate change mitigation and adaptation measures: A literature review. Sci Total Environ 750:141642

Swilling M, Hajer M, Baynes T, Bergesen J, Labbé F, Musango JK, Tabory S (2018) The weight of cities: Resource requirements of future urbanization. International Resource Panel (IRP) Report. https://www.resourcepanel.org/reports/weight-cities Accessed 12 Dec 2022

Talwar N, Holden NM (2022) The limitations of bioeconomy LCA studies for understanding the transition to sustainable bioeconomy. Int J Life Cycle Assess 27:680–703

Thomson G, Newman P (2018) Urban fabrics and urban metabolism–from sustainable to regenerative cities. Resour Conserv Recycl 132:218–229

Tintelecan A, Dobra AC, Marţiş C (2019) LCA indicators in electric vehicles environmental impact assessment. In: 2019 Electric vehicles international conference (EV). IEEE, pp 1–5

Toli AM, Murtagh N (2020) The concept of sustainability in smart city definitions. Front Built Environ 6:77

Tripathi S (2021) Does smart city development promote urbanization in India? In: Kassim AM, Al-Sharif L (eds) Smart cities—Their framework and applications. IntechOpen. https://doi.org/10.5772/intechopen.94568

Tsalis T, Stefanakis AI, Nikolaou I (2022) A framework to evaluate the social life cycle impact of products under the circular economy thinking. Sustainability 14:2196

Ulgiati S, Casazza M, Yang Z (2021) Urban metabolism theory and analysis methods. In: Liu G, Casazza M, Yang Z, Ulgiati S (eds) Urban metabolism and ecological management: vision, tools, practices and beyond. EDP Sciences, pp 41–64. https://doi.org/10.1051/978-2-7598-2520-2.c005

Vandemoortele J (2017) From MDGs to SDGs: critical reflections on global targets and their measurement. In: van Bergeijk PAG (ed) Sustainable development goals and income inequality. Edward Elgar Publishing, pp 32–50

Verma P, Raghubanshi AS (2018) Urban sustainability indicators: challenges and opportunities. Ecol Indic 93:282–291

Verma P, Singh P, Singh R, Raghbanshi AS (2020) Urban ecology: emerging patterns and social-ecological systems. Elsevier. https://doi.org/10.1016/C2018-0-05562-0

Verma P, Kumari T, Raghubanshi AS (2021) Energy emissions, consumption and impact of urban households: a review. Renew Sust Energ Rev 147:111210

Vidal DG, Barros N, Maia RL (2020) Public and green spaces in the context of sustainable development. In: Filho WL, Azul AM, Brandli L, Özuyar PG, Wall T (eds) Sustainable cities and communities. Springer, Cham, pp 479–487

Voukkali I, Zorpas AA (2022) Evaluation of urban metabolism assessment methods through SWOT analysis and analytical hierocracy process. Sci Total Environ 807:150700

Wang HC (2023) Case studies of urban metabolism: what should be addressed next? Urban Aff Rev 59(3):949–968. https://doi.org/10.1177/10780874221080145

Wang Y (2020) New connotation of marx's metabolism concept under the background of information age. Proceedings 47:50

Webb R, Bai X, Smith MS, Costanza R, Griggs D, Moglia M, Neuman M, Newman P, Newton P, Norman B, Ryan C, Schandl H, Steffen W, Tapper N, Thomson G (2018) Sustainable urban systems: Co-design and framing for transformation. Ambio 47:57–77

Wei H, Shenghui C, Yarime M, Hashimoto S, Managi S (2015) Improving urban metabolism study for sustainable urban transformation. Environ Technol Innov 4:62–72

Wolfram M, Frantzeskaki N, Maschmeyer S (2016) Cities, systems and sustainability: status and perspectives of research on urban transformations. Curr Opin Environ Sustain 22:18–25

Wolman A (1965) The metabolism of cities. Sci Am 213:178–193

Wu Y, Que W, Liu YG, Li J, Cao L, Liu SB, Zeng GM, Zhang J (2018) Efficiency estimation of urban metabolism via Emergy, DEA of time-series. Ecol Indic 85:276–284

Xia C, Li Y, Xu T, Ye Y, Shi Z, Peng Y, Liu J (2018) Quantifying the spatial patterns of urban carbon metabolism: a case study of Hangzhou, China. Ecol Indic 95:474–484

Zeng X, Yu Y, Yang S, Lv Y, Sarker MNI (2022) Urban resilience for urban sustainability: concepts, dimensions, and perspectives. Sustainability 14:2481

Zhou J, Chang S, Ma W, Wang D (2021) An unbalance-based evaluation framework on urban resources and environment carrying capacity. Sustain Cities Soc 72:103019

Chapter 3
Urban Metabolism—An Approach for Enhancing Resilience

Ariyaningsih, Riyan Benny Sukmara, Chandrali Sarkar, Dwi Putri Agustianingsih, and Rajib Shaw

Abstract Along with the increase in the size of the city and the high rate of urbanization, unsustainable urban development can cause severe ecological and environmental issues because cities are essential points of contact between natural and socio-economic systems and places where human activity is concentrated. In addition, urban areas are becoming extremely vulnerable to risks caused by a wide range of sources, including the climate change and urban disasters. However, it is not clear how the urban system relates to one another across different scales within a city or how cities compare to one another in terms of their capacity to survive, recover from, and adapt to various sources of stress. In the meantime, the concept of resilience has been around for long time, and it has seen a great benefit in implementation in urban planning and disaster preparedness. Further, in recent years, urban metabolism has been widely implemented as a valuable framework for assessing an urban system's energy efficiency, material recycling, waste management, and infrastructure. The integration of urban resilience and urban metabolism as concepts can be of tremendous value to the development of strategies for sustainability. This chapter discusses the integration of the resilience concept and the urban metabolism framework, and it also provides case studies that illustrate how this integration can be put into practice and its implications. Reengineering cities and urban life to control and balance unsustainable and unequal urban metabolisms must be implemented. The development of theoretical and practical approaches to urban metabolism requires interdisciplinary collaboration.

Ariyaningsih (✉) · C. Sarkar · R. Shaw
Graduate School of Media and Governance, Keio University, Tokyo, Japan
e-mail: ayya19@keio.jp

Ariyaningsih
Urban and Regional Planning, Institut Teknologi Kalimantan, Balikpapan, Indonesia

R. B. Sukmara
Civil Engineering, Institut Teknologi Kalimantan, Indonesia and Department of Civil Engineering, National Central University, Taoyuan, Taiwan

D. P. Agustianingsih
Regional Disaster Relief Agency of Bali Province, Bali, Indonesia

R. Bhadouria et al. (eds.), *Urban Metabolism and Climate Change*,
https://doi.org/10.1007/978-3-031-29422-8_3

Keywords City system · Climate resilience · Integrated resilience and metabolism · Sustainable city · Urban resilience

3.1 Introduction

As per a Japan International Cooperation Agency (JICA) report released in 2013, over the next 40 years, the urban population in developing countries is estimated to increase by 5.5 million people every month (Akukwe and Popejoy 2013). Urban cities are emerging as growth centers and hubs for commercial, social, and cultural activities. However, there is a downside to this growth pattern where rapid urbanization coincides with deteriorating livelihood, air pollution, climate change, and many man-made and natural disasters. Therefore, while promoting such regional and urban development, it is important to manage the impending risk to our ecological environment stemming from rapid urbanization (Jago-on et al. 2009).

Lucertini and Musco (2020) identified cities as the biggest consumer of global resources. Unsustainable consumption of global resources often leads to environmental degradation and ecological imbalance. Therefore, to alleviate environmental pressures, sustainable usage of natural resources is imperative to support citie's development (Ko and Chiu 2020). This intent leads to the emergence of concepts like "safe," "resilient," and "sustainable" metropolises and the inclusion of strategies pushing to build such sustainable cities to reduce over-extraction of and address the consequent degradation of environmental resources. Such direction has been explicitly emphasized in two Sustainable Development Goals: SDG 11 (to make cities and human settlements inclusive, safe, resilient, and sustainable) and SDG 12 (to ensure sustainable consumption and production patterns) (Kaviti Musango et al. 2020).

Following through on these city planning objectives serves as a powerful reminder of the significance of environmentally responsible development and disaster preparedness. Thus, urban metabolism has been widely implemented as a valuable approach for assessing an urban system's energy efficiency, material recycling, waste management, and infrastructure. Lucertini and Musco (2020) highlighted that this is not a new concept but re-emerged with the need of the hour. According to them, the best-known definition of urban metabolism was given by Kennedy et al. (2014) that it is "*the total sum of the technical and socio-economic processes that occur in cities, resulting in growth, production of energy, and elimination of waste.*" The concept of urban metabolism positively impacts strategizing sustainability and ensuring urban resilience, thus correcting the feedback process within the metabolism of the urban system.

This chapter discusses the integration of the resilience concept and the urban metabolism framework; it further discusses how urban metabolism contributes to building a more resilient city with a positive feedback loop. It provides several case studies from many developing and developed countries to illustrate the practical disposition of such integration. In addition, the chapter is divided into five parts; the first part introduces the emergence of urban metabolism as a concept and how it is integrated with the urban system and urban morphology. The second part, authors tries to connect theory from urban metabolism, system, and its morphology. The third part states the contribution or effects of urban metabolism in ensuring urban resilience. The fourth part gives illustration of these theoretical concepts through case studies. Lastly, the fifth part discusses key considerations for integrating urban metabolism and resilience concepts learned from these case studies to influence policy-level implementation during urban planning.

3.2 Connecting the Dots: Urban Metabolism, Urban System, and Urban Morphology

A city's situation becomes more complicated when its populace numbers in the hundreds of thousands or more. In addition, human activities in urban areas can influence the urban metabolism. This section will explain how the city and its elements can interact with one another, as well as how communities can have an impact on urban morphology and comprehension of the metabolic process.

3.2.1 The City as a System Serving the Needs of Its Community

The United Nations assessments show that by 2030, there will be 43 megacities (cities with more than 10 million residents), most of them in developing countries (Kennedy et al. 2014). The rapid urbanization has brought on massive supply utilization and leads to the ecological issues. While urban areas currently make up only 2–3% of Earth's landmass, one-quarter of the Earth's biological assets and two-thirds of its energy are spent in these regions, which also account for more than 70% of the world's carbon dioxide emissions (Jones et al. 2010). There is a critical requirement for comprehensive investigations on urban sustainability and well-being, given that soon, most of the world's population will live in urban area. Significant changes in lifestyles, patterns in the distribution of stocks and services as well as community organization and waste production are examples of supply chains.

Cities and urban systems are complex structures with their metabolic processes that rely heavily on energy and material inputs from beyond their borders (Currie et al. 2017). As a system, the city's complexity is largely due to the interconnected web of feedback relationships among its physical, political, economic, and environmental factors (Ulgiati and Zucaro 2019). It is impossible to separate the human and technical systems because they are so intertwined and interdependent, affecting, transforming, and maintaining one another. Further, sustainable consumption and production can be linked to social processes and ecological process through the urban metabolism concept to change done by the community itself (Broto et al. 2012). In other words, it is important to understand how urban communities self-organize and make decisions to manage better resources and understanding the lifecycle of resources and how to best repurpose them to benefit the local community (Ulgiati and Zucaro 2019).

Essentially, human settlements and cities are about the benefits of combination, which manifest human sociality in their own right (Shennan 2008). "*Cities are places where a certain energized crowding of people takes place*," writes architectural historian (Kostof 2018). In other words, cities are areas where people interact frequently and intensely, and these interactions significantly impact urban behavior and output (Smith and Lobo 2019a, b, c, d). By focusing on the role of community, relationships, and planning, decision-making, and outcomes (Hoernig et al. 2005), the most important part of a community planner's job is probably finding a way to connect knowledge and action. To do this, they must first understand "how and when" knowledge affects decisions. When designing the planning process, some practical problems from the community should be considered. This will help develop knowledge, solve problems, and coordinate decisions and actions (Marasco et al. 2022). In other words, the city cannot stand alone if it does not meet the community's needs.

3.2.2 A Look at the Effects of Urban Morphology on the Metabolism of Cities

The term "urban metabolism" refers to the "exchange processes that produce the urban environment" (Broto et al. 2011, 2012). When we talk about "urban metabolism," we are talking about the processes by which cities convert raw materials, energy, and freshwater into "built environment," "human biomass," and "waste" (Linkov et al. 2014). Cities are not naturally controlled organisms growing and operating by official codes. However, the government constantly enforces regulations and rules to limit excesses and channel the growth of cities (Braham and Lee 2020). Dijst et al. (2018) found that the speed at which urban processes change varies greatly: physical transportation, utility infrastructures, communication, and the distribution of land uses change very slowly, while workplaces, housing, and other non-built-up buildings have long lifecycles, and employment and household composition change relatively quickly compared to the very fast daily processes. This type of change in urban form will affect to the morphology of the city, as well as the urban metabolism (Kostof 2018).

An examination of the effects and ramifications of lifestyle changes such as the widespread adoption of digital technology or cooperative consumption for material product consumption and waste production can influence urban metabolism (Dijst et al. 2018). In the transportation sector, increasing the number of vehicles in a household to make daily life more efficient can increase fossil fuel consumption and air pollution (Alexandros 2018). There will be space demands, air quality effects, and health effects because of urban design on modes of transportation. People who live in or travel through polluted areas will suffer because of high levels of air pollution

(Braham and Lee 2020). When aggregated, these relationships will influence the urban metabolism of a specific area (Demuzere et al. 2014). Urban design and spatial planning, on the other hand, impact resource use in cities and can thus help reduce consumption in those cities (Weisz and Schandl 2008).

Increasing awareness to the development of urban resilience in the perspective of climate change adaptation must be considered similarly (Wilbanks et al. 2007; Demuzere et al. 2014). As an example, Votsis and Perrels (2016) found that housing prices in exposed and safe areas had decreased because of increased transparency regarding natural hazard risks and that this same transparency could have an accumulative effect on future urban morphology (Votsis 2017) and metabolism as a result. In addition, understanding the urban system, including consumption patterns in urban area, and production services can provide useful data for implementing urban metabolism as sustainable urban planning (Fan et al. 2019). Urban metabolism should be studied concerning how various aspects of urban life, lifestyles, and infrastructural landscape influence metabolic variations (Minx et al. 2011). This necessitates the development of new methods for developing metabolic simulations that can analyze the correlation between urban drivers, patterns, urban environmental quality, and lifestyles in metabolic flows (Minx et al. 2011; Dijst et al. 2018).

Spatial planning must consider the existing morphology of cities. This structure of cities (its morphology) is usually not the outcome of a long-term development process, but rather the result of hysterical growth (Weisz and Schandl 2008; Braham and Lee 2020; Hanzl et al. 2021). The density, layout, utility grids such as for electricity, water, transportation, heating and cooling, all have a long-term effect on the regular resources required because infrastructure decisions are long-term investments (Dijst et al. 2018). Davoudi and Sturzaker (2017) also advocate holistic policy formulation in sustainable urban metabolism and urban morphology. A consistent and coherent approach to addressing the root causes of unsustainable practices is advocated by the authors. To that end, policymakers are urged to make decisions that are based on user behavior and social norms, rather than generic ones (Roggema 2014; Davoudi and Sturzaker 2017). In order to overcome the current fragmented and sector-based decision-making, a multilevel and integrated approach to governance is required (Demuzere et al. 2014).

3.2.3 Metabolism in Cities: A Material Flow Analysis Approach

The urban system's boundaries must be established to gather data on the city's material flows. A city's administrative boundary is commonly used to define its system boundary (Wang et al. 2020). Research of urban systems has utilized a variety of economic and biophysical methods. Embodiment energy (embodied), material flow analysis (MFA), life-cycle analysis (LCA), CO_2 emissions, cost/benefit, and economic returns have all been used to develop performance indicators for cities, regions, and countries (Kennedy et al. 2014). Furthermore, when it comes to the interpretation of these types of metrics, the Organization for Economic Cooperation and Development (OECD) suggests using Material Flow Analysis (Liang and Zhang 2011; Bristow and Mohareb 2020). Understanding the system's material basis and identifying inefficiencies in resource use are both made simple with MFA. Using MFA techniques, a city's physical flow of energy and materials is studied. Analyzing the flow of specific substances in and out of a constrained urban system can be described as an accounting approach that employs statistical data to track the movement of specific substances (Allesch and Brunner 2015; Li et al. 2019). Yet, the most substantial constraint of MFA is the requirement for high-quality data (Allesch and Brunner 2015; Dijst et al. 2018; García-Guaita et al. 2018).

Some researchers have attempted regional and urban material flow analysis frameworks. For instance, in Aichi, Japan's 20-year direct material inputs were estimated by Tachibana et al. (2008). To better understand how materials move through Paris, Barles (2009) calculated and analyzed the data. Stocks, which are defined as an accumulation of materials that are currently present in the system, are differentiated from flows within the context of material flow analysis, for example, to enter and exit the system. The urban system's inputs minus its outputs over a specific time period are what determine the stocks using MFA. Conversions may occur in practice (for example, burning imported fuels may result in CO_2 exports), and these conversions can be accounted for using material's balance principles. As production, consumption, and development patterns shift, the accumulation (as a percentage of imported materials) may also shift. The activities that take place within the urban environment have an impact on the flows and stocks that are present there. There is a direct correlation between human activity, its accompanying flows and stocks, and the individual and community needs that drive both of them (Chávez et al. 2018).

Many investigations into the movement of urban materials and energy focus on "accounting exercises" (Kennedy et al. 2007; Chávez et al. 2018), and there are some practical applications based on them as well, such as encouraging the use of urban metabolism in city planning (Barles 2009; Kennedy et al. 2014; Dijst et al. 2018). MFA emphasizes the linearity of urban metabolisms as a specific cause of vulnerability, similar to other urban ecology studies (Chen and Chen 2014; Allesch and Brunner 2015; Rahman et al. 2019). An investigation conducted in Toronto, Canada, found a link between different types of neighborhood's design characteristics and metabolic flows (Kennedy et al. 2007). It is widely believed that one key characteristic of any long-term metabolic system is the ability to function without relying on external sources of resources or waste disposal, as natural systems do (Tseng and Chiueh 2015; Lucertini and Musco 2020; Musango et al. 2020). MFA can also be applied to identify "urban inefficiencies," or instances where resources are being used inefficiently (Li et al. 2019). Regarding environmental issues and urban planning, MFA is seen as an effective tool for problem detection and policy formulation. However, few studies have examined the role that urban morphology can play in helping to achieve sustainable urban metabolism objectives (Allesch and Brunner 2015; García-Guaita et al. 2018; Bristow and Mohareb 2020).

3.3 The Contributions of Urban Metabolism Toward Resilient and Sustainable City

Urban development is causing serious environmental and ecosystem issues because cities are places where human activity is concentrated, making them important points of contact between natural and socio-economic systems. This section will discuss the evolution of urban metabolism research to explain its role in improving both resilience and sustainability. In addition, we will discuss the issues that have been resolved because of this evolution, which will lead to our concluding discussion of some of the issues that have not yet been resolved.

3.3.1 Historical Perspective on the Development of the Idea of the Urban Metabolism

A city's "urban metabolism" (UM) is a reporting framework that aims to quantify the flow of resources (energy and material) in and out of the city (Ulgiati and Zucaro 2019; Bristow and Mohareb 2020). For the first time, Abel Wolman published a paper in Scientific American entitled "The Metabolism of Cities" in 1965. His research focused on pollution because of his position in the sanitary engineering field. He realized that tracking the movement of assets inside and outside an urban system was critical to answering the underlying problem. Wolman's career path led him to become a pollution researcher. During the early stages of its development, this method caught the interest of only a few numbers of scholars. Urban metabolism was first studied using case studies ranging from Miami to Tokyo to Brussels to Hong Kong in early 1970s (Zhang et al. 2015). Measuring the flows of resources and energy through the urban system was accomplished by the authors of this study by employing material and energy flow analysis techniques and using either mass or energy components to measure these flows.

The concept of "urban metabolism" offers a fresh perspective on how cities can be environmentally, socially, and economically sustainable and how to effectively manage these interrelated variables. A great deal of academic study has used the urban metabolism analysis since Wolman's pioneering work. In addition, Odum (1975) was a pioneer in the field of quantitative ecosystem theory with the growth of his urban heterotrophic model, which paved the way for the quantitative assessment of urban metabolism. In the early 2000s, when it came time to start researching sustainable development and how to identify the largest energy consumers and greenhouse gas emitters, the idea received attention. Marxist ecology, industrial ecology, and urban ecology are the three main schools of urban metabolism's understanding (Roggema 2014; Meerow et al. 2016). Ecologists in the disciplines of urban ecology and industrial ecology see urban metabolism as a complex web of interdependent natural and social systems that lead to inequitable outcomes. In a broader sense, urban metabolism can be viewed as part of the socio-metabolism field, which

Haberl et al. (2019) define as "*a systems approach to study society–nature interactions at different spatiotemporal scales.*" Urban sustainability and resilience can be aided by the centrality of UM and its ability to influence city policies and designs directly (Bristow and Mohareb 2020). Because the potential of the urban metabolism approach was so obvious, there has been a surge of research on urban metabolism in recent years.

3.3.2 Integration of Urban Metabolism and Disaster Resilience and How They Can Support Sustainable Development

The phenomenon and development mode of spontaneous urban growth has been widely studied in global cities and megacities, planned and unplanned cities, and formal and informal cities. Its widespread occurrence has been documented (Chen and Chen 2022a, b, c). Ecological and biological concepts have long been a part of sustainable development debates. According to Newman, reducing resource consumption and waste emissions is one way that a city can achieve its long-term sustainability goals. This will help to minimize the negative effects within the capacity of the local or global ecosystem to sustain these collisions, which will also improve human well-being. Newman also noted that this would be an effective way to achieve sustainability goals (Chen and Chen 2022a, b, c). On the United Nations Sustainable Development Goals (SDGs), specifically SDG 11—Sustainable Cities and Communities (SCP), Ecological footprint is discussed (Zhang et al. 2015). Human physical metabolism and Earth's ecosystem are two examples of how the concept can be used to real-world situations (Galli et al. 2016; Meerow et al. 2016).

Attaining and fostering sustainable development are a primary objective of urban planning ideas. As can be seen by looking at India, there is much potential and relevance in today's urban growth theories to deal with India's complexities and diversity. It is a call to rethink, accept, and implement the above concepts in a holistic way to address and solve the existing key urban development concerns of urban sprawl and climate change caused and faced by the Indian megacities (Bhadouria et al. 2022).

The promotion of responsible and balanced use of resources is the focus of an increasing number of policies being enacted at the international level. Developed in response to SDG-11, the New Urban Agenda calls for a "new paradigm" implementation in cities. For urban sustainability transitions to be accelerated, analysis and monitoring must take a holistic approach (Nagy et al. 2018; Attolico and Smaldone 2020). There has also been recent development of a holistic approach to the processes of urban atmospherics (Davoudi and Sturzaker 2017). In more recent times, there has also been the development of an all-broad method for studying the processes of urban atmospherics (Alexandros 2018). Analysis of urban systems using ecological principles and techniques has been proposed in numerous studies (Demuzere et al. 2014;

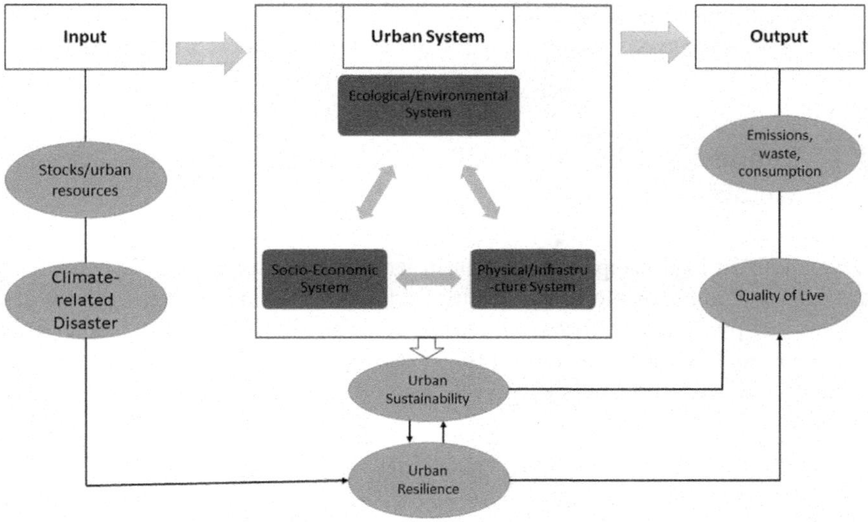

Fig. 3.1 Linkage among urban system, resilience, and sustainability

Abidin et al. 2015; Moraci et al. 2020). Cities can be viewed from this perspective as giant organisms that consume, transform, accumulate, and reduce waste from their surrounding environment. To reduce the likelihood of cascading failures in the urban system, every step in urban metabolism must be organized to allow for consistent energy–material flows (Fig. 3.1).

To determine potential future directions for urban energy metabolism growth, an overview of common resilience system features has been conducted. Urban resilience is not a normative concept like urban sustainability, according to (Hamstead et al. 2021; Morshed et al. 2021). Meanwhile, the planning and design community needs to pay more attention to this issue. As described by the IPCC (Intergovernmental Panel on Climate Change), "*the ability of a social or ecological system to absorb disturbances while maintaining the same basic structure and ways of functioning, capacity for self-organization, and capacity to adapt to stress or change,*" urban resilience is applicable. Cities that have well-designed infrastructure and efficient energy and material flow can better withstand and recover from major disasters, according to McGill (2020). As a result, a gap in urban resilience has been highlighted by this tool's success in promoting sustainability.

Rooted in the marxist–structuralist tradition, urban metabolism offers an integrated and relational way to look at resilience as a process of stabilization and change. To be resilient, a city or community must be able to bounce back from various internal and external threats (McGill 2020). From the standpoint of urban resilience, these (and other) urban services are under threat as cities cope with new anthropogenetic disputes such as climate change, rapid urbanization, technological change, and malicious acts. Considering this shifting risk context, researchers are increasingly concerned about the complexity and risk to cities (Bristow 2020).

3.4 Enhancing Climate Resilience Through Urban Metabolism

Developing an interdisciplinary way of urban metabolism approach is essential to measure and improve the sustainability performance of urban areas. These case studies are presented to focus on a broader discussion about sustainable cities, resilience, and urban metabolism. From these examples, it is hoped to bring new insight regarding urban metabolism that supports resilience, especially in climatic matters.

3.4.1 Case Study of Bali, Indonesia

The province of Bali, which has Denpasar as its capital, is essential to national and regional economic growth. The country's unique culture and tropical setting have drawn tens of millions of visitors worldwide. Environmental concerns appear not to be ignored. On the other hand, Denpasar had a serious waste problem prior to 2011 (Prajnawrdhi et al. 2012). Traditional markets, temporary roadside food stalls, and other commercial activities that support tourism have resulted, that is why the city of Denpasar has a major problem with solid waste that is generated from these tourism activities. Water, rivers, and mangroves are polluted because dumpsite capacity is no longer adequate (De Angeli et al. 2022). Denpasar's air quality and traffic flow have deteriorated to critical levels. Only 3% of Denpasar residents use public transportation, as they prefer to drive around the city (Leung 2016). There has also been a significant rise in the use of private vehicles. It is estimated that private transportation in Denpasar is growing at 11% per year. Further, from November to March, Denpasar is often flooded. Most severe floods happen in the west and south of Denpasar, where there are more people, homes, and business activities (Liu and Tao 2020). Not only does flooding occur because of inadequate drainage, but it is also a result of waste generated by commercial and residential construction not being disposed of in landfills.

By referring to the concept of metabolism city, the flow of material will be smooth, allowing the city to be more efficient in its use of natural resources. On the other hand, the concept of a metabolism city must be supported by a more operational approach, specifically a water sensitive city and a water metabolism city. The idea behind the water-sensitive city approach is that the city is a water reservoir. All water, both rainwater and wastewater, is accommodated, including domestic and non-domestic waste and industrial waste (Newhart et al. 2019; Musango et al. 2020;

Muliana et al. 2021). As a means of promoting tourism, supporting water sensitive city, and implementing urban metabolism, the government has shifted its economic focus from sustainable agriculture (with *subak*—a traditional irrigation system) to tourism (Prajnawrdhi et al. 2012). It has been centuries since the traditional *subak* irrigation systems irrigated rice terraces (Agung et al. 2019). The water source or its descent through a temple to irrigate subak land is marked by terraced paddy fields, rice fields linked by a network of canals, tunnels, and weirs, villages, and temples of various sizes and importance, and these are all components of a *subak* landscape.

Urbanization and shrinking agricultural land have resulted in imbalances in the irrigation system, such as a lack of green space and rice fields that have been converted into housing and commercial properties. The water temple has lost its significance and value due to the city's rapid growth. Denpasar's urban symbolism, such as statues and landmarks, has been lost because of the city's rapid population growth and uncertain development, according to Agung et al. (2019). The Balinese irrigation system has many informal and autonomous water temple networks. Irrigation and temple institutions are examples of how adaptive governance is implemented. She noticed that every farming family in Bali must belong to a *subak* community in their locality (Agung et al. 2019).

City planning in Bali is heavily influenced by the concept of Tri Hita Karana (the harmonious relationship between the gods, humans, and nature). *Subak* is drawn to the "tri hita karana" philosophy, a modern worldview based on ancient Balinese wisdom. The region suffers from major environmental degradation because of unchecked development and excessive petrochemical use. There are three levels of existence in Balinese mythology: the lower world (nista), the middle world (madya), and the upper realm (utama), and this cosmological tenet teaches how to maintain harmony in all three levels of existence. Green areas, public facilities, roads, and open spaces all played an important role in people's daily lives and traditions. Reduced metabolic throughput and improved eco-environment well-being are critical challenges for coming citie's conversion to more sustainable and resilient ways (Meerow et al. 2016; Boin et al. 2017). It is possible to slow down metabolism by looking into all the renewable and secondary energy sources that can be used in cities. In general, to have less of an effect on the well-being of the eco-environment, government should figure out the best ways to use resources by using nexus thinking with water (Fan et al. 2019).

3.4.2 Case Study of Jakarta Province, Indonesia

Jakarta City is the capital of Indonesia, the largest metropolitan city, and the most developed area in Indonesia. As a province, Jakarta has four levels of government, there are Provincial Government (*Provinsi*), City Government (*Kota*), Sub-District Government (*Kecamatan*), and Urban Suburbs Government (*Kelurahan*). Below these four official levels of administration are two different levels, both subsidiaries of Urban Suburbs. The first is Hamlet (*Rukun Warga*) and contained several neighborhoods (*Rukun Tetangga*). These administrative systems were established during the Japanese Occupation (Evans 2020a, b, c, d). Jakarta has five cities, one municipality in Jakarta such as North, West, East, South, and Central Jakarta, and one municipality called Thousand Island, located in Jakarta Bay.

The total population in Jakarta reaches more than 10.6 million inhabitants in 2021 (BPS 2022). This population does not count the people from the surrounding area named Bogor, Depok, Tangerang, and Bekasi (BODETABEK) who commute to Jakarta for work, study, and business. The huge number of population activities forces Jakarta to consume a lot of energy and resources. As the center of economic activity, Jakarta has GPD (at the current price) in the first quarter of 2022 which is Rp. 763.367.734.300 (USD 50.97 million). The activities in Jakarta generate an environmental impact such as an increase in waste, greenhouse gases, and even a decreasing quality of life.

The climate condition shows that Jakarta has an annual rainfall of about 2394.6 mm. The average temperature is 28.5 °C, and the minimum and maximum temperatures are 23.0 °C and 35.3 °C, respectively. The duration of sunshine is about 4.2 h (BPS 2022). The topographical condition of Jakarta is dominated by low land area with slopes ranging between 0 and 2° in the northern and central parts and between 0 and 5° in the southern part, and the most southern part has an elevation of about 50 masl (Abidin et al. 2015). Land use of Jakarta is dominated by the residential area with more than 50% coverage area, 30% of green open space, and roughly 9% of industrial area, and the rest is utilized as an institutional and commercial area as well as water body (Putri et al. 2020).

Regarding the concept of urban metabolism, it represents the city as an ecosystem with an input (energy and materials) while generating an output (waste) (Tseng and Chiueh 2015). We need to consider the technical and socio-economic processes in the city, then analyze the resource consumption, growth, and production of energy and attempt to eliminate the waste (Gunawan and Pusaka 2016). In Jakarta, there are a lot of socio-economic activities, and the natural process occurs as a city dynamic. That obtains the impact of urban metabolism balance, especially for the process and activities which generate waste. Socio-economic activities like transportation, industry, manufacturing, farming, and daily activities of human life will contribute to the total amount of waste and the natural process.

Climate change has become a global issue that affects almost all places, even if it is an indirect impact. Different climatic, demographic, socio-economic, and resource conditions will result in varying energy and material consumption, affecting some citie's urban metabolism performance. Jakarta is the one of cities that should deal with the climate change issue, especially with the increase in sea level and floods (Gunawan and Pusaka 2016). From another point of view as well as the other megacity, Jakarta has several problems, including transportation issues, water resources, air pollution, and domestic waste. For instance, 48% of greenhouse gases (GHGs) in Jakarta are contributed by the transportation sector (C40 CITIES 2020) and also produce more waste heat and enhance the urban heat island index (Farzinmoghadam 2016a, b).

To face the issues and balance the urban metabolism in Jakarta, some strategies need to be implemented to deal with it. The strategy can be a policy, regulation, education, or movement to increase people awareness about urban metabolism and sustainability. For instance, to eliminate waste like pollutants and reduce the consumption of fossil fuels from transportation issues, the Government of Jakarta (GoJ) develops mass transportation. Bus rapid transit, named Transjakarta (Busway), was developed by GoJ and was firstly operated in 2004. GoJ also implemented the *odd/even vehicle license policy* to reduce the number of vehicles on road. The other efforts are creating car-freer day spots and the bicycle line. In 2007, following the national regulation to converting fossil fuel to gas fuel, Jakarta is the one of cities as a pilot project of that regulation, and for the first phase, Taxi and Bajaj (three-wheel vehicles) become a target to be converted to gas fuels. In addition, the Transjakarta was targeted to be converted to gas fuels after Taxi and Bajaj did. After several year's operation, in 2019 Transjakarta stops using gas fuels due to lack of quality of gas fuels and return to use diesel fuels (Republika.co.id 2019). Following the increasing population, the government also expands the development of the railway's systems such as electric train (name KRL) which started operation in 1930 (Hata 2003) to the Mass Rapid Transit (MRT), named Jakarta MRT. Good mass transportation will create efficient mobility that will decrease the fuel consumption and reduce the amount of GHG emissions.

Another government's strategy to eliminating waste and reducing GHG is suggesting people to use induction cookstove. Currently, more than 80% household uses Liquid Petroleum Gas (LPG), although LPG is better than biomass and kerosene, but mostly LPG was imported and subjected to price volatility and supply disruption (UNESCAPE 2021). The use of induction cookstove mostly consumes high electrical energy, which mainly powers plant use coal to produce electricity, so it will be a contradiction (Isla-Cabaraban and Cabaraban 2016).

An extreme flood hit Jakarta in 2007; approximately 70% of Jakarta was flooded with depths up to 4 m. There were 600.000 people displaced and 79 people died. The total cost of this flooding disaster was 890 million USD for direct costs and more than 620 million USD for the indirect cost (Surya et al. 2019). The phase after flood event will produce a lot of solid waste and become an additional burden for landfill area. The impact of flood was damaging the private and public infrastructures, and these will add more material consumption for repairing and rebuilding the infrastructures. Due to the worst flood situation, the National Government prioritized the East Canal Flood and finished the work in 2010. This canal gives additional support to the Jakarta's West Flood Canal that was developed in 1922 by colonial government (Silver 2012).

Regarding the urban metabolism concept for energy input, in 2022 based on the data published by Statistic Indonesia, Jakarta consumes electricity of about 32,875,758.256 KWh, natural gas of about 982,886,312 m^3, gasoline (industry) of about 16,444,293 L, diesel fuel (industry) of about 91,241,519 L. Jakarta also consumes clean water, mainly from the public water company (PDAM) which is 633,306,752 m^3. For biomass input, rice is about 2033.59 tons, shallot about 18.8 tons, spinach about 34,689,95 tons, water spinach about 53,443,79 tons, and Chinese cabbage about 26,840,41 tons, but on the other hand the biomass input will change to be waste. Households generate the most municipal solid trash in Indonesia, followed by traditional markets (Aye and Widjaya 2006). Regarding the enormous population of Jakarta, household waste will increase linearly. Household waste is dominated by solid waste such as food waste, plastic, paper, wood, and even metal. That condition force city to innovate to reduce waste or utilize waste as an energy source. Currently, in Jakarta, there are several companies which are now investing in and operating landfill gas to electricity generation facilities (Aprilia et al. 2012). There are benefits to implementing waste-to-energy projects, such as reduced greenhouse gas (GHG) emissions, improved air quality in landfills, reduced methane emissions through methane capture, reduced passive emissions of landfill gases (LFG), and reduced air pollution from landfill fires and open burning of household waste (Aprilia et al. 2012).

3.4.3 Case Study of Taipei, Taiwan

Taipei is the capital city of the Republic of China, known as Taiwan, and has become the leading city in Southeast Asia (Shmelev et al. 2018). Taipei city is located in Northern Taiwan and becomes the country's most important economic center (Wang et al. 2018). The area of Taipei is about 271.8 km^2 with the total population in 2021 being 2,592,878 inhabitants and the population density in Taipei city is 9732 person/km^2 (Hellmer et al. 2021). In quarter 4 (Q4) 2021, real GDP of Taipei City grew by 6.45% (Invest Taipei Office 2021).

Taipei has a humid subtropical climate, in which summers are lengthy, hot, and humid, with major rainstorms and typhoons on occasion. At the same time, winters are brief, normally warm, and foggy due to the north-easterly winds from the huge Siberian High being enhanced by the pooling of this cooler air in the Taipei Basin. In 2021, Taipei's mean temperature, relative humidity, and sunshine duration are about 24.1 °C, 76%, and 1692.4 h, respectively. The total amount of precipitation is 1908.5 mm with 116 precipitation days (Taipei City Government 2021).

The fast-growing economy of Taipei and the occurrence of mass urbanization have forced Taipei to consume more construction materials such as cement and gravel for transportation, building, residential area, and other construction needs. Furthermore, following the regulation of construction health, some construction needs to be demolished, resulting in construction waste. Most construction waste (such as gravel and cement) would enter the landfills in Taipei Metropolitan city. According to the annual report of the Government of Taipei and New Taipei City, the population density has reached 10.000 per Km^2, which will make finding a new landfill site difficult. Hence, it is critical to manage construction waste from the resource aspects (Wang et al. 2018). According to the analysis of Huang and Chen (2009), land use change in Taipei significantly increases. They mentioned that the urban area is sprawling from 12.7% (296.4 km^2) in 1981 to 22.0% (510.5 km^2) (Huang and Chen 2009).

Ko and Chiu (2020) selected six major strategies in Taipei City for their research (from 2016 to 2018) related to evaluation indicators of urban metabolism concept including natural environment sustainability, improving man-made environment, transportation development, revitalizing urban space, energy sustainability policy, and health (Ko and Chiu 2020). The natural environmental sustainability policy includes installing sewage treatment systems, increasing water resource treatment in the Tamsui River, food waste recycling, furniture recycling, and establishing recycling boxes for used books. Air quality monitoring, environmental quality improvement, and noise reduction were included in the improving man-made environment policy. In the transportation section, the strategy consists of encouraging green vehicles such as bicycle stations and lanes, subsidized bicycle and electric vehicles purchases, installing smart bus stop signs and combining the smart card transaction data to estimate bus operation.

Urban space revitalization, such as increasing the number of sidewalks, uses permeable pavement and green park areas to establish a convenient environment of mountain-friendly leisure and linking the large green space in park with green corridors. Currently, only 47.6% area of Taipei is classified as "*land for development,*" while 52.4% is ranked as a conservation area, wetlands, forestry, and agriculture lands. Just 5.69% of total areas are designated as "park and green space" among "lands for development." However, in the year of 2020, the average amount of green space per person in this basin's urbanized section was 6.18 m^2, which is the range of area about 2.05 m^2 in the central area (Daan district) and about 11.17 m^2 at the outskirt (Hellmer et al. 2021).

In case of a healthy living policy, government was suggested to greening roof and facades for the building to enhance greenery and better air quality. Furthermore, regarding to the energy sustainability policy, the government was promoting renewable energy such as establishment of solar energy for parks, building and green space, improving the efficiency of waste incineration for power generation, and improving efficiency of hydropower generation. The last, health policy strategy was covered several programs, such as building the sports city and optimizing sport complexes, promoting sport activities, establishing an elder-friendly environment, and improving long term for older adults.

Following the natural environment policy, especially for Taipei's river management, could be a challenging sustainability issue. This because of the step hillslope and weak control of land use makes the watershed in Taipei which is associated with frequent floods, debris torrents, and landslides during typhoon (Chou and Li 2012). It will generate more waste and energy consumption during the recovery phase. According to the energy management policy, industrial sector in Taiwan consumes 38.56% of the total energy consumption in 2011. There are ten administrative regions categorized as the concentrated industry area and 12 administrative regions categorized and service concentrated area includes Taipei City. Following Yang et al. (2016), Taipei City got the highest rank of Taiwan's most sustainable administrative region (Yang et al. 2016).

3.4.4 Case Study of Budapest, Hungary

Budapest City is the capital of Hungary and is located along the Danube River in the Carpathian Basin. The total area is 525 km^2 and the entire population reaches 1.7 million people with a density of about 3200/km^2, but if combined with the metropolitan area, the total population will be 2.5 million people with a total area of about 2500 km^2. The city was established in 1872 by the unification of three towns, Pest, Buda, and Obuda (Old Buda), and now becomes one of the largest and most economically dynamic cities in East Europe (Probáld 2014). Budapest will become highly developed with a ratio of about 52% and 60% of built-up area soon (Buzási 2022). A built-up area with regular morphological patterns dominates the eastern part of Budapest, which is renowned as an urban model of concentric zones. The delayed development area covered the west part of Budapest. The irregular pattern of the city is because of the complicated orography and the Budai Hills area which is 400 m above sea level, covered by recreational forest, and becomes a protected area. This area is also found the good quality 4–5 story houses which are spread up in the slope area and surrounded by more and less green spaces. The city center of Budapest is dominated by residential areas, including dilapidated housing from the end of the nineteenth century which started to be renovated in the early twentieth century. Large abandoned industrial districts, a monotonous housing complex, and aging high-rise structures from the socialist era make up most of the transitional zone (Probáld 2014).

As the center of economic and financial resources, Budapest contributes roughly 40% of the national Gross Domestic Product (GDP) Hungary (GDP: 155 billion USD in 2020). There is radical transformation of Budapest's economy and society during the socialist era and after the socialist period (between 1950 and 1990, respectively). This condition also establishes the following impact on the use and efficiency of resources, such as water, food, energy, and land (Pomázi and Szabó 2008).

Budapest has significantly increased material and energy input since early 1950. Electronic devices, such as TV, washing machines, refrigerators, air conditioners, and cars, expand largely. For instance, from 1965 to 2005, the use of passenger cars increased from 39 to 596 thousand stock (Pomázi and Szabó 2008). Roughly, the transportation sector contributes 30–40% of GHG emission of the city and becomes one of the major GHG contributors (Buzási 2014). The increasing number of cars will linearly vary with the gasoline consumption, and the increase of electronic devices increases the total consumption of electricity and heat emission. In June 2013, during heat wave strike results, the increment of national consumption was higher by 20,000 MWh/day than average weekday in May and attributed this to the significant use of air conditioners (Probáld 2014). This condition will also emerge waste and environmental consequences. In the long term, it will also contribute to climate change occurrence. Budapest downtown has annual mean temperature 1.2 °C warmer than outside the city and the peak temperature difference occurred in January (1.5 °C). To face the climate change impact, the Government of Hungary adopted The National Climate Change Strategy program called VAHAVA in 2007 (abbreviation of

Hungarian-language VÁltozás-HAtás-VÁlaszadás, The program ("Change-Impact-Response) developed in 2003 and focused on the decreasing GHGs emission (Farago et al. 2010).

In general, the inputs side of urban metabolism in Budapest are water consumption, natural gas consumption, heat consumption, and food consumption. There are total solid waste, wastewater, and air pollutants for the output side. To balance the urban metabolism of Budapest, The City Government joined the Covenant of Mayor initiative IN 2011 to establish the framework for cutting the city's CO_2 emissions by 20% by 2020, Budapest formed its own Sustainable Energy Action Plan (SEAP). Sectors group several climate-related actions; there are Natural Environment, Transport and Utility systems. The natural environment action covered several programs such as Landscape feature and biodiversity protection, development of green areas and water body, green oriented development for public space, stopping to reduce biologically active areas, and ventilation corridor preservation. Transport action has several programs including increasing the number of park and ride area, developing traffic calming measures in the city center, promoting unmotorized and public transport, developing intermodal points, improving pedestrian and cycling facilities, and last, designing river flood control for utility system action (Buzási 2014).

Energy efficiency of public, private, and commercial building, utilization of renewable energy to local energy production, improving district heating system, promoting unmotorized transport, and improving street lighting efficiency become the main target point of the SEAP of Budapest. Regarding to the energy efficiency action, the Government of Hungary (under EU UIA program) funded the public–private partnership housing program, named "E-Co-Housing," which the City Government of Budapest was included in the partnership scheme. The E-Co-Housing project is the program to tackle the social and environmental issue related to the housing provision by creating a social housing model with high energy efficiency and smart environment solution, especially for underprivileged residents in Budapest. The collaborative planning process is carried out with the residents living in that smart housing. Using prefabricated modular building materials and recycled materials (respecting the circular economy concept) will reduce construction costs and contribute to climate resilience and a greener city environment. Each residential apartment will also be equipped with smart integrative energy and an IT system to support collaborative, intergenerational, and healthy lifestyles (Mcguinn et al. 2020).

Regarding the increment of the population in Budapest, it will indeed affect the land use land cover (LULC) change. In suburban area, the land cover change are frequently caused by extremely complicated socio-economic variables, which also weaken peri-urban agriculture capacity and ecological balance by causing environmental harm, landscape degradation, and wildlife habitat loss (Lennert et al. 2020). Lannert predicts the land use change of built-up area extension (for post-socialist era, 1990–2012) will affect the 117.3 thousand ha with an annual growth rate of about 1055 ha. The biggest change affecting agriculture that converted to artificial areas is about 86% of new built-up zones (Lennert et al. 2020). Conversely, the untechnical problem that may not be included in the Lannert's research is the consistency of the government to protect the change of LULC. In the Hungarian planning system, issues of inefficiency connected to the system's legality and the inadequacy of how current laws and regulations are applied regularly arise (Kauko 2003). Notably, if new uses are found that do not fit the plan, the municipality may amend the plans to suit them. In the case of multinational firms and big "bling" developments financed by institutional investors, this is frequently a pertinent series of events. Additionally, developers try to alter the blueprints to obtain extra building rights (Kauko 2013).

The Changes LULC to built-up areas can be one of the main indicators of increasing infrastructure development, increasing waste and emissions during and after development. Furthermore, the government should consider the future to adapt to the fast-growing development. To better human living space and environment, urban development policy should be regarded as the concept of urban metabolism and sustainability, especially to reduce GHG production due to climate change impacts.

3.5 Conclusion and Future Perspectives

As discussed above, many decades of research from around the world shed light on the changing metabolism of cities. Metabolic differences can be affected by a wide range of factors such as urban forms, lifestyles, infrastructure landscapes, and urban morphology. Urban metabolism from a critical perspective reveals new ways of thinking about the city's production through social relations, resource flows, and discourses. There are differences in how urban metabolism theories are implemented, such as ranging from unusual approaches that focus on developing tools and integrating planning into systems. For instance, reengineering cities and urban life to control and balance unsustainable and unequal urban metabolisms must be implemented. The development of theoretical and practical approaches to urban metabolism requires interdisciplinary collaboration. This chapter has stated that each theme was developed in response to a different set of concerns about urban sustainability and resilience. Thus, the way urban metabolism is conceptualized and approached which serves other functions.

A wide range of data must be gathered for a comprehensive assessment of urban sustainability and resilience. City planners should be encouraged to learn about their citie's urban metabolisms. They would benefit greatly from knowing whether they are making effective use of water, energy, materials, and nutrients, as well as how their efficiency stacks up compared to other cities. Consider the extent to which their immediate surroundings are nearly depleted and devise applicable approaches to reduce utilization if necessary. Metabolic data have been created for just two cities globally; this chapter shows that there is still much work to be done. Resources are typically accounted for at the national level, but this approach may be too wide and miss out on the driving practices in urban areas.

References

Abidin HZ, Andreas H, Gumilar I, Wibowo IRR (2015) On correlation between urban development, land subsidence and flooding phenomena in Jakarta. IAHS-AISH Proc Reports 370:15–20. https://doi.org/10.5194/piahs-370-15-2015

Agung IG, Suryada B, A B St, and R A Ct (2019) The role of power in the existence of Subak cultural landscape in Denpasar, Bali. Int J Curr Adv Res 8(01):16810–16814. https://doi.org/10.24327/ijcar.2019.16814.3122

Akukwe C, Popejoy MW (2013) Millennium development goals. Global Public Health Policy, pp 373–78. https://doi.org/10.1007/978-0-387-78665-0_6106

Alexandros G (2018) Urban metabolism. In: Defining the urban, pp 223–35. https://doi.org/10.4324/9781315576282-19

Allesch A, Brunner PH (2015) Material flow analysis as a decision support tool forwaste management: a literature review. J Ind Ecol 19(5):753–764. https://doi.org/10.1111/jiec.12354

Angeli SD, Malamud BD, Rossi L, Taylor FE, Trasforini E, Rudari R (2022) A multi-hazard framework for spatial-temporal impact analysis. Int J Disaster Risk Reduct 73(July 2021):102829. https://doi.org/10.1016/j.ijdrr.2022.102829

Aprilia A, Tezuka T, Spaargare G (2012) Household solid waste management in Jakarta, Indonesia: a socio-economic evaluation. In: Marmolejo Rebellon LF (ed) Waste management—an integrated vision, pp 73–100. InTech. https://doi.org/10.5772/51464

Attolico A, Smaldone R (2020) The #weResilient strategy for downscaling local resilience and sustainable development: the Potenza Province and municipalities of Potenza and Pignola case. Disaster Preven Managem: an Int J 29(5):793–810. https://doi.org/10.1108/DPM-04-2020-0130

Aye L, Widjaya ER (2006) Environmental and economic analyses of waste disposal options for traditional markets in Indonesia. Waste Manage 26(10):1180–1191. https://doi.org/10.1016/j.wasman.2005.09.010

Barles S (2009) Urban metabolism of Paris and its region. J Ind Ecol 13(6):898–913. https://doi.org/10.1111/j.1530-9290.2009.00169.x

Bhadouria R, Upadhyay S, Tripathi S, Singh P (eds) (2022) Urban Ecology and Global Climate Change. Wiley-Blackwell. https://doi.org/10.1002/9781119807216

Bobby Rahman M, Nurhasanah IS, Nugraha E, Harahap T (2019) Applying the urban political ecology (UPE) framework to re-visit disaster and climate change vulnerability-risk assessments. J Reg City Plann 30(3):224–240. https://doi.org/10.5614/jpwk.2019.30.3.4

Boin A, Comfort LK, Demchak CC (2017) The rise of resilience. In: Designing resilience, pp 1–12. https://doi.org/10.2307/j.ctt5hjq0c.5

BPS (2022) Provinsi DKI Jakarta Dalam Angka 2022. Jakarta, Badan Pusat Statistik Provinsi DKI Jakarta

Braham WW, Lee JM (2020) Metabolism of urban location: travel time and the morphology of cities. Front Sustain Cities 2:1–13. https://doi.org/10.3389/frsc.2020.00004

Bristow DN, Mohareb EA (2020) From the urban metabolism to the urban immune system. J Ind Ecol 24(2):300–312. https://doi.org/10.1111/jiec.12919

Broto VC, Allen A, Eriksson A (2011) Urban metabolism at UCL—a working paper, no. November 13. https://www.bartlett.ucl.ac.uk/dpu/urban-metabolism/project-outputs/Urban_Metabolim_Report.pdf

Broto VC, Allen A, Rapoport E (2012) Interdisciplinary perspectives on urban metabolism. J Ind Ecol 16(6):851–861. https://doi.org/10.1111/j.1530-9290.2012.00556.x

Buzási A (2014) Will Budapest be a climate-resilient city?—Adaptation and mitigation challenges and opportunities in development plans of Budapest. Europ J Sustain Developm 3(4):277–88. https://doi.org/10.14207/ejsd.2014.v3n4p1

Buzási (2022) Comparative assessment of heatwave vulnerability factors for the districts of Budapest, hungary. Urban Climate 42:101127. https://doi.org/10.1016/j.uclim.2022.101127

C40 CITIES (2020) Benefits of urban climate action: C40 cities technical assistance report 2020—Jakarta electric buses. Jakarta

Chávez A, Kennedy C, Chen B, Chertow M, Baynes T, Chen S, Bai X (2018) Understanding, implementing, and tracking urban metabolism is key to urban futures. In: Urban Planet: Knowledge towards sustainable cities. Cambridge University Press, pp 68–91. https://doi.org/10.1017/9781316647554.005

Chen B, Chen S (2014) Eco-indicators for urban metabolism. Ecol Indic 47:5–6. https://doi.org/10.1016/j.ecolind.2014.09.021

Chen F, Chen Y (2022) Urban metabolism and spontaneous architectural growth: a sustainable strategy featuring participatory co-construction by multiple stakeholders. Buildings 12(3):352. https://doi.org/10.3390/buildings12030352

Chou T-L, Li T-C (2012) Sustainability and river management in Taipei City, Taiwan. J Urban Manage 1(1):77–94. https://doi.org/10.1016/S2226-5856(18)30055-4

Currie PK, Musango JK, May ND (2017) Urban metabolism: a review with reference to cape town. Cities 70:91–110. https://doi.org/10.1016/j.cities.2017.06.005

Davoudi S, Sturzaker J (2017) Urban form, policy packaging and sustainable urban metabolism. Resour Conserv Recycl 120:55–64. https://doi.org/10.1016/j.resconrec.2017.01.011

Demuzere M, Faehnle M, Orru K, Olazabal E, Geneletti D, Heidrich O, Orru H, Feliu E, Bhave AG, Mittal N (2014) Understanding cities: advances in integrated assessment of urban sustainability. https://www.dropbox.com/s/adqrx7mfy5eiid7/Dawson_etal_2014_CostTU0902FinalReport.pdf?dl=0

Dijst M, Worrell E, Böcker L, Brunner P, Davoudi S, Geertman S, Harmsen R et al. (2018) Exploring urban metabolism—towards an interdisciplinary perspective. Resour Conserv Recycl 132(October 2017):190–203. https://doi.org/10.1016/j.resconrec.2017.09.014

Evans K (2020) The five levels of government in Indonesia: an AIC briefing note. In: Millott M (ed) The Australia-Indonesia Centre

Fan J-L, Kong L-S, Wang H, Zhang X (2019) A water-energy nexus review from the perspective of urban metabolism. Ecol Model 392:128–136

Fan JL, Kong LS, Wang H, Zhang X (2019) A water-energy nexus review from the perspective of urban metabolism. Ecol Modell 392(November 2018):128–36. https://doi.org/10.1016/j.ecolmodel.2018.11.019

Farago T, Lang I, Csete L (eds) (2010) Climate change and hungary : mitigating the hazard and preparing for the impacts (The 'VAHAVA' Report). VAHAVA Report. Hungarian Academy of Science

Farzinmoghadam M (2016) Urban metabolism and land use modeling for urban designers and planners: a land use model for the integrated urban metabolism analysis tool. University of Massachusetts Amherst

Galli A, Giampietro M, Goldfinger S, Lazarus E, Lin D, Saltelli A, Wackernagel M, Müller F (2016) Questioning the ecological footprint. Ecol Indic 69:224–232. https://doi.org/10.1016/j.ecolind.2016.04.014

García-Guaita F, González-García S, Villanueva-Rey P, Moreira MT, Feijoo G (2018) Integrating urban metabolism, material flow analysis and life cycle assessment in the environmental evaluation of Santiago de Compostela. Sustain Cities Soc 40:569–580. https://doi.org/10.1016/j.scs.2018.04.027

Gunawan J, Pusaka S (2016) Introducing the urban metabolism approach for a sustainable city: a case of Jakarta, Indonesia. J Appl Manag Account Res 14(1):55

Haberl H, Wiedenhofer D, Pauliuk S et al (2019) Contributions of sociometabolic research to sustainability science. Nat Sustain 2:173–184. https://doi.org/10.1038/s41893-019-0225-2

Hamstead Z, Iwaniec D, McPhearson T, Berbés-Blázquez M, Cook EM, Muñoz-Erickson TA (2021). Resilient urban futures. https://link.springer.com/book/10.1007%2F978-3-030-63131-4

Hanzl M, Geerse A, Guschl L, Dewan R (2021) Urban metabolism and land use optimization: in quest for modus operandi for urban resilience. In: Understanding disaster risk. Elsevier Inc. https://doi.org/10.1016/b978-0-12-819047-0.00007-x

Hata T (2003) Improvement of railway system in Jakarta metropolitan area. Japan Railway Transp Rev

Hellmer E, Shih W-Y, Hartung A (2021) Food water energy nexus in Taipei. In: Salehi P (ed) Bonn, ICLEI

Hoernig H, Leahy D, Zhuang ZX, Earley R, Randall L, Whitelaw G (2005) Planning for people: integrating social issues and processes into planning practice. Berkeley Plann J 18:35–55. https://doi.org/10.5070/bp318111500

Huang SL, Chen CW (2009) Urbanization and socioeconomic metabolism in Taipei an emergy synthesis. J Ind Ecol 13(1):75–93. https://doi.org/10.1111/j.1530-9290.2008.00103.x

Invest Taipei Office (2021) Summary of current economic dynamics

Isla-Cabaraban MT, Marc G, Cabaraban S (2016) Generating electricity from spent activated carbon: life cycle environmental benefits. WSEAS Trans Environ Developm 12. http://www.wseas.org/multimedia/journals/environment/2016/a025815-077.pdf

Jago-on KA, Bianet SK, Fujikura R, Fujiwara A, Imai T, Matsumoto T, Zhang J et al (2009) Urbanization and subsurface environmental issues: an attempt at DPSIR model application in Asian Cities. Sci Total Environ 407(9):3089–3104. https://doi.org/10.1016/j.scitotenv.2008.08.004

Jones PD, Parker DE, Osborn TJ, Briffa KR (2010) World development report: climate change. Carbon dioxide information analysis center. vol 1942. https://doi.org/10.3334/CDIAC/cli.002

Kauko T (2003) Planning processes, development potential and house prices: contesting positive and normative argumentation. Hous Theory Soc 20(3):113–126. https://doi.org/10.1080/14036090310016970

Kauko T (2013) On sustainable property development—the case of Budapest. Open Urban Stud J 6(1):9–26. https://doi.org/10.2174/1874942901306010009

Kennedy C, Stewart ID, Ibrahim N, Facchini A, Mele R (2014) Developing a multi-layered indicator set for urban metabolism studies in megacities. Ecol Indic 47:7–15. https://doi.org/10.1016/j.ecolind.2014.07.039

Kennedy C, Cuddihy J, Engel-yan J (2007) "Of Cities" 11(2)

Ko YY, Chiu YH (2020) Empirical study of urban development evaluation indicators based on the urban metabolism concept. Sustainability 12(17):7129. https://doi.org/10.3390/su12177129

Kostof S (2018) Gridded Worlds: an urban anthology. In: Gridded Worlds: an urban anthology, pp 55–73.https://doi.org/10.1007/978-3-319-76490-0

Lennert J, Farkas JZ, Kovács AD, Molnár A, Módos R, Baka D, Kovács Z (2020) Measuring and predicting long-term land cover changes in the functional urban area of budapest. Sustainability 12(8):1–20. https://doi.org/10.3390/SU12083331

Leung KH (2016) Indonesia's summary transport assessment. ADB Papers on Indonesia, vol 15. pp 20. www.adb.org

Li Y, Zhang Y, Hao Y, Wang X (2019) Exploring the processes in an urban material metabolism and interactions among sectors: an experimental study of Beijing, China. Ecol Indic 99(December 2018):214–24. https://doi.org/10.1016/j.ecolind.2018.12.019

Liang S, Zhang T (2011) Urban metabolism in China achieving dematerialization and decarbonization in Suzhou. J Ind Ecol 15(3):420–434. https://doi.org/10.1111/j.1530-9290.2011.00343.x

Linkov I, Bridges T, Creutzig F, Decker J, Fox-Lent C, Kröger W, Lambert JH et al (2014) Changing the resilience paradigm. Nat Clim Chang 4(6):407–409. https://doi.org/10.1038/nclimate2227

Liu Y, Tao H (2020) Strategic sustainability plan in Denpasar 2040. In: IOP conference series: earth and environmental science, vol 526(1). https://doi.org/10.1088/1755-1315/526/1/012009

Lucertini G, Musco F (2020) Circular urban metabolism framework. One Earth 2(2):138–142. https://doi.org/10.1016/j.oneear.2020.02.004

Marasco S, Kammouh O, Cimellaro GP (2022) Disaster resilience quantification of communities: a risk-based approach. Int J Disaster Risk Red 70:102778. https://doi.org/10.1016/j.ijdrr.2021.102778

McGill R (2020) Urban resilience—an urban management perspective. J Urban Manag 9(3):372–381. https://doi.org/10.1016/j.jum.2020.04.004
Mcguinn J, Fries-Tersch E, Jones M, Crepaldi C, Masso M, Kadarik I, Lodovici MS, Drufuca S, Gancheva M, Geny B (2020) Social sustainability. Luxembourg, Study for the committee on employment and social affairs, policy department for economic, scientific and quality of life policies, European Parliament
Meerow S, Newell JP, Stults M (2016) Defining urban resilience: a review. Landsc Urban Plan 147:38–49. https://doi.org/10.1016/j.landurbplan.2015.11.011
Minx JC, Baiocchi G, Peters GP, Weber CL, Guan D, Hubacek K (2011) A 'Carbonizing Dragon': China's fast growing CO_2 emissions revisited. Environ Sci Technol 45(21):9144–9153. https://doi.org/10.1021/es201497m
Moraci F, Errigo MF, Fazia C, Campisi T, Castelli F (2020) Cities under pressure: strategies and tools to face climate change and pandemic. Sustainability 12(18):1–31. https://doi.org/10.3390/su12187743
Morshed SA, Arafat M, Mokhtarimousavi S, Khan SS, Amine K (2021) 8R resilience model: a stakeholder-centered approach of disaster resilience for transportation infrastructure and network. Transport Eng 4:100058. https://doi.org/10.1016/j.treng.2021.100058
Muliana R, Ghozali A, Ariyaningsih (2021) Factors influencing adequate sanitation on domestic wastewater management of black water type in Margasari village. In: IOP conference series: earth and environmental science, vol 778 (1). https://doi.org/10.1088/1755-1315/778/1/012015
Musango JK, Currie P, Smit S, Kovacic Z (2020) Urban metabolism of the informal city: probing and measuring the 'unmeasurable' to monitor sustainable development goal 11 indicators. Ecol Indic 119:106746. https://doi.org/10.1016/j.ecolind.2020.106746
Nagy JA, Benedek J, Ivan K (2018) Measuring sustainable development goals at a local level: a case of a metropolitan area in Romania. Sustainability 10(11):3962. https://doi.org/10.3390/su10113962
Newhart KB, Holloway RW, Hering AS, Cath TY (2019) Data-driven performance analyses of wastewater treatment plants: a review. Water Res 157:498–513. https://doi.org/10.1016/j.watres.2019.03.030
Odum EP (1975) Ecology, the link between the natural and the social sciences. Holt, Rinehart and Winston
Pomázi I, Szabó E (2008) Urban metabolism: the case of budapest. In: ConAccount 2008: urban metabolism, measuring the ecological city, September, pp 352–75. https://www.researchgate.net/publication/251231474%5CnUrban
Prajnawrdhi TA, Sivam A, Karuppannan S (2012) Is urbanised city sustainable ? A case study of Denpasar. In: 6th international conference and workshop on the built environment in developing Countries, December, 4–5
Probáld F (2014) The urban climate of budapest: past, present and future. Hungarian Geograph Bullet 63(1):69–79. https://doi.org/10.15201/hungeobull.63.1.6
Putri RF, Abadi AW, Kafafa U (2020) The correlation analysis between urbanization phenomena and landuse change in Jakarta special Province. In: E3S web of conferences—the 1st geosciences and environmental sciences symposium (ICST 2020). vol 200. EDP Science, pp 01002. https://doi.org/10.1051/e3sconf/202020005003
Rahman MA, Moser A, Rötzer T et al (2019) Comparing the transpirational and shading effects of two contrasting urban tree species. Urban Ecosyst 22:683–697. https://doi.org/10.1007/s11252-019-00853-x
Republika.co.id. (2019) Ini Tiga Alasan Transjakarta Tak Utamakan Armada BBG, July 2019
Roggema R (2014) Towards enhanced resilience in city design: a proposition. Land 3(2):460–481. https://doi.org/10.3390/land3020460
Shennan S (2008) Not by genes alone: how culture transformed human evolution, by Peter J. Richerson and Robert Boyd. Biol Philos 23(2):293–299. https://doi.org/10.1007/s10539-005-9007-5

Shmelev SE, Sagiyeva RK, Kadyrkhanova ZM, Chzhan YY, Shmeleva IA (2018) Comparative sustainability analysis of two Asian cities: a multidimensional assessment of Taipei and Almaty. J Asian Finan Econ Business 5(3):143–55. https://doi.org/10.13106/jafeb.2018.vol5.no3.143

Silver C (2012) A case for ecological planning: urban flooding in old and New Jakarta. Citygreen 1:164. https://doi.org/10.3850/s2010098115000778

Smith ME, Lobo J (2019) Cities through the ages: one thing or many? Front Digital Human 6:00012. https://doi.org/10.3389/fdigh.2019.00012

Surya MY, He Z, Xia Y, Li L (2019) Impacts of Sea level rise and river discharge on the hydrodynamics characteristics of Jakarta Bay (Indonesia). Water (Switzerland) 11(7):1384. https://doi.org/10.3390/w11071384

Tachibana M, Matsumura Y, Fukuda M, Kimura H, Shinkai Y (2008) G9a/GLP complexes independently mediate H3K9 and DNA methylation to silence transcription. EMBO J 27(20):2681–2690. https://doi.org/10.1038/emboj.2008.192

Taipei City Government (2021) Taipei City Statistical Abstract 2021. Taipei City, Taipei City Government

Tseng WL, Te Chiueh P (2015) Urban metabolism of recycling and reusing food waste: a case study in Taipei city. Proc Eng 118:992–999. https://doi.org/10.1016/j.proeng.2015.08.540

Ulgiati S, Zucaro A (2019) Challenges in urban metabolism: sustainability and well-being in Cities. Front Sustain Cities 1:1–3. https://doi.org/10.3389/frsc.2019.00001

UNESCAPE (2021) Energy trasition pathway for the 2030 agenda—sustainable energy transition roadmap for Jakarta. United Nations Publications, New York

Votsis A (2017) Utilizing a cellular automaton model to explore the influence of coastal flood adaptation strategies on Helsinki's urbanization patterns. Comput Environ Urban Syst 64:344–55. https://doi.org/10.1016/j.compenvurbsys.2017.04.005

Votsis A, Perrels A (2015) Housing prices and the public disclosure of flood risk: a difference-in-differences analysis in Finland. J Real Estate Financ Econ 53(4):450–471. https://doi.org/10.1007/s11146-015-9530-3

Wang X, Li Y, Liu N, Zhang Y (2020) An urban material flow analysis framework and measurement method from the perspective of urban metabolism. J Cleaner Prod 257:120564. https://doi.org/10.1016/j.jclepro.2020.120564

Wang Y, Chen PC, Ma HW, Cheng KL, Chang CY (2018) Socio-economic metabolism of urban construction materials: a case study of the Taipei metropolitan area. Resour Conserv Recycl 128:563–71. https://doi.org/10.1016/j.resconrec.2016.08.019

Weisz H, Schandl H (2008) Materials use across World Regions: inevitable pasts and possible futures. J Ind Ecol 12(5–6):629–636. https://doi.org/10.1111/j.1530-9290.2008.00097.x

Wilbanks TJ, Romero Lankao P, Bao M, Berkhout FGH, Cairncross S, Ceron J-P, Kapshe M, Muir-Wood R, Zapata-Marti R (2007) Industry, settlement and society. In: Climate change 2007: impacts, adaptation and vulnerability, contribution of working group II to the fourth assessment report of the intergovernmental panel on climate change, Cambridge University Press, pp 357–90

Yang WC, Lee YM, Jin Li H (2016) Urban sustainability assessment of taiwan based on data envelopment analysis. Renew Sustain Energy Rev 61:341–353. https://doi.org/10.1016/j.rser.2016.04.015

Zhang Y, Yang Z, Xiangyi Y (2015) Urban metabolism: a review of current knowledge and directions for future study. Environ Sci Technol 49(19):11247–11263. https://doi.org/10.1021/acs.est.5b03060

Chapter 4
Urban Metabolism to Understand Changes in Urban Ecology: A Case of Bengaluru

N. S. Nalini and Neesha Dutt

Abstract The impact of the interaction between socio-economic and political processes with natural ecosystems and the built environment has been well captured by the urban metabolism framework in recent decades. In this concept, cities are equated to living organisms in terms of the consumption of natural resources and the excretion of waste products. This concept refers to cities as biological entities having internal processes continuously exchange matter and energy with their surrounding environment in order to grow. The growth pattern induced by metabolic processes may be uneven spatially and socio-economically. Master plans were prepared by the concerned authorities to control such uneven development. However, as these metabolic processes diverged, the impact was on natural resources such as vegetation, temperature, and water bodies. This chapter analyzes the changes that are seen in vegetation patterns in Bengaluru. The results showed that the metabolic processes of the city transformed vegetation to a large extent spatially. There was either a shrinking or proliferation of green cover converting the indigenous greenery to fast-growing varieties during different phases of urbanization. Resource-intensive lawns became important in corporate and residential landscaping. Finally, these changes in vegetation invariably affected temperature patterns, also showing an interconnection between these natural elements.

Keywords IT phase · Lawns · Spatial · Uneven · Urban metabolism · Vegetation

N. S. Nalini (✉)
Nitte School of Architecture, Planning and Design, NMIT Campus, Bengaluru, Karnataka 560064, India
e-mail: arnalinins@gmail.com

N. Dutt
National Institute of Advanced Studies, IISc Campus, Bengaluru 560 012, India

Manipal Academy of Higher Education, Manipal 576 104, India

R. Bhadouria et al. (eds.), *Urban Metabolism and Climate Change*,
https://doi.org/10.1007/978-3-031-29422-8_4

4.1 Introduction

With more than half of the world being urban, cities in today's world are complex entities within which a myriad of social, environmental, and economic processes takes place (Ulgiati and Zucaro 2019; UN World Urbanization Prospects (revision) 2022). Urban metabolism is a useful framework to understand these processes and their interactions in such complex milieus (Broto et al. 2012; Dijst et al. 2018). It provides a multidisciplinary framework through which to study urban areas wherein dynamic flows and interactions between interrelated biophysical, social, and built systems constantly build, destroy, create, and recreate urban environments (Swyngedouw 2006a, b). If applied effectively, the urban metabolism framework can help plan cities better and steer them toward more sustainable pathways (Kennedy et al. 2011; Dijst et al. 2018; Ulgiati and Zucaro 2019). The recent attention afforded to the urban metabolism framework for its ability to provide effective solutions for urban sustainability, health, and well-being is indicative of its potential as a multidisciplinary framework (Dijst et al. 2018).

Urban metabolic processes, including socio-economic and political processes, are interlinked with each other and the environment (Swyngedouw 2006a, b). Natural elements like air, water, soil, temperature, and vegetation change over space and time due to these metabolic processes. Although the natural environment also influences metabolic processes—the reverse—i.e., the impact of urban metabolic processes on the environment is very prominent. For example, urbanization gives rise to various environmental problems such as urban heat islands, flooding, land degradation, loss of vegetation, and so on; this kind of environmental change and degradation, in turn, impacts human health and well-being (Ulgiati and Zucaro 2019; Bhadouria et al. 2022). However, such effects are not borne equally by all urban inhabitants (Dijst et al. 2018). For instance, metabolic flows of matter and energy are socially mobilized through networks of power structures embedded within the complexities of urban settings (Swyngedouw 2006a, b). Thus, metabolic processes lead to uneven outcomes whether it be social, geographical (or spatial), economic, material, or environmental (Broto et al. 2012). In order to address issues of sustainable and equitable living and to build more just and inclusive cities in the future, the urban metabolism framework can help unravel some of the complexities underlying this unevenness (Swyngedouw 2006a, b).

The chapter uses the urban metabolism framework to understand changes in urban ecology in the south Indian city of Bengaluru. Specifically, it studies changes in vegetation during different phases of urbanization starting from the colonial phase to the more recent information technology (IT) phase. The first section elaborates on the concept of urban metabolism in some detail with a focus on various dimensions including social, political, economic, and spatial metabolism. The second section shows the transformation of urban vegetation in the city during three urbanization phases, namely, the public sector phase, the garment sector phase, and the information technology (IT) phase. The latter section focuses on understanding contemporary lawn landscapes in the city during the last two decades, a hitherto understudied yet important component of urban green space. And finally, transformation in temperature patterns is also discussed in relation to the city's urbanization. The chapter concludes with an overview of the transformation in vegetation experienced by Bengaluru due to urban metabolic processes.

4.2 Urban Metabolism

Metabolism is a process observed in living organisms in which one form of energy gets transformed into another and in the process produces waste. This concept dates back to when Karl Marx used this approach to explain the exchange of energy and material between nature and the city (Swyngedouw 2006a, b). Thereafter, urban areas are equated to living organisms since various activities consume and store resources and produce energy and waste in the process. Urban metabolism as an analytical framework helps in understanding these interactions between human beings, natural resources, and the manmade environment (Wolman 1965; Kennedy et al. 2007; Pincetl et al. 2012). The exchange of resources and energy that is observed within the city can also be quantified using the urban metabolism framework. Urbanization has gradually increased human activities with the increase in the rate of metabolism (Pincetl et al. 2012).

Metabolic processes are inherently more circular in nature, whereas, in urban areas, these processes are more linear (Girardet 1990; Chrysoulokis 2013). In circular metabolic processes, waste produced is recycled and the same is used as input material for other activities (Karakeiwicz 2011). For sustainable development of urban processes, cities should devise efficient planning methods to reduce waste materials by employing proper recycling techniques (Kennedy et al. 2011). Recent urban studies stress the need to envisage the cities as made up of interrelated systems, wherein all the services like transportation, water supply, sanitation, and zoning are considered as systems. Understanding the interrelationship between these various services is very essential in the preparation of master plans and the useful application of the urban metabolism framework (Karakeiwicz 2011; Musango et al. 2020).

Energy consumption, resource inflow, and waste outflow of the city are essential information required for efficient urban planning and master plan preparation. The urban metabolism framework enables such quantification and therefore can be adopted by urban planners as a new model and method for urban planning (Chrysoulokis 2013). Several studies have used urban metabolism models for quantifying greenhouse gas emissions, heat waves, groundwater depletion, and so on (Kennedy et al. 2011; Dousset et al. 2011). The exchange of energy and pollutants in urban areas needs quantification for better urban planning (Grimmond et al. 2010; Chrysoulakis et al. 2013). Mitigation of these issues requires circular metabolism for the sustainable development of cities (Karakeiwicz 2011).

Driving forces like socio-economic, political, and environmental processes shape the resources and energy flows inside the city. The urban metabolism framework being multidisciplinary is a useful platform to understand the influence of such forces. Each of these forces along with other infrastructural services is considered a separate system in urban ecology and industrial ecology. As a result, the city was considered to be made up of several systems that drive metabolic processes. In this chapter, we consider social, economic, political, and spatial metabolic processes to understand the changes in urban ecology. Urban ecological processes involve the changes in the biophysical aspect of cities as a result of urbanization, which also form an integral part of urban metabolic processes that transform natural elements like air, soil, water, green space, etc. It is imperative to understand the interactions between these systems in order to better understand the complexity of urban socio-ecological systems to plan cities effectively for sustainable development using the urban metabolism framework.

4.2.1 Social Metabolism

The total exchange of all resources and materials that takes place between society, the biosphere, and the geosphere is captured in social metabolism (Schott 2004). Society brings about changes in nature by consuming resources. The consumption of resources is organized in such a way that the flow of materials and energy can derive the maximum benefit from resources (Alier 2009). Within the city, these metabolic processes are driven completely by social activities rather than by interaction with nature (Geizen and Roemers 2015). Nevertheless, one can observe the uneven distribution of resources creating uneven socio-spatial spaces as a result of socio-metabolic processes. Uneven material flow is observed between societies and different classes of people in cities (Heynen 2014). In addition to resource consumption, other social factors such as employment rate, literacy rate, mobility, and health conditions can also provide an estimation of social metabolism in cities (Alier 2009; Rapoport 2011).

4.2.2 Political Metabolism

Politics plays an important role in controlling the flow of materials and energy. Politics along with social factors determine the extent of access to resources by different classes of people and control and shape the environment of the city (Geizen and Roemers 2015). Policy decisions, whether intentional or unintentional, also shape the energy flows within the city and thereby produce the urban environment as per the needs and aspirations of the people. Thus, political metabolism has the power to urbanize the social and physical environment of the city (Pincetl et al. 2012).

4.2.3 Economic Metabolism

Social and political metabolism mobilizes capital and labor power through the exchange of resources and energy. Economic metabolism thus produces commodities by operating through certain social control, appropriation, and possession (Niza et al. 2009). Natural resources are used as inputs and are transformed to produce energy which is used in the fabrication of products. The waste thus produced can cause environmental damage (Niza et al. 2009). In order to reduce waste production, the quantity of natural resources used as input in economic activities should be minimized. Consumption and production patterns also require modification (Rapoport 2011). Economic metabolism can be made sustainable by minimizing environmental degradation through innovations in technology and thereby reducing natural resource input (Fisher and Freudenburg 2001).

4.2.4 Spatial Metabolism

Interaction between social, political, and economic metabolisms initiates spatial development and brings about changes that are generally uneven with uneven distribution of resources and infrastructure (Gandy 2006). Metabolic processes give a specific spatial form to the city, and it is important to connect these processes to understand the problems of the city. Social activities define and occupy their own spaces based on the characteristics of the activity (Harvey 1970). The spatial property of metabolism helps in identifying the development pattern of livable and vulnerable spaces within the city (Idrus et al. 2008; Bahers et al. 2022). In this chapter, spatial is referred to as a built-up area that consumes resources like land, materials, and energy (Schremmer 2009). Therefore, spatial metabolism of urban areas also depends on the intensity of these resource consumption. Spatial metabolism is measured with spatial metrics which estimate the percentage of pervious and non-pervious areas, urban sprawl, and fragmentation and coherence of urban spaces (Asian Development Bank 2014).

Uneven resource management has in a way manifested in spatially heterogeneous socio-cultural development (Pickett et al. 2001). Political power influences the spatial development of the city creating a hierarchy of towns and cities by means of socio-economic transformations (Allen et al. 2016). Spatial transformation observed in urban and peri-urban areas and their interrelationships are due to the urban sprawls of cities. Master plans identify the direction where maximum capitalizing is available for spatial development (Mukherjee 2016). Recent development in automobile industries has restructured the socio-economic spaces spatially, thereby even the planning projects of cities (Hajer 2016). Consequently, gentrification in urban areas due to escalation in land values and the development of slums in peri-urban regions is the process generally observed in all cities. Along with this, the location of gated communities in peri-urban regions has added to the spatial disparity of urban cities (Lawton 2016).

Master plans delineate spatial boundaries in order to facilitate equitable distribution of infrastructure provision (Vanka 2014). However, boundaries thus produced more often follow concentric circle patterns encouraging symmetric development. The rationale for such planning patterns is the conceptualization of development of cities during the pre-independence era, when urban planning theories followed the Chicago School of thought taking forward the concentric ring theory of EW Burgess (Das 1981; Dossal 1989; King 1989; Drakakis 2000; Dear 2002). Conversely, spatial metabolism of cities more often followed the indigenous socio-economic and political processes than urban planning theories. Thereby, this chapter discusses the impact of metabolic processes on the transformation of vegetation in the city of Bengaluru.

4.3 Urban Vegetation Metabolism

Bengaluru is one of the fastest-growing urban agglomerations in India (UN World Urbanization Prospects (revision) 2022). In terms of area, it has increased from 501.21 km^2 in 1961 to 748.42 km^2 in 2011, with a decadal population growth rate of 49.44% in 2011 (Census of India 2011). Spatial analysis of the city shows that its urban expansion is characterized by different phases of urbanization, namely the public sector, garment industries (those which were established as small scale industries), and IT sector industries (Nalini 2016). Most of this expansion has been attributed to the region's economic growth during each phase. The 1950s to 1970s was the public sector phase, and 1980s to 1990s and later period is contiguous with the rise of garment industries and the information technology industry, a surge in foreign investments, and growth of the real estate market (Ramachandra and Kumar 2010; Sudhira and Nagendra 2013; Nalini 2016).

The rapid expansion of the city during this period provides an opportunity to understand the linkages between urban metabolic processes and vegetation. Urbanization processes directly affect the biophysical environment of a city including its green spaces, land, and water (Swyngedouw 2006a, b). The changes in vegetation due to the socioeconomic processes that accompany urban expansion can be complex. For example, urbanization can destroy or pollute natural elements like air, soil, and water; however, socioeconomic processes of urban spatial expansion and development can also build and re-shape urban environments. This, in turn, creates modified natural environments including urban vegetation. Thus, political, social, and economic processes together influence vegetation change patterns during city development due to the interlinkages and metabolic interactions among all such processes (Heynen et al. 2006).

Bengaluru's urbanization processes in general have destroyed greenery to a large extent, but recent studies have shown that there is an increase in vegetation during particular phases of urbanization (Nagendra et al. 2012). It is imperative to understand the impact of urban metabolic processes on vegetation which influences the percentage of green cover and distribution pattern of vegetation. To understand these transformations, satellite images of different time intervals were analyzed.

4.3.1 Transformation of Urban Vegetation

4.3.1.1 Extraction of Vegetation from Satellite Images

Remotely sensed images from LandSat Satellite were downloaded from the Earth Explorer website supported by USGS, for 1976, 1992 and 2013. Satellite images for Bengaluru region are available only from 1973 onward. These images corresponded with different phases of urbanization, namely the public sector industries, garment industries, and IT sector industries, respectively. Since lush green cover is found in the months of November and December, after monsoon rains in Bengaluru, satellite images of these months were downloaded for Normalized Difference Vegetation Index (NDVI) analysis. NDVI extracts vegetation pixels using near-infrared and red bands of satellite images (NIR-R/NIR + R). An index value more than 0.2 was selected to pick the vegetation pixels. The extent of urban growth was extracted from satellite images and overlaid on NDVI images to understand the changing patterns of vegetation. The total area of the green cover was calculated using the zonal statistics tool of QGIS software.

The green cover within the built-up area in 1976 was 36%, that of 1992 was 30%, and in 2014, it was 48% (Nalini 2020). It is apparent from the above analysis that urbanization does not necessarily reduce vegetation, instead can even enhance the green cover. The different phases of urbanization had a diverse impact on greenery.

4.3.1.2 Vegetation During Colonial Period

During the colonial phase, the region had agricultural lands, vegetable gardens, and eucalyptus plantations (Nair 2005). It was the British who introduced vegetation inside the residences and along the streets. New extensions like Malleshwaram and Basavanagudi also had streets with trees, parks, and large houses with large trees (Nagendra 2016). About 80% of the city had open spaces and 15% of the area comprised parks, playgrounds, and water bodies. There were two large parks in the city, namely Lal Bagh and Cubbon Park (Nair 2005).

Lal Bagh is an important landmark of the city even to this day. It was created by Hyder Ali and his son Tippu Sultan. They encouraged Mughal gardens and brought many different varieties of plants and trees from other countries (Udvardi 2014). Eventually, Lal Bagh was turned into a horticulture center by the British giving training to gardening enthusiasts (Iyer et al. 2012; Nagendra 2016).

Cubbon Park on the other hand was created as a buffer area between the Pete and Cantonment to cordon off the movement of native people to the cantonment. The design was prepared by Major General Richard Sankey, chief engineer of Mysore. It was named as Meade's Park initially, named after John Meade, the acting commissioner of Mysore in 1870 (Gowda 2003). It covers an area of 300 acres and houses 6000 different varieties of trees and plants (Gowda 2003; Vanka 2014).

4.3.1.3 Vegetation Before IT Phase (1980s)

Green Spaces of the Public Sector Townships

During the 1940s–1950s, central public sector industries were established, namely HAL and ITI in the east, and HMT and BEL in the north. They had a substantial impact on the social and spatial metabolism of the city. These industries occupied agricultural lands and were aptly referred to as green-field townships (Shivaramakrishnan 2014). These townships were well planned with gardens, parks, and trees flanking the streets, following the urban planning principles of colonial cities. While large shady trees were part of big houses, small houses had small gardens in the front and back. 'Durvaninagar township' established by ITI was the best-designed township among all (Nair 2005).

Residential layouts established during this period followed the grid-iron pattern, with roads running in north–south and east–west directions. These extensions too had parks and open spaces in between the houses as seen in the toposheets of 1948 and 1960. Major roads had an avenue of large shade-giving trees (Nagendra 2016). By 1955–56, there were many parks and gardens, including Lal Bagh and Cubbon Park (Rao 1985). There were ornamental plants and lawns in the traffic roundabouts and smaller open spaces (Mani 1985). Trees during this period had large canopies and comprised different varieties of species (Nagendra and Gopal 2011). The trees were large and the park sizes were also large during this phase (Nagendra and Gopal 2011). While every house generally had gardens and trees, the city was surrounded by vineyards, vegetable gardens, and plantations (Nair 2005). The agricultural tract was demarcated in 1963 to control urban sprawl (Ravindra 1995).

Analysis of satellite images of the year 1976 shows 36% of green cover within the built-up area during that year, whereas inside the public sector townships, percentage of the green cover was slightly higher. Total green cover inside the HAL township was 56%, ITI township was 47%, and HMT and BEL townships were 41% and 34%, respectively. In general, the city had a substantial amount of natural green cover inside and outside the city limits conferring the coveted status of 'Garden City' (Nair 2005).

4.3.1.4 Vegetation During IT Phase

The emergence of small-scale and IT industries during the 1980 and 1990s provided jobs for unskilled, semi-skilled, and skilled laborers. The large number of people migrated in search of jobs, including women from economically weaker sections. The growth rate was 76.7% during this period (Ravindra 1995). Built-up area increased in 1980 due to urbanization, leading to the deterioration of green spaces and open spaces to 5% and 40%, respectively (Nair 2005). The green belt policy was introduced in 1984 Comprehensive Development Plan (CDP) to control the urban sprawl, which covered around 65% of the planning area (Ravindra 1995). Green belt policy restricted the type of construction inside this zone. This zone resembled a 'fortification' provided to control the urban sprawl (Nair 2005).

Restriction of urban growth due to the green belt policy led to encroachment of open spaces inside the city to meet the housing demands of the growing population. Many spacious bungalows were converted into residential apartments (Swamy 2013). Even though Bangalore Development Authority (BDA), Karnataka Town and Country Planning (KTCP), and Karnataka Parks, Playgrounds, and Open Spaces Acts undertook several policy measures to encourage the development of green spaces, they dwindled inside the city area as well as in the peripheral regions. Bangalore Development Authority (BDA) implemented a policy of planting one or two trees for a site area of 2400 sq ft or 4000 sq ft, respectively (BDA 2007; Swamy 2013). KTCP act laid restrictions on the conversion of parks into other types of land use (KTCPA 2005). Regardless of all the measures taken, urban vegetation was reduced to an alarming extent. While several parks were converted into civic amenities (Vanka 2014), gardens inside the house almost disappeared due to a lack of space (Swamy 2013). Industrialization and the rapid explosion of the built-up area were detrimental to the greenery inside the city (Adkin 2009). The skyline of the city soon changed from low-rise to high-rise buildings (Gopinath 2014) with minimum or no space for greenery.

Analysis of the 1992 NDVI image showed a total green cover of 30% within the built-up area of 1992. Overall vegetation cover was reduced when compared to the 1976 vegetation cover. Vegetation metabolism inside the industrial townships was slightly different; for example, the green cover inside HAL reduced to 51%, and that inside HMT, HMT watch factory, BEL, and ITI increased to 64%, 32%, 49%, and 52%, respectively, compared to the green cover in 1976. Vegetation metabolism within the green belt zone underwent a noticeable transformation. The location of IT industries inside this zone led to encroachment of green spaces by the built-up area, particularly in the eastern and southeastern directions. The anticipation of a similar development in other directions, particularly in the periphery, inspired landowners to convert agricultural lands either into revenue layouts or plantations ready to sell when the demand arrives (Nagendra et al. 2012; Madalasa 2014).

IT development in the city skyrocketed land prices (Stallmeyer 2006; Vanka 2014), and high demand for land led to encroachment of many green spaces including parks and playgrounds within the city (Vanka 2014). The majority of IT parks were located in the periphery, reducing the total area of the green belt to 20.7% (Ravindra 1995).

These socio-economic processes decreased the green cover of the city during this phase of urbanization.

4.3.1.5 Vegetation During Later IT Phase

During this phase, the green cover was 48% within the respective year built-up area. This improvement was revealed by the analysis of the 2013 NDVI image, both inside and in the periphery of the city. The reason for this change is the growth of plantations in the periphery and the development of gardens and small parks inside the city. Awareness was brought among the public about the significance of greenery in improving the quality of environmental conditions. Participation of the public and NGOs has resulted in the development and maintenance of neighborhood parks and gardens (Swamy 2013). Most importantly, lawns and landscaped gardens became an integral part of luxury apartments and villas, which were used as promotional elements by real estate developers and builders to sell their property and also to meet the aspirations of the urban elite, particularly IT professionals (Rai 2005).

The following sections discuss the metabolism of lawns in the city of Bengaluru.

4.4 Urban Lawn Metabolism

As mentioned earlier, urban metabolic processes alter city environments to give rise to new forms of nature (Swyngedouw 2006a, b). A part of these urbanization processes involves producing built environments with managed turfgrass areas in the form of parks, landscaped gardens, sports fields, etc. In an urban setting, lawn spaces are produced, maintained, and utilized through complex interactions and metabolic processes including socio-economic processes as well as political and environmental processes (Robbins 2007). Globalizing cities such as Bengaluru are a hub of economic activities that are spatially distributed to form contemporary urban landscapes that include well-maintained lawn spaces (Stallmeyer 2006). In spite of lawns being a common urban green space in cities around the world, most studies, till now, have focused on lawns in the Global North (Ignatieva et al. 2015). In the cities of the Global South such as Bengaluru, lawns have been an understudied component of urban green space (Dutt 2017).

This section examines the spatial distribution of lawns at a city-wide scale in Bengaluru. Being an important component of urban green space, understanding lawn space distribution contributes to the overall aim of the chapter in understanding the effects of urban metabolic processes on the changes in vegetation across the city. Specifically, this section presents a survey of lawns across all wards in the city to find out the average proportions of lawns and their spatial distribution across the city's center and periphery. The results allowed for an understanding of lawn space transformation due to urbanization in the city during the last two decades.

4.4.1 Detection of Lawns—First Survey

Turfgrass lawn areas in the city require high-resolution imagery for detection and analysis (Zhou et al. 2008). This is mainly due to the nature of their physical disposition in the form of small parcel-sized patches except sports lawns such as golf fields which can span several kilometers. For the city of Bengaluru, in the absence of any previous investigation on the spatial distribution of lawns, and with the resources that were available, freely available satellite imagery, online factual information, and site visits were used to study lawns spatially. A multi-stage cluster sampling approach was used to first select three zones (consisting of 67 wards) randomly from a total of eight zones (198 wards) within the administrative ward boundary of the Bruhat Bengaluru Mahanagara Palike (BBMP). In these three zones, there was a selection of a proportionate random sample of wards based on an overall sample size of 20 wards (which is approximately 30% of the total of 67 wards in the three selected zones). For random selection, QGIS application software was used. Next, after being georeferenced and corrected for projection, the Bruhat Bengaluru Mahanagara Palike (BBMP) ward map was added as a vector layer over Google Earth Pro (GEP) satellite imagery. Within each of the 20 wards that were selected, all lawn spaces were identified by applying 275 × 275 m sampling grids making it possible to cover every area of a ward. The lawn sites that were detected in GEP imagery were cautiously verified again through Google Maps imagery, official website information, and Google images in a period from March 2020 to June 2020. Information from prior field visits in a period before 2020 was also used wherever available.

4.4.2 Lawn Spaces in the City

Lawns are ubiquitous in the city of Bengaluru and are present in various establishments. The first survey showed that the number of lawns in a single ward can range from zero to as many as forty lawn spaces in different forms and sizes. Thus, certain wards have no lawn space at all while other wards have a considerable amount of lawn space within them. The wards without any lawn space appear to have a higher degree of compaction and urban density and lesser vegetation, in general, than the wards which have some degree of green space—whether this is in the form of lawns, or any other vegetation like trees. Lawn sizes across the city vary from small patches like street medians covering a few meter squares to large spaces like golf courses covering several kilometers. Further, lawns in the city can be categorized according to their utility. For example, there are institutional lawn spaces constituted by establishments such as parks, educational and research institutions, government administrative buildings, and defense and health establishments. There are also residential lawn spaces constituted by individual houses, apartments, and villa complexes. Then, there are office lawn spaces composed of corporate campuses and office buildings of companies. And then, there are commercial lawn spaces constituted by businesses encompassing hospitality, retail, and so on. Finally, there are sports lawns mainly comprised cricket, golf, and football grounds. Each of these different lawn spaces and establishments is connected to how the city has developed over the years through its urbanization process. For instance, in the more recent phase of IT development, there has been a proliferation of well-managed lawn spaces in corporate office campuses and gated residential complexes and mixed-use development projects constituting office, residential, as well as commercial spaces (Dutt 2017). Also, different types of lawns entail different levels of resource use and management (Townsend-Small et al. 2011). For example, sports lawns like golf and cricket require high maintenance levels, and hence, use a higher amount of resources such as water and fertilizers for their maintenance and upkeep. In contrast, individual home lawns may not require the same level of intensive resource use and management as that seen in sports lawns.

The intensity of lawn management has the ability to affect environmental processes like oxygen regeneration, soil fertility, carbon sequestration, and in general, surrounding air and water quality (Robbins et al. 2001; Dutt 2017; Dutt and Tanwar 2020). This points toward to the need to make responsible decisions during urban planning and policy deliberations in order to optimize their environmental benefits (for example, carbon sequestration) and reduce adverse impacts (for example, groundwater pollution), being mindful of the various metabolic processes that underlie the production of such spaces in the city (Bhadouria et al. 2022). Overall, the first survey characterized lawn spaces in Bengaluru as a carefully managed form of urban green space present in various forms and sizes and exhibiting varying intensities of resource use depending on their utility. The production of such lawn spaces in the city can be associated with the type of establishments that have emerged at various intervals during the urbanization process in Bengaluru brought about by various metabolic processes that included socio-economic, political as well as spatial metabolic processes.

4.4.3 Surveying Lawn Proportions Across the City—Second Survey

Using the results from the first survey in 20 wards, random sampling points were generated across all 198 wards within the BBMP ward boundary using QGIS random points tool (Figs. 4.1 and 4.2). From each random point, the closest lawn space within a radius of 600 m was selected according to the identification process described earlier. Lawn selection was done according to the five major categories of lawn establishments identified in the first sampling stage—institutional, residential, office, commercial, and sports. Within the residential category, individual home lawns were excluded from the second survey as they were difficult to verify with absolute certainty due to the absence of secondary information from online sources and official websites. Not all random points generated in a ward were found to have a lawn space nearby. Random sampling points with no lawns within the 600 m radius

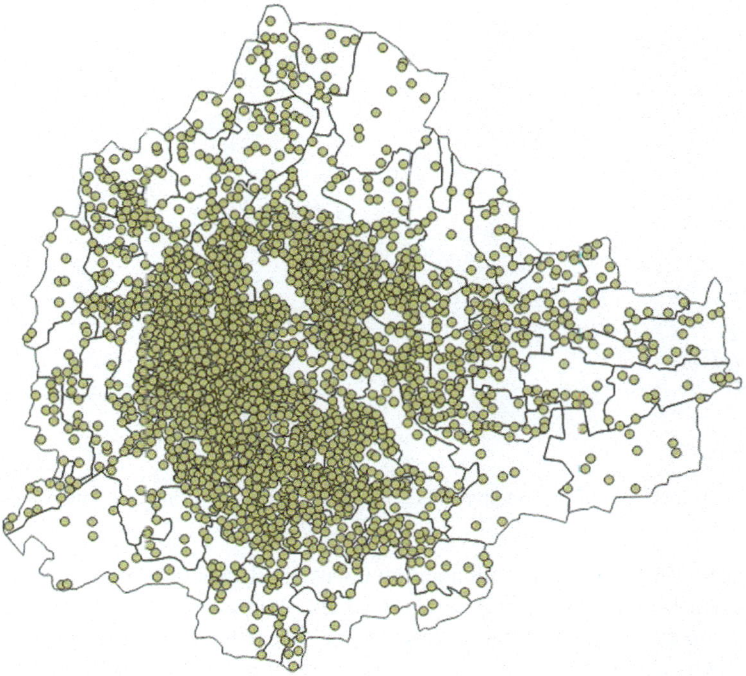

Fig. 4.1 Random point generation across all wards in the BBMP ward boundary using QGIS mapping software

were marked as such in order to note down no-lawn point locations. Thus, the final dataset comprised 2369 random sampling points is divided into sampled lawn points and no-lawn points. This allowed for a spatial study of average proportions of all five lawn categories across the 198 wards in the city in terms of when they had emerged (year of lawn establishment) and where they were located (spatial coordinates).

4.4.4 Spatial Distribution of Lawns in the City

Urban growth patterns have influenced the spatial distribution of lawns in Bengaluru. Studies show that the spatial expansion of the city is defined by unplanned and uneven sprawl patterns with an increase in built-up area from the core toward the periphery (Sudhira et al. 2007; Ramachandra et al. 2012). Post-liberalization, there has been rapid urbanization in the periphery, particularly in the last two decades, due to a lack of space and high land prices in the center (Goldman 2011; Nagendra et al. 2012; Sudhira and Nagendra 2013). The lawn spaces found in the city center and periphery reflect this pattern of urban spatial expansion. Lawns were found in 157 wards out of

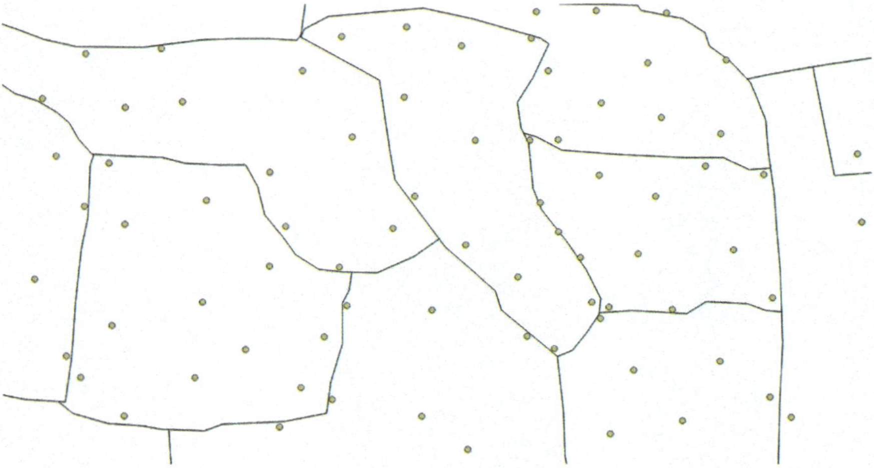

Fig. 4.2 Zoomed-in view of random points generated within wards using QGIS mapping software

198 wards, while 41 wards contained no lawns at all. About 65% of the wards that did contain lawns were central wards. Of these central wards, nearly fourth-fifths contained lawns in proportions not exceeding 30% (Fig. 4.3). Thus, more than half of the lawns that were found were in the center; these lawns constituted not more than 30% of the individual wards that they were found in.

Among the lawn categories that were identified, lawns in the institutional category were spread across 115 wards; this was followed by residential (72 wards), office (40 wards), commercial (35 wards), and sports (9 wards) lawn categories. About 67% of the wards containing institutional lawns were found to be central wards. Of these central wards that contained institutional lawns, almost half of these wards contained institutional lawns in proportions >90–100% of the lawns found in the ward (Fig. 4.4). That is, almost all the lawns found in these wards were in campuses that were part of institutions. There are many institutional spaces in the central wards of the city that have managed to preserve their lawn spaces over time even as urbanization started to gradually shift to the periphery; these include spaces such as educational establishments, parks, defense and military establishments, administrative buildings, heritage sites. Thus, institutional lawns are the largest category of lawns that are mostly located in the center and are also the most dominant type of lawns found at the center.

In the case of the residential lawn category, 61% of the wards containing residential lawns were found to be peripheral wards. Of these, nearly two-fifths contained residential lawns in proportions >30–50% of the lawns are found in the ward (Fig. 4.5). Peripheral urbanization in the last two decades has affected peripheral vegetation change patterns in such a way that they reflect the kind of development extolled by real estate residential and commercial properties and IT development, which mostly comprises water-hungry ‘exotic’ species such as the *Eucalyptus* plantations

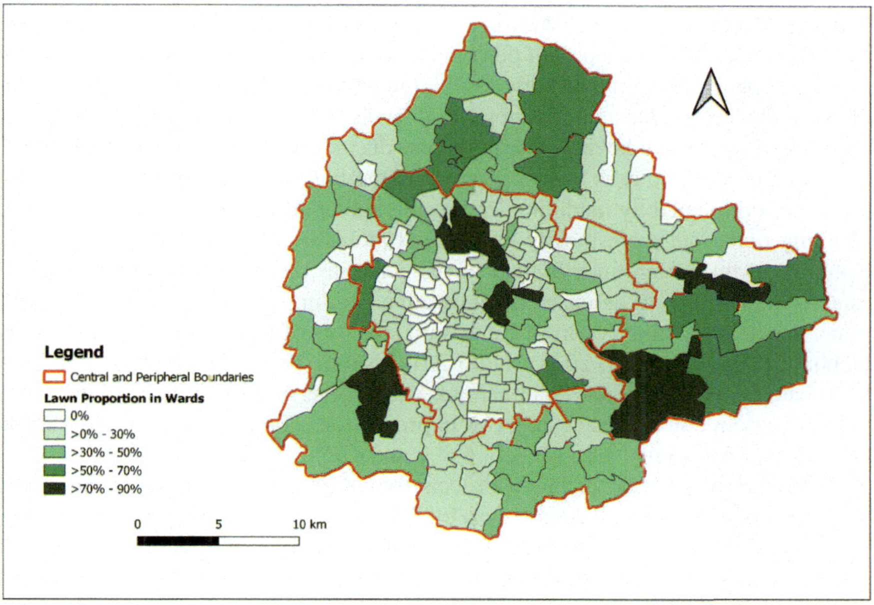

Fig. 4.3 Proportions (%) of lawns in all wards within the BBMP ward boundary

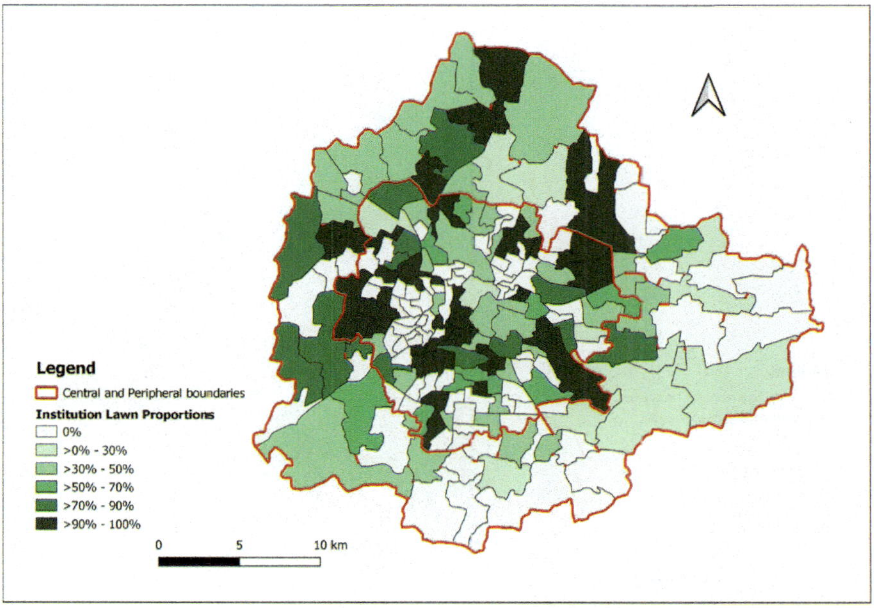

Fig. 4.4 Proportions of institutional lawn category within lawns in a ward

(Nagendra et al. 2012; Nalini 2020). Therefore, in the survey, it was found that residential lawns, which are a part of this kind of urban development, are mostly located on the periphery. And, in their turn, residential lawns are the second largest category of lawns that are the most dominant form of lawns found in the periphery. The mushrooming of residential lawns in the periphery due to urbanization associated with IT development in the city is also supported by the finding that peripheral residential lawns are dominated by medium-sized residential apartments with 100–500 family dwelling units that were mostly established after 2012 by real estate developers. Contrary to this, many institutional lawns came up before 2000 and were part of establishments such as defense, administrative buildings, parks, educational institutions, which were built as part of Bengaluru's phases of industrialization before the IT phase (Nagendra et al. 2012; Nalini 2016; Nalini 2020). In the case of all other lawn categories, i.e., office, commercial, and sports, more than half of the wards did not contain any of these categories in proportions greater than 30%–50% of the lawns found within the ward (Figs. 4.6, 4.7 and 4.8).

Thus, the distribution of lawns reflects the urban growth pattern of the city. It becomes important to understand, therefore, the type of establishments' lawn spaces which are associated with in order to understand the type of greening they extoll, and by extension, the type of urban green space that gets manifested on the ground. It is worth noting that many of the lawns in the survey were in the form of clusters of lawn spaces belonging to different establishments. Further, citing the limitations of the survey, a significant void in the dataset was the exclusion of major software

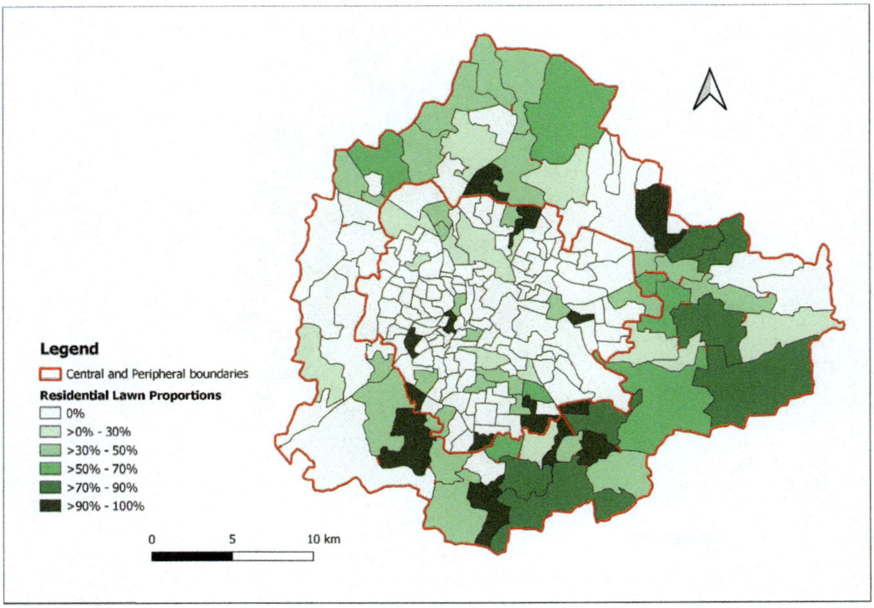

Fig. 4.5 Proportions of residential lawn category within lawns in a ward

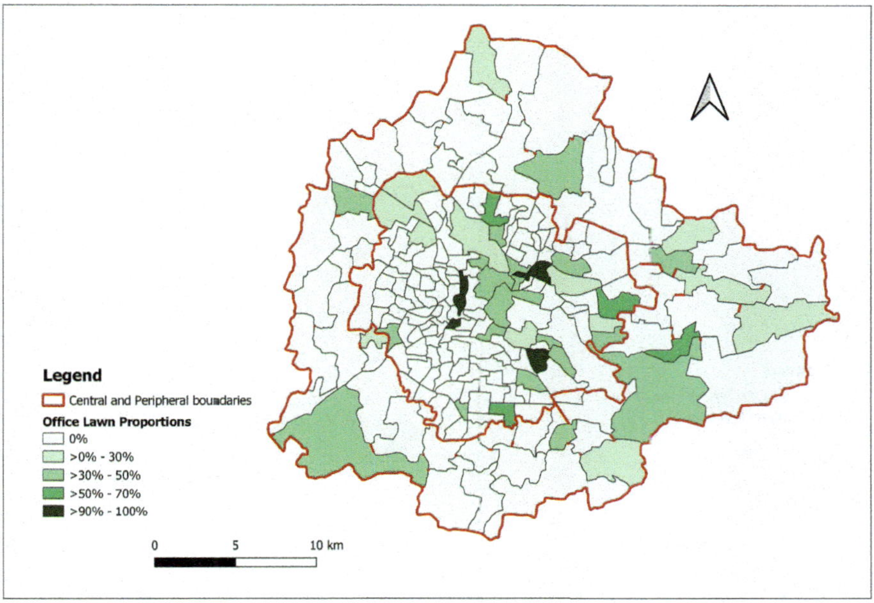

Fig. 4.6 Proportions of office lawn category within lawns in a ward

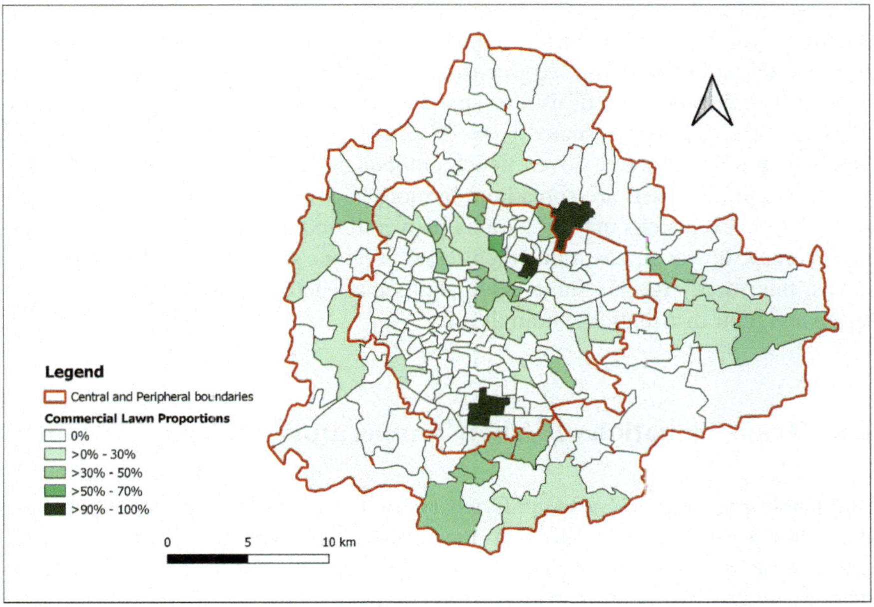

Fig. 4.7 Proportions of commercial lawn category within lawns in a ward

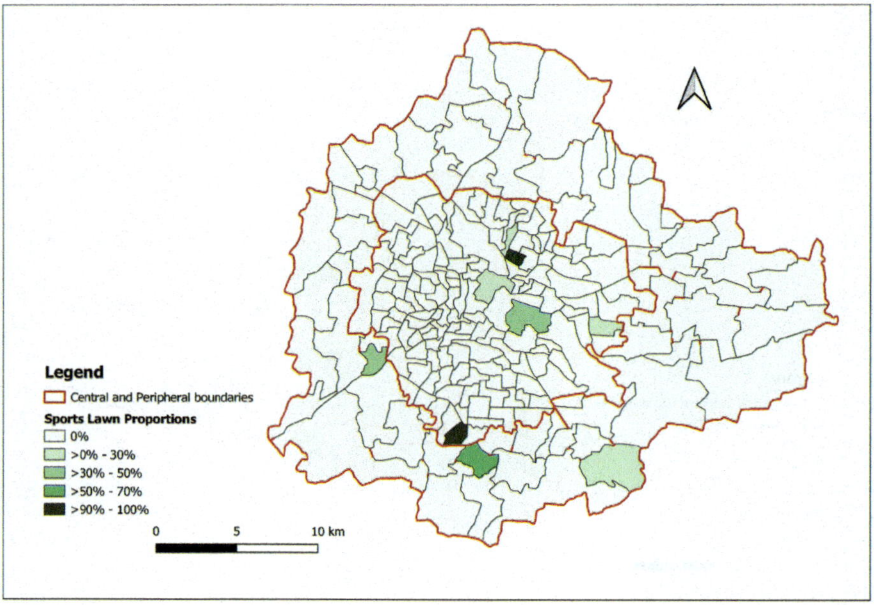

Fig. 4.8 Proportions of sports lawn category within lawns in a ward

development zones such as Electronics City which fall outside of the BBMP ward boundary. Since the IT industry signifies Bengaluru's urban growth trajectory in recent times harboring definitive green landscapes mirroring global-scale enterprises, it points out the necessity of using a spatial analysis frame that covers a larger area of study than the one presented here. Also, the exclusionary criteria of sampling lawns with no visible tree cover, which automatically filters out lawns under trees could also imply lawns are spread out in more locations than were detected in this study. These exclusions may moderate the overall findings from the survey, limiting its accuracy; however, given that there is no previous study on the spatial aspect of lawns, this survey reveals some key insights that shed light on the nature of lawn space and their distribution within the city.

4.5 Transformation of Urban Temperature Pattern

The name 'Garden City' was conferred to Bengaluru due to the presence of large trees, home gardens and gardens in the institutions, open spaces, parks, roadside trees, and greenery inside industrial townships until the 1970s–1980s. These green patches along with ponds and lakes helped in maintaining the soil moisture. Studies have shown that soil moisture is a very important factor to control the temperature patterns of the city (Nalini 2020). Industrialization and rapid urbanization due to garment industries and the IT boom during the 1980 and 1990s resulted in the loss of the majestic green spaces of the city. Many water bodies were encroached on to meet the demand for housing and real estate. Many other water bodies got polluted due to the uncontrolled release of sewage water and industrial effluents into them. The loss of green spaces and water bodies reduced the soil moisture, and thereby, the temperature of the city increased considerably (Nalini 2020). However, the public was made aware of the need for lung space and the aspiration of the citizens to live in close proximity to nature inspired builders and authorities to develop small parks, gardens, lawns, and terrace gardens. Economic benefits inspired the development of plantations in the peripheral regions. Though these initiatives improved greenery, the species of plants and trees that developed were more detrimental to the ecosystem than they were in providing benefits. The species were water-intensive and fast-growing varieties which added to the groundwater depletion that was already observed due to socio-economic activities. Thus, improvement in green cover was not beneficial to increasing the soil moisture content. Thus, the salubrious climate of the city which conferred the name 'Air-conditioned city' is now experiencing hot summers with high temperatures recorded every year (Citizen Matters 2021).

4.6 Conclusion and Future Recommendations

Urbanization has transformed Bengaluru's urban landscape. Over the years, Bengaluru has grown into a burgeoning metropolis with an ever-expanding boundary enabling a host of socio-economic, political, and spatioenvironmental interactions to take place. As urbanization occurred, the city's growth became influenced by metabolic processes of social, political, spatial, and economic development. Urban planning and policy implementation became challenging in the face of uneven development experienced by urban inhabitants, impacting their health and well-being. The gradual expansion of human activities across spatial boundaries has simultaneously affected urban vegetation in accordance to the phases of the urbanization experienced by the city, namely, the public sector, garment sector, and IT sector phase. Together, these urbanization phases had a diverse impact on the city's green spaces.

While the sociopolitical climate leading up to and during the colonial phase encouraged greenery inside residences and along streets, exclusionary buffer regions like Cubbon Park also existed and shaped the city's environment. These are some of the most definitive green spaces in Bengaluru till date. The period of the public sector industries, in turn, contributed to establishing lush gardens and parks meticulously planned and laid out into orderly neighborhoods embedded within industrial townships. The recognition of Bengaluru as the 'Garden City' of India is thus emerged alongside its industrial expansion. The influence of metabolic processes on the city's changing vegetation cover and its patterns were thus already apparent in these built-up processes before the IT phase. With the liberalization of the economy and the emergence of Bengaluru as an IT hub, globalizing processes rendered another phase of vegetation change and transformation during the 1980s–1990s. Urban planners and policymakers favored foreign investments and real estate development. The coming together of social, political, and economic processes meant that the city's vegetation changed according to the nature of development that was encouraged. In this period, urbanization led to a considerable transformation in vegetation with an overall reduction in greenery, conversion of peripheral areas into urban developments, and increased concretization in the center with rising land prices. A prominent trend during the later IT phase was the development of lush green lawn landscapes in coveted real estate development projects like gated residential communities and business parks.

The metabolism of lawns in the city is an important aspect of urban vegetation change and transformation. Bengaluru's lawn spaces are many and varied, representing a managed form of urban green space that has until recently been neglected in scholarly inquiry. The management of lush green lawn spaces in public and private establishments forms an integral part of urban planning and policy decisions that have the ability to affect the biophysical environment both beneficially and adversely. Thus, lawn management decisions bring together both human and non-human actors at the intersection of social and political metabolic processes that also influence their spatial distribution across the city. Urban growth strategies that are favored by urban planners and policymakers determine green space development in accordance with global trends in the production of contemporary urban landscapes; lawn landscapes are a part of such trends. That is, they are mostly part of landscaping efforts that help to create esthetically pleasing environments. Being an understudied component of urban green space in Bengaluru, the complexities of different metabolic processes that produce lawns as an end-product in homes, offices, and other designed landscapes need to be understood better. These overall transformations in greenery have indirectly brought about transformations in temperature pattern, which also needs to be looked at thoughtfully.

As part of future endeavors to build more just and sustainable cities, urban policymakers, decision-makers, researchers, planners, and other relevant stakeholders should keep in mind that:

- Urban metabolic processes and urban ecological processes are interlinked, and that urban planning processes should look at these linkages to understand their complex interactions to plan cities better.
- Apart from economic processes, urban metabolisms include social, spatial, and environmental processes which should be considered together while developing blueprints for urban expansion. There should be focused efforts on ensuring that the benefits of urban green space planning and architecture are accessible to all sections of society.
- The urban metabolism framework provides an opportunity to understand and apply integrated knowledge gained from studying different socioeconomic processes that enable a holistic approach toward achieving the health and well-being of both humans and ecosystems.

Acknowledgements The study on lawn spatial distribution is based on research carried out with funding support from the National Institute of Advanced Studies, Bangalore, as part of a doctoral research project.

References

Adkin RK (2009) India's 'Garden City'? Bangalore's Disappearing Greenbelt: Green To Gone. Thesis in Master of Arts in Geography. University of Delaware.
Alier JM (2009) Social metabolism, ecological distribution conflicts, and languages of valuation. Capital Nat Soc 20(1):58–87
Allen A, Lampis A, Swilling M (2016) Why untamed urbanisms? In: Allen A, Lampis A, Swilling M (eds) Untamed urbanisms. London and New York, Routledge, pp 1–15
Asian Development Bank (2014) Urban metabolism of six Asian cities. Metro Manila, Asian Development Bank, pp 1–78
BDA (2007) Bangalore master plan—2015. Vol. I. Government of Karnataka, Bangalore
Bahers JB, Athanassiadis A, Perrotti D, Kampelmann S (2022) The place of space in urban metabolism research: towards a spatial turn? A review and future agenda. Landsc Urban Plan 221:104376
Bhadouria R, Upadhyay S, Tripathi S, Singh P (eds) (2022) Urban ecology and global climate change. Wiley-Blackwell.
Broto VC, Allen A, Rapoport E (2012) Interdisciplinary perspectives on urban metabolism. J Indus Ecol 16(6):851–861
Census of India (2011) Provisional population totals: urban agglomerations and cities. New Delhi, Ministry of Home Affairs, Government of India
Chrysoulakis et al. (2013). Sustainable urban metabolism as a link between bio-physical sciences and urban planning: the BRIDGE project. Landscape and Urban Plann 112:100–17
Citizen Matters (2021) Available at https://bengaluru.citizenmatters.in/urbanisation-high-land-surface-temperature-65541#:~:text=India's%20urban%20population%20went%20up,%E2%80%94%20a%20whopping%2024%25%20increase!. Accessed on 5 Aug 2022
Das B (1981) Urban planning in India. Social Scientist. 9(12):53–67
Dear M (2002) Los Angeles and the Chicago School. City and Commun 1(1):5–32
Dijst M, Worrell E, Böcker L, Brunner P, Davoudi S, Geertman S, Zeyringer M (2018) Exploring urban metabolism—towards an interdisciplinary perspective. Resour Conserv Recycl 132:190–203
Dossal M (1989) Limits of colonial urban planning: a study of mid-nineteenth century Bombay. Int J Urban Reg Res 13(1):19–31
Dousset B, Gourmelon F, Laaidi K, Zeghnoun A, Giraudet E, Bretin P, Vandentorren S (2011) Satellite monitoring of summer heat waves in the Paris metropolitan area. Int J Climatol 31(2):313–323
Drakakis-Smith DW (2000) Third world cities. Routledge, London
Dutt N (2017) Being green?. Seminar 694:50–54
Dutt N, Tanwar T (2020) Nitrous oxide emissions from turfgrass lawns as a result of fertilizer application: a meta-analysis of available literature. Curr Sci 118(8):1219–1226
Fisher DR, Freudenburg WR (2001) Ecological modernization and its critics: assessing the past and looking toward the future. Soc Nat Resour 14:701–709
Gandy M (2006) Urban nature and the ecological imaginary. In: Heynen N, Kaika M, Swyngedouw E (eds) In the nature of cities: urban political ecology and the politics of urban metabolism. Routledge, London and New York, pp 62–72

Giezen M, Roemers G (2015) The metabolic planner: reflection on urban planning from the perspective of urban metabolism. In: Savini S, Verschuuren S, Salet WGM, Raats K (eds) Urban metabolism. Universiteit van Amsterdam, pp 14–27

Girardet H (1990) The metabolism of cities. In: Cadman D, Payne G (eds) The living city: towards a sustainable future, 1st edn. Routledge. https://doi.org/10.4324/9780429197307. eBook ISBN9780429197307

Goldman M (2011) Speculative urbanism and the making of the next world city. Int J Urban Reg Res 35(3):555–581

Gopinath R (2014) Bangalore .... a baseline study. Chapter 2. In: Parametric study of heat Island effect due to urban sprawl in Bangalore. PhD thesis, Jawaharalal Nehru Technological University, Anantpur

Gowda R (2003) Working plan for Bangalore urban forest division. Karnataka Forest Department, Bangalore

Grimmond CSB, Blackett M, Best MJ, Barlow J, Baik J-J, Belcher SE et al (2010) The international urban energy balance models comparison project: first results from phase-1. J Appl Meteorol Climatol 49:1268–1292. https://doi.org/10.1175/2010JAMC2354.1

Heynen N, Kaika M, Swyngedouw E (2006) Urban political ecology: politicizing the production of urban natures. In: Heynen N, Kaika M, Swyngedouw E (eds) In the nature of cities: urban political ecology and the politics of urban metabolism. Routledge, London, pp 1–15

Heynen N (2014) Urban political ecology I: the urban century. Prog Hum Geogr 38(4):598–604

Harvey D (1970) Social processes and spatial form: an analysis of the problems of urban planning. Papers of the Reg Sci Assoc 25(1):47–69

Hajer MA (2016) On being smart about cities: seven considerations for a new urban planning and design. In: Allen A, Lampis A, Swilling M (eds) Untamed urbanisms. London and New York, Routledge, pp 50–63

Idrus S, Hadi AS, Shah AHH, Mohamed AF (2008) Spatial urban metabolism for livable city. In: Blueprints for sustainable infrastructure conference. Auckland, NZ. 9–12 December, pp 1–11

Ignatieva M, Ahrné K, Wissman J, Eriksson T, Tidåker P, Hedblom M, Bengtsson (2015) Lawn as a cultural and ecological phenomenon: a conceptual framework for transdisciplinary research. Urb Fores Urb Green 14(2):383–387

Iyer M, Nagendra H, Rajani MB (2012) Using satellite imagery and historical maps to investigate the contours of Lalbagh. Curr Sci 102:507–509

Karakeiwicz J (2011) Urban metabolism of low carbon cities. In: Conference proceedings of 47th ISOCARP congress 2011. Wuhan, China. 24–28 October, pp 1–13

Kennedy C, Cuddihy J, Engel-Yan J (2007) The changing metabolism of cities. J Ind Ecol 11:43–59. https://doi.org/10.1162/jie.2007.1107

Kennedy C, Pincetl S, Bunje P (2011) The study of urban metabolism and its applications to urban planning and design. Environ Pollut 159(8–9):1965–1973

Kennedy C, Baker L, Dhakal S, Ramaswami A (2012) Sustainable urban systems an integrated approach. J Ind Ecol 16(6):775–779

King AD (1989) Colonialism, urbanism and the capitalist world economy. Int J Urban Reg Res 13:1–18

KTCPA (2005) The Karnataka town and Country planning (Amendment) Act, 2004. Available at http://dpal.kar.nic.in/ao2005/01%20of%202005%20(E).pdf. Accessed on 19 Dec 2017

Lawton P (2016) Beyond an imaginary of power? Governance, supranational organisations and 'just' urbanisation. In: Allen A, Lampis A, Swilling M (eds) Untamed urbanisms. London and New York, Routledge, pp 204–215

Madalasa V (2014) Analysing urban growth boundary effects on the city of Bengaluru. Econ Pol Wkly 49(48):54–61

Mani A (1985) A study of climate of Bangalore. In: Vyasulu V, Reddy A (eds) Essays on Bangalore. Karnataka State Council for Science and Technology, IISc, Bangalore, vol 2. pp 1–17

Mukherjee J (2016) Sustainable flows between Kolkata and its peri-urban interface: challenges and opportunities. In: Allen A, Lampis A, Swilling M (eds) Untamed urbanisms. London and New York, Routledge, pp 33–49
Musango JK, Currie P, Smit S, Kovacic Z (2020) Urban metabolism of the informal city: probing and measuring the 'unmeasurable' to monitor sustainable development goal 11 indicators. Ecol Indic 119:106746
Nalini NS (2020) Urbanisation and changing patterns of temperature in the city of Bengaluru. Environ Developm Sustain 23:9090–9109 (2021). https://doi.org/10.1007/s10668-020-01014-3, ISSN 1387–585X Springer Link
Nagendra H, Gopal D (2011) Tree diversity, distribution, history and change in urban parks : studies in Bangalore India. Urban Ecosyst 14(2):211–223
Nagendra H, Nagendran S, Paul S, Pareeth S (2012) Graying, greening and fragmentation in the rapidly expanding Indian city of Bangalore. Landsc Urban Plan 105(4):400–406
Nagendra H (2016) Nature in the City: Bengaluru in the past, present and future. New Delhi, Oxford University Press
Nair J (2005) The promise of the metropolis Bangalore's twentieth century. Oxford University Press, New Delhi, New Delhi, India
Nalini NS (2016) The great divide: exploring the divergence between urban metabolism in theory and practice in water supply system in Bengaluru. Int J Urban Sustain Developm 9(1):1–20
Niza S, Rosado L, Ferr P (2009) Urban metabolism: methodological advances in urban material flow accounting based on the lisbon case study. J Ind Ecol 13(3):384–405
Pickett STA, Cadenasso ML, Grove JM, Nilon CH, Pouyat RV, Zipperer WC, Costanza R (2001) Urban ecological systems: linking terrestrial ecological, physical, and socioeconomic components of metropolitan areas. Annu Rev Ecol Syst 32:127–157
Pincetl S, Bunje P, Holmes T (2012) An expanded urban metabolism method : toward a systems approach for assessing urban energy processes and causes. Landsc Urban Plan 107:193–202
Rai S (2005) Indians find they can go home again. The New York Times, December 26. Available at http://www.nytimes.com/2005/12/26/business/worldbusiness/indians-find-they-cango-home-again.html. Accessed on 20 Nov 2016
Ramachandra TV, Kumar U (2010) Greater Bangalore : emerging urban heat Island. GIS Development, pp 1–16
Ramachandra TV, Aithal BH, Sanna DD (2012) Insights to urban dynamics through landscape spatial pattern analysis. Int J Appl Earth Obs Geoinf 18:329–343
Rao Sundara BN (1985) Bengalurina Itihasa. Bangalore, Vasantha Sahitya Grathalaya
Rapoport E (2011) Interdisciplinary perspective on urban metabolism: a review of the literature. UCL Environmental Institute working paper. London, Development Planning Unit, UCL
Ravindra A (1995) In: Metropolitan Bangalore: a management perspective. Kolkata, Times Research Foundation
Robbins P (2007) Lawn people: how grasses weeds, and chemicals make us who we are. Temple University Press, Philadelphia
Robbins P, Polderman A, Birkenholtz T (2001) Lawns and toxins: an ecology of the city. Cities 18(6):369–380
Schott D (2004) Urban environmental history : what lessons are there to be learnt ? Boreal Environ Res 9:519–528
Schremmer C (2009) Urban development and urban metabolism: a spatial approach. SUME-Sustainable Urban Metabolism in Europe (FP7), Deliverable D, 1.1, Vienna
Shivaramakrishnan KC (2014) Governance of megacities: fractured thinking, fragmented setup. Oxford University Press, New Delhi
Stallmeyer JC (2006) Architecture and urban form in India's silicon valley: a case study of Bangalore. University of California, Berkeley
Sudhira HS, Nagendra H (2013) Local assessment of bangalore: graying and greening in Bangalore–impacts of urbanization on ecosystems, ecosystem services and biodiversity. In: Urbanization,

biodiversity and ecosystem services: challenges and opportunities, Springer, Netherlands, pp 75–91

Sudhira HS, Ramachandra TV, Subrahmanya MB (2007) Bangalore. Cities 24(5):379–390

Swamy S (2013) Reshaping neighbourhood parks for biodiversity and people: a study on pocket green spaces in Bangalore, India. PhD thesis, Manipal University, Manipal, Karnataka

Swyngedouw E (2006a) Metabolic urbanization. The making of cyborg cities. In: Heynen N, Kaika M, Swyngedouw E (eds) In the nature of cities-urban political ecology and the politics of urban metabolism, pp 20–39

Swyngedouw E (2006b) Circulations and metabolisms: (Hybrid) Natures and (Cyborg) cities. Sci Culture 15(2):105–121

Townsend-Small A, Pataki DE, Czimczik CI, Tyler SC (2011) Nitrous oxide emissions and isotopic composition in urban and agricultural systems in southern California. J Geophys Res: Biogeosci 116(G1)

Udvardi M (2014) Bangalore : urban development and environmental injustice. Wellesley College. Massachussets, USA, Honors Thesis Collection, Paper 176

Ulgiati S, Zucaro A (2019) Challenges in urban metabolism: sustainability and well-being in cities. Front Sustain Cities 1:1

Vanka PS (2014) In: Public space and life in an Indian City : the politics of space in Bangalore. PhD thesis submitted to Urban and Regional Planning in University of Michigan

Wolman A (1965) The metabolism of cities. Sci Am 213:179–190

United Nations (2022) Department of Economic and Social Affairs, Population Division: World Urbanisation Prospects, the 2022 revision. https://population.un.org/wpp/. Accessed on 5 Aug 2022

Zhang Y, Yiyun C, Qing D, Jiang P (2012) Study on urban heat island effect based on normalized difference vegetated index: a case study of Wuhan City. Proc Environ Sci 13:574–581

Zhou W, Troy A, Grove M (2008) Modeling residential lawn fertilization practices: integrating high resolution remote sensing with socioeconomic data. Environ Manage 41(5):742–752

Chapter 5
City Core and Urban Sprawl

Shivangi Singh Parmar, Vidhu Bansal, Sunny Bansal, Nazish Abid, Joy Sen, and Mazharul Haque

Abstract Cities around the world occupy 2% of the area of the globe, and are accountable for about 80% of the consumption of all the energy produced by Earth. Within a few decades, this number will exponentially increase. These expanding city cores lead to a consequent demand for the service sector, which generally exists in the city's peripheral areas. Also, this phenomenon of rapid urbanization leads to huge out-flows and in-flows of energy in terms of materials, services, and waste. This in-flow and out-flow of energy is concentrated locationally and spills over the period of time. In other words, the subsequent growth of cities and its spill-over development leads to uncontrolled growth on the periphery of the city (peri-urban areas), which can be termed urban sprawl. The concept of urban metabolism is crucial to understand the in-flows and out-flows in the uncontrolled development of the urban sprawl. The core idea of urban metabolism is to understand a city as an organism and react to the various material flows for efficient resource utilization. The symbiosis of 'urban core' and 'peri-urban areas' lies in the complementarity of dependence on material resources. An example of such can be: the out-flow of waste material from the dense city center used to be toward outer boundaries of the city or urban sprawl, whereas the in-flow of food materials from urban sprawl area to the city core. Hence, it becomes pertinent and of utmost importance to understand the networks of

S. S. Parmar (✉)
RCG School of Infrastructure Design and Management, IIT Kharagpur, Kharagpur 721302, India
e-mail: shivangisp9@gmail.com

School of Architecture, VIT Vellore, Vellore 632014, India

V. Bansal · J. Sen
Department of Architecture and Regional Planning, IIT Kharagpur, Kharagpur 721302, India

V. Bansal · S. Bansal
School of Architecture, Anant National University, Ahmedabad, Gujarat, India

N. Abid
Department of Architecture and Interior Design, College of Engineering, University of Bahrain, Zallaq, Bahrain

N. Abid · M. Haque
Department of Architecture and Planning, NIT Patna, Patna 800005, India
e-mail: mazharul@nitp.ac.in

R. Bhadouria et al. (eds.), *Urban Metabolism and Climate Change*,
https://doi.org/10.1007/978-3-031-29422-8_5

heterogeneous material flows in a city that is its 'Urban Metabolism'. These networks of heterogeneous material flows are intertwined with not only the energy flows but also the economics involved in it, hence bestowing the key role to urban metabolism as a catalyst of various processes responsible for the efficient working of the city. Urban metabolism also becomes a prerequisite to understanding resource allocation, which is the basis for sustainable development. Hence, the present study will examine the understanding of urban metabolism in the context of peri-urban development or urban sprawl. The study would look into multiple cases to base its understanding and develop recommendations for the sustainable development of an urban ecosystem.

Keywords Climate change · Heterogeneous flow · Peri-urbanization · Sustainable development · Urban metabolism · Urban sprawl

5.1 Introduction

The lure of opportunity and prosperity brings people to cities and leads to rapid urbanization. More than half of the world's population is urban in nature. This share will be two-thirds by 2050 (Bansal et al. 2017). Rapid urbanization has a lot of ill effects associated with it, like congestion, poverty, environmental degradation, etc. However, urbanization has many underlying benefits. These benefits include economic activities, better medical and education facilities, technological and infrastructural advancements, improved standard of living, etc. But these benefits may not balance against the costs, especially the environmental ones. Climate change, depletion of fossil fuels, and rising pollution strongly suggest the need for a more efficient and sustainable way of management of resources (Ahuja and Tatsutani 2009).

The solution to the urban problems being faced may not be merely a technological one. Urban planning, design, and infrastructure management have a lot to do with these problems and sustainable resource management. A major concern for planning in urban areas is the uncontrolled growth on the city's periphery, known as peri-urban area or urban sprawl. These urban or suburban sprawls have a unique relationship of mutual dependency with the urban center. Several material flows, service linkages, and networks exist between both. A new approach is needed to understand the concept of urban metabolism for the holistic sustainable development (Mensah 2019).

The ability to utilize resources efficiently and tackle environmental degradation due to human activities at the local level shall be considered first. Further, how the resources and materials flow within the urban sprawl and between the urban center and the sprawl can reveal the influence urban dwellers have on resource and material metabolism. Thus, to decipher the relationship between urban sprawl and urban center, the concept of urban metabolism is relevant for the attainment of resources and waste release with the provision of social services and economic outputs (Kennedy et al. 2011).

It is important to understand the concept and functioning of urban metabolism for the uncontrolled development of the urban sprawl, and its linkage with urban centers is quite important (Tan et al. 2018). While most research focuses on the urban center or city, the least attention is given to the peri-urban or sprawl. Decoding its metabolism is highly relevant as many interactions occur between the city and its peri-city. The current research attempts to address this gap and highlights the urban metabolism in light of urban sprawl. It attempts to study the relationship between the urban center and urban sprawl by applying the metabolism concept.

The present chapter is divided into seven broad sections. The first section introduces the research by presenting a general background framework and the need for the research. The second section explores and defines the concept of urban metabolism. The third section describes urban sprawl and highlights the evolutional relation of urban sprawl and its urban center. In the fourth section, the heterogeneity in the flow of the systems in urban sprawl is discussed with the help of case studies of the Delhi Metropolitan region. The understanding of the urban sprawl metabolism for sustainable development is portrayed in the fifth section with the help of urban infrastructure and urban sprawl metabolism framework. Based on the research, certain recommendations are presented in the sixth section. Finally, concluding remarks and the future scope of the study are discussed in the seventh section.

5.2 Concept of Urban Metabolism

Various disciplines have used the notion of urban metabolism to describe urban phenomena over the past and present centuries. The spatial dimension appears to be gaining importance in the interdisciplinary topic of urban metabolism. The analogy of urban center as living organism is present in literature for more than ten decades. The antecedents of the idea of urban metabolism are contested in various literature. According to Lederer and Kral (2015), the concept dates to Theodore Weyl's 'Essay on the Metabolism of Berlin,' published in 1894. However, (Wolman 1965) seminal article on a hypothetical US million city is associated with its larger applicability to understanding and monitoring city metabolism. The concept of urban metabolism has recently been employed as an analytical tool to understand the in-flows and out-flows of energy (Fischer-Kowalski 2002). The urban metabolism concept is widely used to understand ecology of chemical industry (Bai 2008) and nature-based design process (Benyus 2002), life cycle assessment of materials (McDonough and Braungart 2002), and design process of regeneration (Lyle 1994). The premise would be that if we needed sustainable ecosystem, ecology must be the paradigm for technological growth.

Though metabolism was initially implied to the life processes, it was expanded in 1935 by pioneer ecologist Arthur Tansley to describe the in-flows and out-flows of the cities. As if the city were an ecosystem, urban metabolism is a modeling concept for various in-flows and out-flows. This technique enables the study of city dynamics beyond the conventional settings of built-unbuilt environment (Newman et al. 2009). Each element of the city and its region are the factors of the urban metabolism modeling (Reynolds et al. 2017).

Golubiewski (2012) contrasted the metabolism of cities to that of life science processes. The analogy of organism reflects the age-old and unidirectional concept of metabolism. Most materials are transported to the city core from the urban sprawls (Bai et al. 2017). Urban sprawls are extremely exploited to fulfill the demands of the city core (Agudelo-Vera et al. 2012). City core transports waste to the urban sprawl, majorly used as service backyard of the core. Urban sprawls are thus vulnerable due to unbalanced and unidirectional transfer of resources and wastes to-and form the core. As defined by Doughty and Hammond (2004), circular metabolism efficiently distributes the resources for the sustainable development. Thus, the concept of circular metabolism sketches the optimistic picture of urban sustainability. However, certain social scientists have challenged it since it overlooks the pre-conditioning of living organisms to the change in social environment than natural (McDonald and Patterson 2007). Rather than unchangeable natural laws, societal standards have the greatest influence on human behavior. It is critical to plan city as a living organism because of complexity as shown in Fig. 5.1 (Lucertini and Musco 2020).

Although sociological theories of urban metabolism have exposed societies' irrationality regarding critical resources like water, nutrients, and others, there is only one hope that human settlements can adapt to changing state of the environment. Humans, unlike all other organisms, are conscious of their behaviors and can adapt their behavior. Users' participation in the environment's design, construction, and administration help society attain 'flexibility' (Timmeren 2014).

Modern life has been based on the luxuries than the actual needs which challenges the concept of sustainability of the resources and systems (Timmeren 2014). The following approach of development is having unidirectional flows of the materials and energy. Most people do not want forced, top-down societal control as

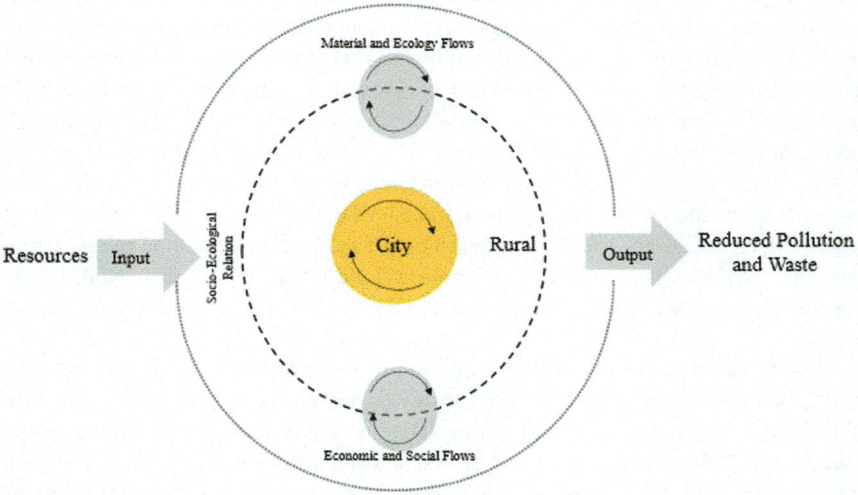

Fig. 5.1 Circular Metabolism Concept (Redrawn from Lucertini and Musco 2020)

an environmental technocracy/ dictatorship. Everyday practices will have to shift. Users' engagement in this development process. The energy crisis of 1973 ignited the concept of sustainability and awareness toward better use of energy (Timmeren 2013).

Meanwhile, there is a sharp distinction between the multiple stakeholders, as well as the different segments, when it comes to solutions for issues such as renewable energy. The energy policies emerge around the globe after the 1973 energy crisis. Many energy-based initiatives started, particularly in the developed nations than developing through inclusion of technologies (Timmeren 2014).

The quality of urban metabolism is inextricably tied to infrastructural morphology. Without any changes in behavior, the city's infrastructure will substantially impact residents' environmental footprint. In the Netherlands, for example, power is primarily derived from fossil fuels, whereas in Norway, electricity is primarily derived from hydroelectric sources. As a result, infrastructure suppliers, not customers, define the ecological footprint and match the 'environmental assessment' criteria. The consequent design of infrastructure influences citizen behavior, increasing the quality of urban metabolism, and the form of transportation affects the city's density. The geographical location of a resource and material supply to the city core determines the energy out-flow or in-flow or urban metabolism (Nelson 2010).

5.3 Evolution and Relationship of Urban Sprawl and the Accompanying Urban Center

Urban or suburban sprawl is generally defined as a low-density and unplanned development, dependent on automobiles, spiraling out from the city or the urban center (10.5176/2301-394X_ACE18.68). It is generally caused due to weak planning laws and to provide accommodation to the ever-growing urban population. It is also an outcome of affordable land prices, improved infrastructure, and the desire for increased living space and other leisure amenities that the urban center cannot provide. Generally, sprawl leads to several negative impacts like pollution, inequality, congestion, and a decline in community cohesion. It also destroys natural areas and wildlife habitats and threatens agricultural lands (Zhang 2021). However, there are advantages to urban sprawl, too, such as economic growth, employment opportunities, and better living conditions and lifestyles. Some researchers argue that urban sprawl is not completely unsustainable; rather, it is a dynamic process of deconcentrating the urban center. Through this process, a new urban structure evolves within the metropolitan area. Thus, it becomes extremely important to understand how the relationship between the urban sprawl and its accompanying urban center evolves and slowly attains a form of mutual dependency (Paramasivam and Arumugavelu 2020).

While the urban centers receive the major attention of planning and development, there is an urgent need to plan and develop the urban sprawl with the same effort.

Also, a mutual dependence exists between the city center and its urban sprawl. The sprawled suburban areas provide accommodation facilities to the people who keep the city centers working and alive (Paramasivam and Arumugavelu 2020). These suburbs provide residences not only to the high-end employees who desire to live in big houses with green spaces and less pollution but also to the low-income workers who cannot afford the high housing costs in the city cores. Thus, these sprawled suburbs become highly important for the urban center's income flows and economic viability (Gemenetzi 2017).

Similarly, the city centers provide job opportunities and social networks to the metropolitan area. The current technological advancements in transportation and communications have led to increased urbanization in the suburban region leading to sprawl. These urban sprawls or 'edge cities' are not only limited to residential spaces but have large clusters of corporate offices, retail, medical, education, and entertainment facilities. Much of the suburban population also gets employed in suburban jobs and services and loses direct connections to the city center. However, a strong linkage exists between the urban center and its sprawled suburb, as described below (Hill et al. 1995; Thomson and Newman 2018):

- The image of the urban center directly influences the prospective choice of future investments for the urban or suburban sprawls. The urban center's past and present performance decides the sprawl's future potential. Thus, the prospects of the urban sprawl's economic prosperity directly link with the urban center's reputation.
- Urban centers often provide metropolitan areas with many crucial amenities that cannot be duplicated easily, such as museums, government offices, civic centers, older universities, medical centers. For such important and irreplaceable amenities, urban sprawl has to depend completely on the urban center.
- Sprawled urban and suburban areas provide accommodation to the high-end and low-income workers working in the urban centers and are home to many low-wage workers working in the urban sprawls. Most of such workers are either slum dwellers or tenants in the older parts of the urban center. Thus, urban sprawl is home to a major share of the urban center working population.
- Urban centers are home to key nodal and switching facilities for a number of utilities and amenities that eventually serve the suburban and urban sprawl, like electric power, railway stations, water and sanitation systems, older airports, highways, and telecommunication services.
- For their smooth functioning, many suburban firms rely on the products and services provided by the urban central firms where agglomeration economies exist. Similarly, the urban center depends upon the urban sprawl for its logistical and warehousing needs (Aljohani and Thompson 2016).

Urban centers perform vital functions of key importance to the urban sprawl and the entire country. Similarly, urban sprawl performs few significant functions like residential, modern amenities, leisure, new offices, warehousing, etc. Thus, it becomes highly significant to understand the degree of social and economic linkages and interdependence between the urban center and urban sprawl in the whole metropolitan area. Subsequently, a regional approach shall be taken in the form of integrated planning with the help of the concept of urban metabolism between the urban center and the urban sprawl to develop the metropolis.

5.4 Heterogeneity in the Flow of the Systems in Urban Sprawl-Case Studies

Urban sprawl and environmental quality are the two critical issues of rapid urban growth around the globe (Chadchan and Shankar 2012). The flow of systems in urban sprawl helps analyze the energy-matter flow within and outside the system and determines its sustainability. The process of urban metabolism is a product of out-flows and in-flows of the materials, resources, energy and wastes between core and unplanned boundaries of the city known as urban sprawl (Kennedy et al. 2014).

In 1970, Doxiadis presented the concept of energy consumption in the growth and evolution of cities in terms of energy consumption with following two cases: (i) a compact village with a per capita energy consumption of 8000 cal and (ii) sprawling cities having 33,000 to 100,000 cal (Braham and Lee 2020). The multiscale approach of urban metabolism indicates that energy consumption is function of the density and distribution of the material and resources. The urban sprawl demands high energy consumption due to agricultural practices and high construction materials and ground levels for waste disposal. There is a mutual energy dialogue between the urban centers and their sprawl. In the next section, two case studies have been selected to understand the heterogeneity in the flow of the systems in urban sprawl.

5.4.1 The Case of Megacity Urban Sprawl: Delhi Metropolitan Region

The Delhi Metropolitan Region is one of the fast-growing urban centers in the world as shown in Fig. 5.2. Unprecedented changes in the outer boundaries of the fastest growing metropolitan region are taking place. The Delhi Metropolitan region's expansion has occurred on its periphery, leading to the growth of urban sprawl. Due to this unplanned urbanization of rural areas, New Delhi's outskirts have seen the majority of Delhi's urban sprawl growth (Jain et al. 2016). Between 1991 and 2011, Delhi's geographic area nearly doubled, doubling urban families and halving rural housing. The sprawling urban areas of the Delhi Metropolitan region have been facing several administrative, social, political, and environmental metabolic issues. New urban centers of the Delhi Metropolitan region, like Bahadurgarh, Ghaziabad, Noida, Faridabad, and Gurugram, are also facing the unplanned growth of urban sprawls. Urban sprawl formation is the real challenge for growing cities.

The urban metabolism of Delhi Metropolitan region involves heterogeneous energy flows between environmental, social, and economic factors as shown in

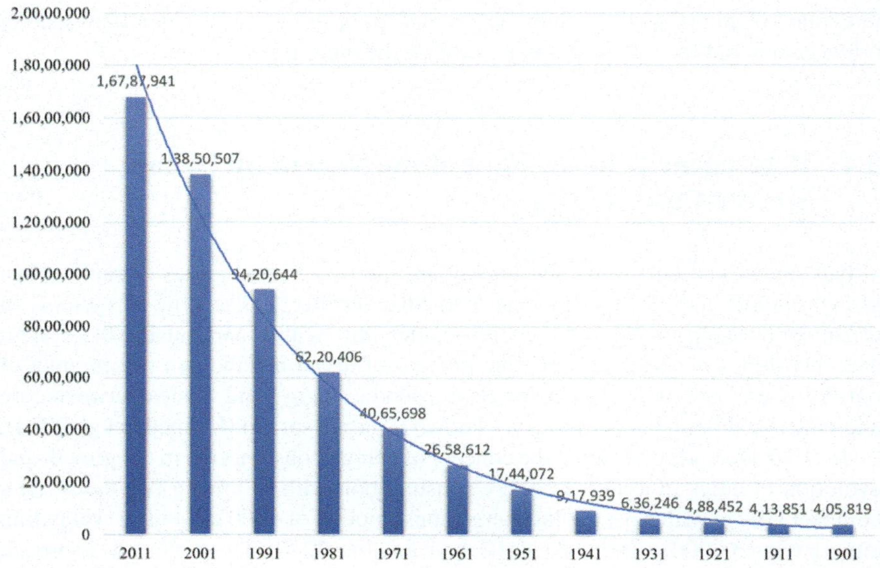

Fig. 5.2 Delhi Metropolitan Region population growth (Redrawn from Census of India 2011)

Fig. 5.3. It involves cross-disciplinary interventions for various urban systems of the region and thus contributes to urban sustainability (Jain et al. 2016). The urban sprawl of the Delhi Metropolitan region is dependent on the capital, New Delhi, for jobs and resources. The opportunity for transit-oriented development is eminent in the growth of the sprawls. Cities as ecosystems reflect a complex and extensive system of social, economic, and environmental activities, necessitating an inter- and multidisciplinary approach to studying urban metabolism (István and Elemér 2008). The linear flow of the urban metabolic system affects the development process of urban sprawl. This effect of the linear flow can be seen in the sprawls of the Delhi Metropolitan Region.

The linear flow of inputs and outputs from the core of the Delhi Metropolitan region to its core area to the sprawling growth is leading to various environmental, traffic, quality of life, and institutional and administrative issues. Delhi Metropolitan region is a megacity with a swelling sprawling population exceeding 15 million (Kennedy et al. 2014). Drivers of metropolitan region growth, a SEIRS of secondary growth centers are developing at the edge which is further engulfing farmlands and villages. The phenomenon is known as the 'polarization-reversal process' (Villa and Rodriguez 1996). However, the polarization-reversal process could be advantageous for reducing the load to the core city, but without proper planning, it will lead to the formation of sprawls. The heterogeneous flow of materials and energy within cities and urban sprawl lead to understanding the sustainable developmental issues and challenges as follows:

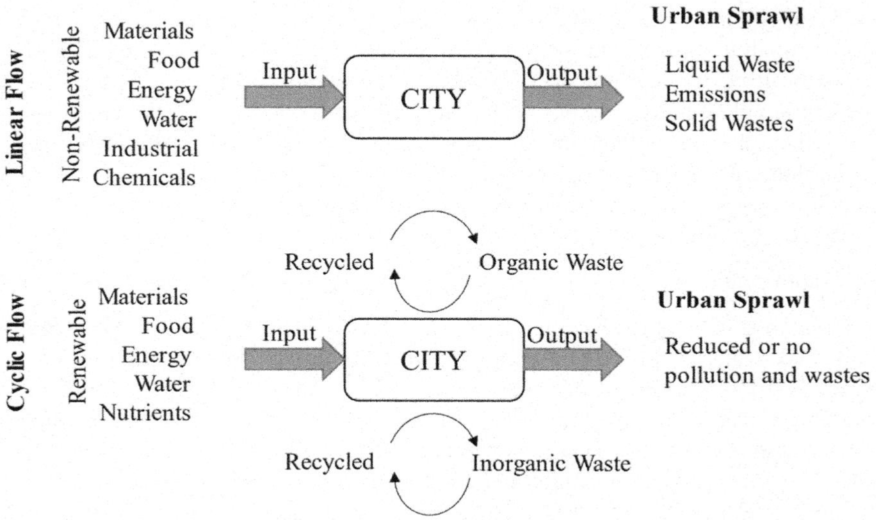

Fig. 5.3 Urban metabolic relationship of city core and sprawl (Redrawn from Novotny 2011)

- **Urban sprawl and agricultural lands**

The pressure of the land use change is high on the boundaries of the Delhi Metropolitan region. Most of the land resources belong to private ownership leading to challenges for the local bodies to prepare the land resources for future development (Davies and Wright 2014). Further, this situation leads to urban growth in fertile agricultural lands (Duarte et al. 2022). Thus, change in land use leads to the threat to the flow of food supply to the Delhi Metropolitan region. Most of the agricultural lands around the core metropolitan region are in the production of vegetables. The loss of agricultural land leading to reduced production will pressure transporting the required food supply. Hence, more energy flow will be required in this unsustainable process.

- **Traffic congestion**

The urban sprawl growth of the Delhi metropolitan region is based on better transportation modes and facilities. Housing prices and taxes are high in the metropolitan region, whereas sprawl has affordable housing facilities (Sun et al. 2016). The Delhi metropolitan region's road connectivity and sprawl are also quite good, leading to the increased traffic of private vehicles, and thus, high energy consumption.

- **Urban water cycle and sprawls**

The case of the Delhi metropolitan region provides the opportunity to understand the institutional role in sustainable and efficient water in-flow and out-flow (Maria 2008). The poor quality of water and supply services are provided by the municipal bodies on the edges of the metropolitan region, subsequently led to the ground water extraction through tube-wells and pumps by the high-income residents. These individual groundwater extraction practices have led to the uneven water table of the sprawl. Some individual houses do not receive water for a week or more. Water resources are vulnerable to contamination from soiled groundwater and surface runoff because of the limited number of water system plants. This scenario highlights the heterogeneity of urban water cycles in the core metropolitan region and accompanying sprawls (Dixit 2022).

- **Waste cycle**

In urban metabolism studies, the out-flow of energy happens through consumption and waste (Shao et al. 2014). The rapid growth of urban sprawl in the case area is limited, and the traditional drainage system pollutes the downstream. Waste disposal is a major issue in metropolitan edges due to the absence of efficient waste management institutions. Urban sprawls are the service backyard of the main city core, and thus, it resulted in the additional high generation of waste at the edges. Proper waste disposal is needed for sanitation, good health, and uncontaminated groundwater and air (Parihar et al. 2022). The absence or inadequacy of the existing sewage and waste disposal system is evident in the case area's urban sprawls. The existing waste disposal of trash and other debris in an uncovered region on the outskirts of communities poses a threat.

- **High land prices**

The growth of urban sprawls in the Delhi metropolitan region causes a dramatic rise in land and real estate prices. Rapid population growth is focused in areas outside of the core of the Delhi metropolitan region may lead to the future growth of urban sprawls. This rapid change in urban population leads to the changes in land use patterns of the sprawls of Delhi which lead to the unsustainable increase in land prices (Sharma and Abhay 2022).

Further, the key issues due to the heterogeneity of the energy flows in the urban sprawl of the Delhi metropolitan region are:

(1) Determining the edges of the Delhi metropolitan region is a challenge leading to urban sprawl growth. Determination of boundaries for the city core is important to quantify the total energy input or output flow of energy.
(2) The absence of data on the parameters to assess heterogeneity of urban metabolisms due to unclear boundaries of administrative units.
(3) The parameters to assess the heterogeneity of urban metabolism are also needed to study the quality of life in sprawl development.
(4) The energy flow analysis will help assess the efficiency of urban metabolism in the Delhi metropolitan region and facilitate sustainable development.

Hence, diverse energy exchanges between environmental, social, and economic factors are the crucial element of the urban metabolism of the Delhi Metropolitan Region and its sprawl development. The most important issue related to urban sprawl is defining its geographic boundaries and tracking its growth through the evolution of the sprawl (Katsoulakos et al. 2016). The urban sprawls of Delhi depend on the capital city of New Delhi for employment opportunities. While on the contrary, urban sprawls play a role of the service backyard for the Delhi metropolitan region, leading to the heterogeneous in-flow and out-flow of energy.

5.5 Understanding of the Urban Sprawl Metabolism for Sustainable Development

Urban centers and their sprawls are responsible for approximately 80% of the total greenhouse gas (GHG) emissions worldwide. As the urban population is increasing rapidly, the contribution of urban areas to GHG emissions is increasing too. While most of the studies focus on the sustainable planning and development of urban centers with the application of urban metabolism, a holistic approach of the urban metabolism for the sustainable development of urban sprawl is required.

The most commonly accepted definition of sustainable development was given by Brundtland Commission (UN 1987): '*Development that meets the needs of the present generation without compromising the ability of future generations to meet their own needs.*'

In the case of urban areas, it includes the progress of the urban region's economic, social, and environmental aspects. Similarly, urban metabolism talks about the healthy usage of urban resources accounting for the material, technical, and human interactions. Ecological footprints of the urban centers extend far from their boundaries. Thus, to create a real, sustainable ecosystem, the concept of sustainable planning and development shall also be integrated into urban sprawl areas. Urban sprawl metabolism quantifies the resources, energy, and waste exchange within and between the urban center and urban sector to achieve sustainable development. Decoding the relationship between the consumption of energy, natural resources, and other materials in urban sprawl areas and the locations where these resources are produced or extracted, and the final waste products are decomposed in understanding the area's ecological footprint.

5.5.1 Urban Infrastructure and Urban Sprawl Metabolism

To link the urban sprawl metabolism with the urban center metabolism and to mobilize resources and people, urban infrastructure is required. The urban infrastructure linking urban centers and sprawl consists of roads, metro rails, pedestrian walkways, bicycle paths, water and sanitation lines, electric cables, gas pipelines, etc. The urban infrastructure massively exploits natural resources. It comprises large-scale volumes, financing, and timelines. However, transforming urban infrastructure with the help of a stakeholder co-creation process is possible with many added benefits such as increased ownership, more inclusivity, innovative solutions. Here, urban infrastructure facilitates an efficient flow of resources through urban core and urban sprawl and leads to sustainable development (Chen and Chen 2022).

Sustainable development requires an efficient technical urban sprawl metabolism and infrastructure through human activity and design. Human intervention is also of prime importance when it comes to urban sprawl metabolism. Here, the importance of inclusive production methods and tools through a cooperative approach is highlighted. One way to achieve this is by focusing on new and innovative approaches for community-driven sustainable mobility at the metropolitan or regional level. Reducing the consumption of resources from the city edges falls under the ambit of urban sprawl metabolism. Methods like compact neighborhoods, improvement in connectivity, safeguarding open spaces, transport-oriented development (TOD) helps in controlling urban sprawl to a certain extent.

5.5.2 *Urban Sprawl Metabolism Framework*

With the help of the urban sprawl metabolism framework, the technical and socio-economic processes of an urban sprawl can be analyzed. It includes the assessment of the inputs, stores, and outputs of urban sprawl areas' energy, materials, and water (Codoban and Kennedy 2008). As discussed in Sect. 5.2, urban sprawl, analogous to living organisms, transforms materials into infrastructure and waste. Similarly, they are analyzed as ecosystems to decipher the relationship and dependency between urban centers and sprawl. These can be compared with natural ecosystems (the most sustainable systems) to develop sustainable regions.

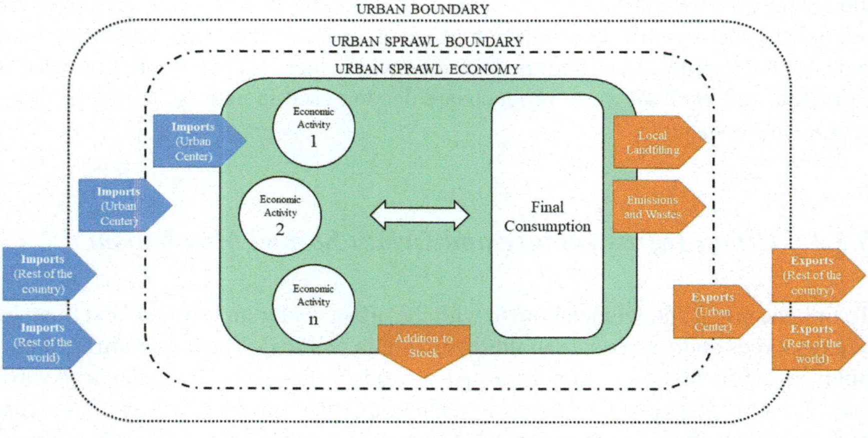

Fig. 5.4 Material flows in an urban sprawl (Redrawn from Asian Development Bank 2014)

Studying urban sprawl metabolism is a complex task. It requires capturing the multifaceted and cross-scale relationship and dependencies among nature, infrastructure, society, and institutions shaping the interactions within urban sprawl and between the urban center and sprawl. However, these studies are generally difficult in the real world due to a lack of data and defined metrics (Palme and Salvati 2020). On the other hand, interactions between materials within the urban sprawl and between the urban center and sprawl provide an opportunity for analysis. The production and consumption of materials in an urban area are crucial for assessing sustainability in terms of the availability of resources, efficient functioning, and environmental protection (Braham and Lee 2020). Material flows in an urban sprawl can be assessed as illustrated in the figure and include the following as shown in Fig. 5.4 (Asian Development Bank 2014):

- **Inputs**: Extraction of resources and import of raw materials and finished products. They may come from the accompanying urban center, any other city or region of the country, or abroad.
- **Outputs**: Exports of raw materials and finished products to the accompanying urban center, any other city or region of the country or abroad; wastes and emissions to the surroundings and beyond.
- **Internal processes**: Include intermediate consumption due to various urban economic activities and final consumption at households and services.
- **Stock**: Share of consumption unused for the time being and accumulates in the system in the form of buildings, urban infrastructure, and other goods.

Once the urban sprawl metabolism is analyzed, designing policies promoting sustainable development in urban centers and sprawls is not easy (Fróes and Lasthein 2020). Depending on consumption levels and targets, such policies are specific to its region. Urban sprawl metabolism can contribute to designing and implementing sustainable development policies in three ways (Asian Development Bank 2014):

1. Comparing quantitative data on the usage of resources and generation of waste and identifying best practices amongst different urban sprawls.
2. Enhancing clustering of urban areas based on demography, climate, morphology, governance, economic conditions, etc., to develop sprawl typologies.
3. Providing a framework to assess and weigh various policy alternatives and their impact on resource consumption and waste generation.

5.6 Recommendations for Future Perspectives

The study of urban sprawl metabolism has not only scientific contributions but also leads to practical recommendations. Some of the major practical applications for urban planners, urban designers, and infrastructure experts are listed below:

- **Urban Design**: The urban sprawl metabolism perspective can be used to redesign urban areas or to design the urban infrastructure according to sustainability principles. Best practices like green building design, alternative energy, TOD, etc., can be analyzed using urban sprawl metabolism (Pistoni and Bonin 2017).
- **Dynamic mathematical models**: Urban sprawl metabolism models are used to study specific substances within the process, like metals or nutrients. It also simulates future changes through policy or technological interventions in the urban process (Maranghi et al. 2020).
- **Sustainability**: Urban sprawl metabolism includes appropriate in-flows and outflows of material and waste recycling, urban management, and energy efficiency, which is relevant to urban planners and managers. Such information can assist in describing and analyzing environmental conditions, which can be a predecessor to policy-making (Cui et al. 2019).
- **GHG accounting**: With the help of urban sprawl metabolism, GHG emissions can be quantifiable and policies can be designed to reduce them (Restrepo and Morales-Pinzón 2018).

At the metropolitan level, understanding patterns or spatial flows in an urban center-sprawl system through urban metabolism can help decide future projects' location (Wang 2022). Such projects can attempt to improve the metabolism of the whole metropolitan area.

5.7 Conclusion

Urban sprawl is a product and catalyst of economic growth and development in the city core. In contrast to the urban sprawl, its core used to have a high concentration of resources, materials, technology, and human resources. Urban cores have become the drivers of productivity while urban sprawls as its service backyard with unplanned development. Urban sprawl is the most common effect of rapid urbanization, and the relationship between urban center and urban sprawl is complex. Urban sprawl brings a few negative externalities like congestion, pollution, etc. The key concern is how to maximize the benefits of the growing urbanization and minimize the dis-benefits to the people and environment. A deep understanding of the urban dynamics of the economy, resources, and wastes within urban sprawl and between the urban center and its sprawl are required. And to decipher these unique dynamics, the application of urban sprawl metabolism is suggested.

With the application of urban sprawl metabolism, a detailed quantitative analysis of the patterns of urbanization is possible. A unique understanding of the correlation between resources and urban economic activities can also be made. Not only that, several practical recommendations related to sustainability indicators, GHG emission quantification, mathematical models for policy making, and urban design are achievable. Urban metabolism shall eventually alter the approach to practice sustainability rather than just a tool for some added information for urban designers, planners, and infrastructure experts. Urban metabolism shall encourage them to think and work in terms of synergies and dependencies across flows, scales, locations, and technologies. Further detailed and deep knowledge is required in terms of the management of various resources. A participatory approach to implementing the urban metabolism concept may be valuable at the urban sprawl level. The research strongly emphasizes the stakeholders' involvement in the sustainable development of urban sprawl. The study has also highlighted the improvement of urban infrastructure to achieve green mobility in the urban sprawl area by designing transit-friendly neighborhoods.

Certain ongoing and future research directions related to urban metabolism include understanding the relationship between urban poor and urban metabolism. Deeper integration of parameters of quality of life into the urban metabolism framework is crucial. Also, a universal standard classification system for flows and stocks is required for the concept of urban sprawl metabolism. Research and practical understanding are essential to apply the concept of urban metabolism in urban planning to understand the complexity of city-sprawl relationships. Currently, the tracking of material and energy flows is taking place at the building scale. It is the need of the hour to take this to higher levels like neighborhood, city, metropolitan, or regional levels, as real sustainability cannot be achieved in isolation. The current challenge is 'designing sustainable cities' urban metabolism.'

References

Agudelo-Vera DM, Mels A, Keesman K, Rijnaarts H (2012) The urban harvest approach as an aid for sustainable urban resource planning. J Ind Ecol 16:839–850

Ahuja D, Tatsutani M (2009) Sustainable energy for developing countries. Surv Perspect Integrating Environ Soc 2(1):1–16. http://journals.openedition.org/sapiens/823

Aljohani K, Thompson RG (2016) Impacts of logistics sprawl on the urban environment and logistics: taxonomy and review of literature. J Transp Geogr 57:255–263

Asian Development Bank (2014) Urban metabolism of six Asian cities. Asian Development Bank, Metro Manila

Bai X, McPhearson T, Cleugh H, Nagendra H, Tong X, Zhu T, Zhu YG (2017) Linking urbanization and the environment: conceptual and empirical advances. Annu Rev Environ Resour: 215–240

Bai X (2008) Industrial ecology and the global impacts of cities. J Ind Ecol 11:1–6

Bansal S, Pandey V, Sen J (2017) Redefining and exploring the smart city concept in Indian perspective: case study of Varanasi. In: From poverty, inequality to smart city: proceedings of the national conference on sustainable built environment 2015. Springer, Singapore, pp 93–107

Benyus MJ (2002) Biomimicry: innovation inspired by nature. Perennial, New York

Braham WW, Lee JM (2020) Metabolism of urban location: travel time and the morphology of the cities. Front Sustain Cities 2:1–13. https://doi.org/10.3389/frsc.2020.00004

Chadchan J, Shankar R (2012) An analysis of urban growth trends in the post-economic reforms period in India. Int J Sustain Built Environ 1(1):36–49. https://doi.org/10.1016/J.IJSBE.2012.05.001

Chen F, Chen Y (2022) Urban metabolism and spontaneous architectural growth: a sustainable strategy featuring participatory co-construction by multiple stakeholders. Buildings 12:352

Codoban N, Kennedy CA (2008) Metabolism of Neighborhoods. J Urban Planning Develop 134:21–31. https://doi.org/10.1061/(ASCE)0733-9488(2008)134:1(21)

Cui X, Wang X, Feng Y (2019) Examining urban metabolism: a material flow perspective on cities and their sustainability. J Cleaner Prod 214:767–781

Davies PJ, Wright IA (2014) A review of policy, legal, land use and social change in the management of urban water resources in Sydney, Australia: a brief reflection of challenges and lessons from the last 200 years. Land Use Policy 36:450–460. https://doi.org/10.1016/J.LANDUSEPOL.2013.09.009

Dixit K (2022) Groundwater rises in some areas, but Delhi loses overall. Times of India, Delhi

Doughty MRC, Hammond GP (2004) Sustainability and the built environment at and beyond the city scale. Build Environ 39:1223–1233

Duarte AC, Ferreira C, Vitali G (2022) Use of simulation models to aid soil and water conservation actions for sustainable agro-forested systems. Natural resources conservation and advances for sustainability, pp 389–412. https://doi.org/10.1016/B978-0-12-822976-7.00006-5

Fischer-Kowalski M (2002) Exploring the history of industrial metabolism. In: Ayres RU, Ayres LW (eds) A Handbook of Industrial Ecology. Edward Elgar, p 11. https://doi.org/10.4324/9781351282086-3

Fróes I, Lasthein MK (2020) Co-creating sustainable urban metabolism towards healthier cities. Urban Transformations 2:5(1–19)

Gemenetzi G (2017) Exploring the relationship between urban sprawl and the urban system. Evidence from Thessaloniki, 1991–2011. Spatium 2017:1–11. https://doi.org/10.2298/SPAT1737001G

Golubiewski (2012) Is there a metabolism of an urban ecosystem? An ecological critique. AMBIO: J Hum Environ 41(7):751–764. https://doi.org/10.1007/s13280-011-0232-7

Hill EW, Wolman HL, Cook Ford C (1995) Can suburbs survive without their central cities?: Examining the suburban dependence hypothesis. Urban Aff Rev 31:147–174

István P, Elemér S (2008) (PDF) urban metabolism: the case of Budapest. In: Urban metabolism: measuring the ecological city. In: Urban metabolism: measuring the ecological city conference. https://www.researchgate.net/publication/251231474_Urban_Metabolism_The_Case_of_Budapest

Jain M, Dimri AP, Niyogi D (2016) Urban sprawl patterns and processes in Delhi from 1977 to 2014 Based on remote sensing and spatial metrics approaches. Earth Interact 20(14):1–29. https://doi.org/10.1175/EI-D-15-0040.1

Katsoulakos NM, Misthos LM, Doulos IG, Kotsios VS (2016) Environment and Development. In: Poulopoulos SG, Inglezakis VJ (eds) Environment and Development: Basic Principles, Human Activities, and Environmental Implications. Elsevier, pp 499–569. https://doi.org/10.1016/B978-0-444-62733-9.00008-3

Kennedy C, Stewart ID, Ibrahim N, Facchini A, Mele R (2014) Developing a multi-layered indicator set for urban metabolism studies in megacities. Ecol Indic 47:7–15. https://doi.org/10.1016/J.ECOLIND.2014.07.039

Kennedy C, Pincetl S, Bunje P (2011) The study of urban metabolism and its applications to urban planning and design. Environ Pollut 159:1965–1973

Lederer J, Kral U (2015) Theodor Weyl: A Pioneer Of Urban Metabolism Studies. J Ind Ecol 19:695–702

Lucertini G, Musco F (2020) Circular urban metabolism framework. One Earth 2(2):138–142. https://doi.org/10.1016/j.oneear.2020.02.004

Lyle JT (1994) Regenerative design for sustainable development. Wiley, New York

Maranghi S, Parisi ML, Facchini A, Rubino A, Kordas O, Basosi R (2020) Integrating urban metabolism and life cycle assessment to analyse urban sustainability. Ecol Indic 112:106074

Maria A (2008) Institut Du Développement Durable et Des Relations Internationales-Adresse Urban Water Crisis in Delhi. Stakeholders Responses and Potential Scenarios of Evolution. Institut Du Développement Durable et Des Relations Internationales

McDonald GW, Patterson MG (2007) Bridging the divide in urban sustainability: from human exeptionalism to the new ecological paradigm. Urban Ecosyst 10:169–192

Mc.Donough W, Braungart M (2002) Cradle to cradle: remaking the way we make things. North Point Press, New York

Mensah J (2019) Sustainable development: meaning, history, principles, pillars, and implications for human action: literature review. Cogent Soc Sci 5(1):1653531

Nelson N (2010) Investigation into a sustainable implant in the Erasmusveld neighborhood, the Hague. Master Thesis, WUR Wageningen/TU Delft, DSO, Gemeente Den Haag

Newman P, Beatley T, Boyer H (2009) Resilient cities: responding to peak oil and climate change. Island Press, Washington, DC

Novotny V (2011) Water and energy link in the cities of the future—achieving net zero carbon and pollution emissions footprint. Water Sci Technol: A J Int Assoc Water Pollut Res 63:184–190. https://doi.org/10.2166/wst.2011.031

Palme M, Agnese S (2020) Sustainability and urban metabolism. Sustainability 12(1):353

Paramasivam B, Illanthirayan A (2020) Urban sprawl. In: Almusaed A, Almssad A, Truong L (eds) Sustainability in urban planning and design. Intechopen, Hong, London

Parihar DS, Singh M, Deepak (2022) Impact of urban sprawl in waste disposal management: need actions required managing in Almora town. EPRA Int J Econ Growth Environ Issues 10:1–13. https://doi.org/10.36713/EPRA9728

Pistoni R, Bonin S (2017) Urban metabolism planning and designing approaches between quantitative analysis and urban landscape. City Territory Architect 4:20

Restrepo JDC, Morales-Pinzón T (2018) Urban metabolism and sustainability: precedents, genesis and research perspectives. Resour Conserv Recycl 131:216–224

Reynolds J, Rezgui Y, Hippolyte JL (2017) Upscaling energy control from building to districts: current limitations and future perspectives. Sustain Cities Soc 35:816–829. https://doi.org/10.1016/j.scs.2017.05.012

Shao L, Chen GQ, Hayat T, Alsaedi A (2014) Systems ecological accounting for wastewater treatment engineering: method, indicator and application. Ecol Indic 47:32–42. https://doi.org/10.1016/J.ECOLIND.2014.04.026

Sharma M, Abhay RK (2022) Urban growth and quality of life: inter-district and intra-district analysis of housing in NCT-Delhi, 2001–2011–2020. GeoJournal 87(4):797–819

Sun L, Dong H, Geng Y, Li Z, Liu Z, Fujita T, Ohnishi S, Fujii M (2016) Uncovering driving forces on urban metabolism—a case of Shenyang. J Cleaner Prod 114:171–179. https://doi.org/10.1016/J.JCLEPRO.2015.05.053

Tan LM, Arbabi H, Li Q, Sheng Y, Tingley DD, Mayfield M, Coca D (2018) Ecological network analysis on intra-city metabolism of functional urban areas in England and Wales. Resour Conserv Recycl 138:172–182

Thomson G, Newman P (2018) Urban fabrics and urban metabolism—from sustainable to regenerative cities. Resour Conserv Recycl 132:218–229

Timmeren (2013) Sustainable development; a retrospective. SUET, Delft University of Technology, Delft

Timmeren (2014) The concept of the urban metabolism (UM). SUET

UN (1987) Brundtland commission. United Nations

Villa M, Rodriguez R (1996) Demographic tendences in Latin America's Metrpolises, 1950–1990. In: Gilbert A (ed) The Mega-City in Latin America, pp 25–52. https://books.google.com/books/about/The_Mega_city_in_Latin_America.html?id=TySzAAAAIAAJ

Wang HC (2022) Case studies of urban metabolism: what should be addressed next? Urban Aff Rev 59(3):949–968

Wolman A (1965) The metabolism of cities. Sci Am 213:179–190

Zhang H (2021) The impact of urban sprawl on environmental pollution: empirical analysis from large and medium-sized cities of China. Int J Environ Res Public Health 18(16):8650

Chapter 6
Adaptive Reuse of Historic Buildings: An Ecological Indicator

Ayesha Agha Shah, Tania Ali Soomro, Naji Akbar, Asifa Iqbal, and Ismaila Rimi Abubakar

Abstract Urban metabolism (UM) is the sum of processes for which cities mobilize, consume, and transform their resources for built environments to function effectively. It consists of interrelated processes working at various urban levels forming an intricate socio-environmental network to achieve urban sustainability, including adaptive reuse of heritage buildings. As significant components of historic urban areas, heritage buildings are adapted for appropriate functions to prolong their lifespans. This standard practice is an ideal solution for reducing adverse environmental impacts of the construction industry given that new buildings consume energy and resources and have large carbon footprints. Moreover, the adaptive reuse of heritage buildings increases socio-cultural viability and promotes eco-friendly environments. This practice of rehabilitation and reuse has been studied vastly. However, the relationship between urban metabolism and heritage buildings from an ecological perspective has not largely gained scholarly attention. Hence, this chapter explores this relationship by advocating that reuse and conservation are ecological principles, and that adopting existing infrastructure such as historical buildings and increasing their life cycles contributes toward desirable urban metabolism. This brings down

A. A. Shah (✉)
Department of Architecture and Interior Design, University of Bahrain, Zallaq, Bahrain
e-mail: aashah@uob.edu.bh

T. A. Soomro
Department of Architecture and Urban Studies, Politecnic Di Milano, Milano, Italy
e-mail: taniaali.soomro@polimi.It

N. Akbar
College of Architecture, Art and Design, Ajman University, Ajman, UAE
e-mail: n.akber@ajman.ac.ae

A. Iqbal
Department of Computer and Geospatial Science, Faculty of Engineering and Sustainable Development, University of Gavle, Gavle, Sweden
e-mail: asifa.iqbal@hig.se

I. R. Abubakar
Department of Urban and Regional Planning, College of Architecture and Planning, Imam Abdulrahman Bin Faisal University, Dammam 31441, Saudi Arabia
e-mail: irabubakar@iau.edu.sa

R. Bhadouria et al. (eds.), *Urban Metabolism and Climate Change*,
https://doi.org/10.1007/978-3-031-29422-8_6

cost and time of construction, requires less resources, retains a set of materials for longer periods for metabolic processes, and promotes environmental and social sustainability. The chapter reviews relevant literature and specific adaptive reuse case studies in urban areas around the globe including restoration and redevelopment of singular or multiple heritage buildings that display significant positive environmental impacts such as addressing resource depletion and reducing greenhouse gas (GHG) emissions. The chapter also highlights the ecological aspect of a built environment concerning sustainable supply of materials and energy required for a desired urban metabolism. Based on this, the research builds a case for conservation and adaptive reuse to be employed as an important indicator for urban metabolism through maintenance and management of historical urban built environments.

Keywords Adaptive reuse · Ecological wisdom · Historic urban areas · Social · Economic · And environmental indicators · Urban metabolism

6.1 Introduction

Urban metabolism (UM) applies the analogy of metabolic processes carried out by living organisms to the way cities function so that their sustainability can be ascertained (Zhang et al. 2015; Restrpo and Morales-Pinzón 2018). UM studies material and energy inflows, processing of these inflows, and generating desired products along with the resultant waste. UM processes, relationships, and interactions can be environmental, social-cultural, and economic (D'Amico et al. 2020). Given this, UM is not just a reproduction process; it is more of a transformation (Burgess 1925). Newman (1999) rephrases the same thought by saying that UM is a changeover of nature to society; the city is the locus of this activity where nature is metabolized to accommodate human existence (Girardet 1996; Washsmuth 2012).

Urban metabolism is of two types. Circular UM relates to nature, where the by-product from an organism is food for another. On the other hand, linear UM characterizes cities, where resources enter from one side and throwing out waste from the other. This distinction is significant as it relates to the current sustainability crisis, which is abundant in linear UM (Shaw 1977; Girardet 1996; Charruadas 2012; Feldhoff 2013; Hartel et al. 2013; Bagheri 2015; Zacharias and Lei 2016; Kapoor et al. 2020; Lucero and Cruz 2020; Bhadouria et al. 2022). Also, cities are majorly assessed using monetary functions although bio-physical factors closely related to human well-being (Raffestin and Lawrence 1990; Tomasetti 2016; Tan et al. 2019) may not always be given due importance in urban development. As a consequence, researchers have difficulty balancing and integrating urban metabolism and smart city aspects (Thomson and Newman 2018). Given the biological nature of the metaphor, UM shares similarities with ecological wisdom (EW) and its guidelines. Generally defined as contextual knowledge gained, processed, and internalized over large expanses of time (Wang et al. 2016), and EW is emerging as a benchmark, domain, and framework for various knowledge and planning inputs. However, human action guided by EW is an age-old process that fosters ecological embeddedness and harmonious social conditions in many notable past civilizations, which guarantees the co-evolution of humans and nature (Hornborg 1998; Patten 2016; Wang et al. 2016; Kakoty 2018; Xiang 2019; Young 2019; Akbar et al. 2021).

According to EW, entities within an environment are involved in continuous interaction and dialogue with culture, place, and history (Young 2019). EW does not see nature as a problem to be addressed but rather as a relationship to be studied, elaborated, understood, and refined (Xiang 2019). Given this, EW seeks problem resolution through harmony between culture, environment, and practice promoting sustained innovation with the meaningful integration and coherence of policies (Michael 2020). People naturally associate with EW as contextual, whose application is not optional; it is a principled covenant that humans have made with nature to promote beneficial co-existence by marrying moral knowledge and righteous actions (Xiang 2019). EW can be obtained through the following four principles (Yang et al. 2019).

- **Respect for nature**
  Engagement with nature engenders respect for nature and taking actions that are detrimental to nature tantamount to environmental disrespect.
- **Sustained relevance**
  Temporal dimension is of significant value for EW. A wealth of relevant information is available in past records that can be tapped into to guide solutions for contemporary sustainability challenges.
- **Holism**
  Complex systems are planned as organic and coherent wholes. This means that the property sum of components is larger than the simple sum. This is due to the interdependent and interrelated nature of components.
- **Practicality**
  Faced with the continuous demand for making choices in the social system, making the right decisions becomes extremely important. Given this, ecological wisdom cannot just be abstract; it has to be practical to inform mature and wise decisions. For this, EW integrates wide-ranging knowledge areas and goes beyond the philosophical level.

In light of the above discussion, certain key indicators emerge through EW.

1. Following nature, there is a possibility of doing real and permanent good to the environment.
2. Temporal dimension is paramount in developing EW. Traditional practices are revered; the past is viewed as "the long present" or deep time; experience gained by previous generations is replete with ecological and social knowledge.
3. Inclusion of the diverse knowledge of an environment's interconnected and interdependent bio-physical and socio-cultural aspects will give rise to ecologically wise policy making.
4. By considering the constraints and opportunities of a context, ethical decisions result in symmetrical enhancement in social and ecological systems.

5. Apart from material benefits, non-material outputs such as culture and recreation also emerge from ecosystem services.
6. Circularity of resources plays a significant role in environmental sustainability.

Researchers believe that by following the stability of EW and its principles, UM can further effectively mimic natural systems (Zhang et al. 2015). For instance, both UM and EW study the human-nature relationship (Xiang 2019; Cárdenas-Mamani and Perrotti 2022). Comparable to EW, UM functions at multiple scales (Sallis et al. 2006). Therefore, it would be reasonable to conceive of a city as an open ecosystem demanding a holistic and multidimensional view (D'Amico et al. 2020). These open systems can promote the circularity of a resource within the system. Supply-chain footprint can then be minimized by retaining useful resources in the systems for maximum time (Tan et al. 2019).

Many built environments today have drawn planning inputs from the EW gathered over past millennia. In this respect, various case studies have been carried out about towns in countries such as Iran (Mazraeh and Pazhouhanafar 2018), Japan (Okubo 2016), China (Shaowen and Hanqi 2017; Zheng et al. 2018), and India (Dhingra and Chattopadhyay 2016). The following key planning decisions that can be learned from these case studies:

- Regions ecological limits and carrying capacity must be well understood
- Blue-green infrastructure should be incorporated for urban services as well as flood mitigation
- Paved surfaces should be limited
- Urban agriculture should be a part of urban set up
- If applicable, forest cover must be maintained
- Local materials should be used in construction
- Existing topography, sun movement, and wind direction should be respected while laying out a settlement.

Since this chapter promotes adaptive reuse of heritage buildings, the adoption and contribution of this practice provide ecological indicators toward desirable urban metabolism. A heritage building is a piece of architecture that the society has a shared respect for and is strongly associated with its history. It is a built environment component that contributes actively and creatively to the cultural landscape. It retains the material originality, structural integrity, functional continuity, and experiential value of a historical site as an ecological space (González 2020). In other words, a heritage building grounds a shaken identity in a rapidly changing world by storing and displaying intimate and respected local culture, sense of place, and valuable memories (Grammenos and Russel 1997). Even though heritage buildings save values and transfer them to the next generation, a community can appropriate these building making them liveable by adding another layer of meaning, character, and identity-making. These, therefore, bring together social setup and cultural landscapes to create an authentic urban metabolism (González 2020). This respect and association help people to relate and feel connected to the city, which is one of the aspects of ecological thinking.

The fact that historic buildings have survived up to the present is testimony to the fact that these were designed to be sustainable and live well in their location (Chang et al. 2021). This fact demands that this knowledge of sustainability and endurance be transferred by preserving historic buildings. The adaptive reuse of historic buildings also offers numerous other benefits (EBB 1977–1984; Mundaca et al. 2010), as expounded by Grammenos and Russel (1997):

1. Adaptive reuse of the building is the most environmentally friendly way of mitigating adverse environmental impacts.
2. Culturally sensitive use of heritage aligns well with the Global Plan of Action, 2003.
3. Reusing old buildings has enormous economic value in reducing the physical embodied energy required to construct new buildings.
4. Adaptive reuse may change some of the features of a valued historic buildings. By such change, a building becomes useable while conveying a sense of place, historical association, and its evolution over time.
5. Traditional techniques and materials have a low environmental impact due to local production, usage, and retention for future generations.
6. Adaptive reuse will keep old buildings out of landfills.

Based on the discussion above on EW and its relation to the significance of adaptive reuse, the authors of this study deduce eight ecological principles. These are interdependence, coherence, embeddedness, functioning of a system at multiple scales, nuanced approach, circularity, conservation, and adaptability. Table 6.1 shows the major implications of these principles and to what degree the selected case studies for this chapter satisfy these principles.

6.2 Methodology

The methodology followed for this research is based on a deductive approach of the qualitative analysis of facts and is divided into two sections; literature review, and fieldwork based on case studies. The literature review has provided an account of the knowledge and ideas that have been established by researchers on the topics of ecological wisdom, adaptive reuse, and urban metabolism. Eight EW principles have been identified that form the basis for identifying specific criteria for assessing the relationship of urban metabolism with historic buildings from an ecological perspective. The second section deals with fieldwork, where a case study is taken as a tool to understand the application of the knowledge gathered via section 1. A case study is both a research design and a method of analysis that provides a rich starting point that helps advance theory (Yin 2014). Three case studies in three different regions, namely, Pakistan (Southeast Asia), Sweden (Western Europe), and Bahrain (Middle East), are carried out (Table 6.1). The reason for taking case studies from three very different locations is to understand the link between the adaptive reuse of historic buildings to the city's UM in different environments. By investigating these case studies, this chapter explores how these buildings address UM and ecological concerns and present a high potential for adaptive reuse. An inventory form was developed to record information about buildings' history and current status. Buildings are analyzed through site surveys, personal observations, and meetings with conservation architects and building owners. The evaluation process comprised the criteria of EW principles representing different levels of relationship with the strong, medium, and low classification. Finally, a comparative analysis is performed

Table 6.1 Major implications of ecological principles of heritage building reuse

#	Ecological principle	Implications	Authors
1	Interdependence	Adaptive reuse of heritage engages several stakeholders for an effective social and cultural interdependence The following beneficial functions are demonstrated through the adaptive reuse of heritage buildings 1. Storing information in the built form that is collected over many decades, centuries, or millennia, 2. Serving as a vehicle for a harmonious weaving of past and present, 3. Becoming symbols of contextual identity, 4. Imparting a sense of communal ownership to the community, 5. Promoting tourism, 6. Allowing reuse to foster environmental stewardship It might be reasonable to consider heritage conservation as an actor in wider urban metabolism that achieves maximum desired results in certain contexts. In this respect, heritage aligns well with environmental and social topics covered by urban metabolism and the international standards on smart urban metabolism indicators, especially ITUT-T Y.4903/L.1603 and ITU-T.4901/L.1601?	Tomasetti (2016), D'Amico et al. (2020)

(continued)

Table 6.1 (continued)

#	Ecological principle	Implications	Authors
2	Coherence	Heritage site demonstrates the following coherent ecological dimensions that are valuable for society 1. Material originality of the object/building 2. Coherence of the technical-spatial structure with which the object has an association, that is, the space layout is enmeshed with and does justice to technology 3. Functional continuity of the object/building 4. The inherent value of the object 5. Appropriation of a heritage site by the community depending upon the four dimensions mentioned above	González (2020)
3	Embeddedness	A heritage building is well nested/embedded in the wider environment. It competes with urban pressures to stay relevant, or expressed in Darwinian terms, survives as it is the fittest amongst other edifices by creating a niche for itself through timely appropriations An ecological approach seeks alignment between culture and practice. Adaptive reuse of heritage buildings aids this alignment and harmony	Michael (2020), Akbar et al. (2021)
4	Functioning at multiple scales	Multidisciplinarity is a core ecological principle. Heritage buildings function at multiple scales, including social and environmental sustainability	Sallis et al. (2006)
5	Nuance approach (multiplicity of meanings)	Heritage buildings automatically become nuanced given that these offer multiple benefits through addressing social, spatial, and cultural environments	Tomasetti (2016)

(continued)

Table 6.1 (continued)

#	Ecological principle	Implications	Authors
6	Circularity	Heritage buildings promote the circularity of knowledge and materials at a social and physical scale	Tan et al. (2019)
7	Conservation	Adaptive reuse of heritage buildings points toward conservation of and respect for the environment addressing the adverse effects of the "metabolic rift" that Karl Marx famously discussed	Grammenos and Russel (1997), Washsmuth (2012)
8	Adaptability	Declared heritage buildings have stood the test of time by adapting to geographical conditions. Therefore, there are lessons for the current generation to learn from these buildings	Chang et al. (2021)

by simplifying the data associated with the eight proposed EW principles to get a conclusive assessment.

6.3 Adaptive Reuse of Historic Buildings for Urban Metabolism

In heritage management, conservation and ecological sustainability go hand in hand. While the topic might not have been a pivotal point in the past, it is becoming increasingly relevant in today's fast-paced world threatened by global warming. One of the smart approaches toward sustainable development is the preservation of heritage buildings. Measuring and managing the impact of these buildings in the context of global environmental crises is paramount as their embedded energy can be retained (Aksamija 2016; Foster and Kreinin 2020).

It has been argued that heritage envelops human lives and shapes who we are as individuals and a society. As a gift from the previous generation, subsisting nature of heritage demands us to pass it on to the next generation. However, to do so, it is crucial to properly assess and evaluate the opportunity costs in terms of ecological sustenance. One consideration that makes this transfer significantly simpler is the idea of rehabilitation and adaptive reuse. This consideration is corroborated by the 2003 Global Plan of Action (2003), which states, "Conservation, rehabilitation, and culturally sensitive adaptive reuse of urban, rural, and architectural heritage are also in accordance with the sustainable use of natural and human-made resources.

Furthermore, access to culture and the cultural dimension of development is of the utmost importance, and all people should be able to benefit from such access".

Historic buildings are constructed according to the conditions of site, local materials, and climate; their embedded energy is a "quantifiable indicator" to assess the goal of sustainable development (Foster and Kreinin 2020). By establishing ecological and environmental awareness and preservation, there is much to be gained in terms of economic value through reusing historic buildings. For example, Heinonen et al. (2011) highlighted how replacing late twentieth century housing incurs a long payoff time regarding Carbon emissions. Also, if broader impacts are considered, the gains to be made in the long run far outweigh the minor discomfort to be endured today (Heinonen et al. 2011).

One of the problems that the conservation of historic buildings faces these days is a lack of scientific forethought and procedure regarding decisions on environmental benefits. Not only should rational methods be employed, but an analytic approach to these projects can also bring better results than the traditional method of only paying attention to aesthetics. Absence of this mature and rational planning can harm the originality and authenticity of the heritage building, not to mention the enormous financial burden and environmental concerns it may raise. Furthermore, activities such as demolition are now frowned upon by current societies that focus more on ecological issues. This reinforcement by media and community awareness has helped architectural conservation to actively endorse adaptive reuse. Simply put, giving new life to a building allows for more societal benefits for the environment and community (Langston and Shen 2007; Foster and Kreinin 2020).

Research shows that heritage conservation utilizes the retention and reuse of preexisting resources. It also saves capital and energy as far as concept of circularity is concerned (Foster 2019). In turn, this allows for a more transparent and effective procedure that has payoffs for society and the environment since there is minimal need for external agents. However, this is heavily dependent on suitable practices when it comes to maintenance and operation. Key examples include promoting appreciation of repairing instead of replacing, intervening only when necessary, and focusing more on minimal interference.

It is interesting to mention that traditional techniques and materials have a lower environmental impact as studied by Wilson (2007). Furthermore, these techniques also add stature and pride to the work done since labor and materials are local. Due to this, specific skills and expertise evolve organically in a community contributing to economic growth and creating a network that allows for a better quality of life. Keeping these factors in mind, it has been suggested that traditional methodologies are more sustainable for our environment in the long run (Gültekin and Arzuhan 2011).

6.4 Case Studies of Adaptive Reuse of Historic Buildings

"The most environmentally benign building is the one that need not be built because it already exists" (Grammenos and Russel 1997; Elefante 2007). Existing buildings are a sustainable source to maximize the use of embodied energy and building materials; hence reuse of these buildings allows for a low-carbon economy (Abubakar and Bununu 2020; Foster and Kreinin 2020). Heritage buildings are literal as well as symbolic signs that remind us about the past that, if utilized properly, can give direction toward the future. Remnants of culture and their evolution provide a quick glimpse into history. Therefore, it is not only essential but also mandatory for heritage buildings to be conserved and reused, depending on size, location, and potential (Gültekin and Arzuhan 2011).

Adaptive reuse with new functions is a valuable solution that should be considered. However, it should be opted after factoring in cultural significance, context, and synchronization with the historic fabric. Furthermore, the new adaptation of a building should align with its identifying architectural character so that the design intervention does not seem out of place or affect the longevity and relevance of architectural history for the future generations. Therefore, adaptive reuse implies literal reuse of a heritage building that conforms to societal and historical expectations and is honest to the architectural significance that the building presents (Misirlisoy and Günçe 2016). There are many examples around the world to elucidate this point clearly. For instance, the *Roscommon Museum* in Dublin, a former Presbyterian church building, now serves as a museum. Its reuse as a museum keeping original character intact allowing us to utilize heritage better (Dimitrijevic et al. 2000). If one contemplates that such a building was to be demolished or significantly altered in a manner that influenced its integrity, then the externalized cost would be relatively high. However, the process is not as straightforward as it may seem and poses many problems and concerns. Langston and Shen (2007) state that changing the functions of a building can introduce new regulatory conditions. This is so as many planning and code variables are involved requiring forethought and exposure to pertinent existing literature.

To stress the importance of adaptive reuse of heritage buildings as an ecological tool for the UM, three case studies of residential buildings in three different climatic zones, Pakistan (Southeast Asia), Sweden (Europe), and Bahrain (Middle East), were carried out (Table 6.2). These case studies reflect the tangible and intangible attributes of SDG 11-Sustainable Cities, and emphasize adaptive reuse for sustaining ecosystems. Other criteria for the selection of case studies were their typology. In this respect, residential buildings were selected as the general public can easily relate to them. Residences also have a low scale, which made this study manageable and focused. Besides this, the selected cases prominently exist in their location and are celebrated by the city's masses. This offered accessibility to the buildings as well as information.

Table 6.2 Summary of case studies

Name of the building	Aman House, Pakistan (case study 1)	Clergy House, Sweden (case study 2)	Ismail House, Bahrain (case study 3)
Location	CAN-237, Fatima Jinnah (Bonus) Road, Karachi, Pakistan	West 33:1, Little Water Alley, Gävle, Sweden	House 773, Road 935, Block 209, Muharraq, Kingdom of Bahrain
Area	6738.50 m^2	2600 m^2	170 m^2
No. of floors	Ground + 1	Ground + 1	Ground + 1
Construction year	1927	1764	1900
Renovation year	The 1970s: Subsequent minor alterations were done over the following years	1914–15: Major renovation 1952: façade renovation 1975: Current color scheme	The major transformation from 1970 to 1990 Conservation 2021–2022
Current occupancy	Occasionally used as a film shooting studio	residence	NGO head office
Status of the building	Partly maintained	Fully maintained	Fully restored and maintained
Parts of the building rehabilitated/altered	Interiors, exterior	Interior, toilets, new kitchen, and water lines	Interior, Exterior, Major alteration with the addition of a new toilet

6.4.1 Case Study 1: Aman House, Pakistan

The building shortlisted as a case study from Pakistan is Aman House (AH) (Fig. 6.1). It is a residential bungalow in Cantonment Quarter in Karachi. Since its construction, the building has retained its original function and remained under use as a residence, which is why it required minimum architectural design interventions. Although a few minor alterations are observed within the building, they are reversible, having the least impact on the overall status of the building. Another reason for shortlisting this specific building as a case study is based on its present condition, which is fairly maintained, thus requiring minimum expenditure. AH is a notified listed heritage property protected by the Government of Sindh, Pakistan, under the Sindh Cultural Heritage Preservation Act 1994, enlistment no: 1997–179. AH was built in 1926 by the Dinshaw family, who were local Parsi merchants. The building was later leased in 1985 to Mr. Ayaz Fakiruddin. After some time, the Pakistan national army took over a section of this property when the lease ended and promptly built housing for soldiers. Finally, the Fakiruddin family regained the building's possession and sold it to Aman Foundation. Currently, the building is empty and is available for the videography of feature films and other similar functions. Although the building has been maintained periodically, Aman Foundation plans to further the restoration of the building and use it as its headquarters.

The structural advancement adopted by the British in the construction of AH is distinctive. They combined their knowledge with the local attributes developing AH as a well-suited structure to the soil and climatic conditions. The main building material is stone which has a longer lifecycle compared to the current construction materials. Even though assembly of the stone masonry involves high initial capital investment, this investment makes more financial if the cost is distributed over long lifecycle of stone. Sources suggest that stone is the only material with the least embodied energy of 0.1 MJ/kg. On the other hand, cement has the highest embodied

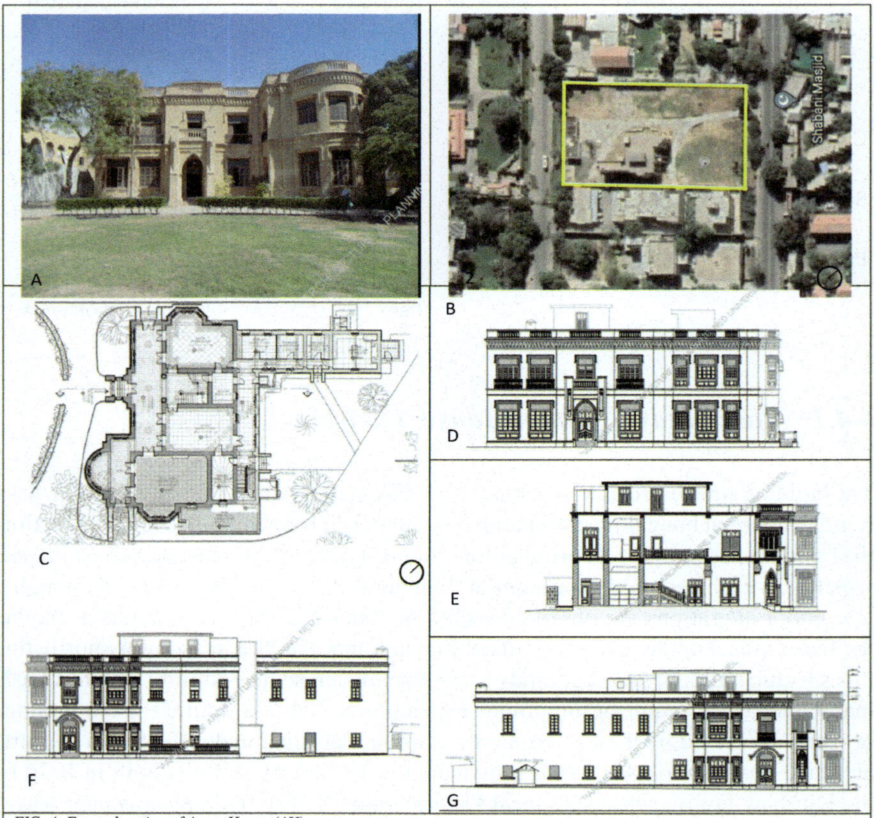

FIG. A. Front elevation of Aman House (AH)
FIG. B. Google earth top view of AH
FIG. C. Ground floor plan of AH
FIG. D to G. Respectively front, rare, right, and left elevation drawing of AH.
© Images & drawings at Heritage Cell – DAPNED, Karachi

Fig. 6.1 A detailed illustrative view of Aman House, Pakistan

energy of 5.6 MJ/kg. Additionally, the development of a patina[1] causes external stone surfaces of the building to age in a more decent manner, especially when looked at from an aesthetic point of view.

The building is well nested in the wider environment. The large open space (front courtyard) with thick foliage and a water body provides a barrier from the city's hustle and bustle creating a cozy green space. The landscape within the property boundary minimizes the urban heat island effect, thus allowing space to be used as a habitat for the local birds and insects that assist the existing ecosystems. It also allows the rain to be absorbed by the land which replenishes the underground water table in an urban area that is greatly covered by impervious land surfaces such as pavements and roads. The operational performance of the building is partly maintained through sustainable design and energy usage. The long corridors adjacent to the rooms ensure passive ventilation. The building has energy-saving features, including 20-inch-thick load-bearing masonry walls, smaller recessed windows, and exterior openings with awnings and shutters. These reduce the amount of heat gained by the building in the summers. The use of vestibules, porches, and other buffer spaces is also mentionable in this regard. The external façade of the building is laid in Ashlar masonry, which is obtained locally and commonly known as *Gizri* limestone. The internal surfaces of the walls are plastered with lime that keeps the building cool.

The building functions promote circular economy through least material wastage (Almulhim and Abubakar 2021). Reusing Aman House is retaining maximum embodied energies required for the building till the present time. On the other hand, a significant amount of energy would be required if the building were to be demolished. The principle of minimum intervention is the inspiration behind the present conservation state of the building. Existing large internal spaces are restored and temporarily subdivided into smaller units, given the need of the adopted function to maintain the cultural heritage value of AH. Overall urban elements and the landscape existing on the site are well maintained not only to create a bond between human beings and the environment but also to balance the ratio between the built to non-built area within the property bounds.

6.4.2 *Case Study 2: Clergy House (CH), Sweden*

Clergy House (Swedish name: Kyrkoherdebostaden) is a residential bungalow built in Gävle, Sweden, in 1764 by the Church of Sweden. It underwent several renovations receiving its current appearance during the twentieth century (Municipality of Gävle 2020). Shortlisting this specific building as a case study from Sweden is based upon its historical and present-day cultural worth and conservation status that is highly maintained (Municipality of Gävle 2020). CH was sold in 2017 by the Church of Sweden to a private owner because of its maintenance cost. The new owner decided

---

[1] An organic film formation on the surface of stone caused by exposure to the environment generally refers to the age of the structure/monument.

to divide the building into three smaller properties for residential purposes creating a common outdoor garden facility in the backyard. CH is a listed heritage property protected according to the Cultural Environment Act declared under Law number RAÄ 51. According to the law, plot separation is not allowed, which means it is not permitted to make visible property division on the outside. The conservation process adopted zero structural design intervention that retains the building's original character.

CH is well embedded in its surrounding medieval town's environment by promoting contextual relevance through interdependence on history and current times. It has retained its original identity as well as met the current needs of the owner. CH respects the ecology of the area by being low-rise that allow sun to penetrate the ground, having a large open backyard that is a multi-functional green space, and providing a natural access to the Gävle river that celebrates the context and exerts little pressure on the ecosystems. The building layout addresses climatic concerns such as rising temperature through having high ceilings, entrance halls, porches, and recessed windows and doors. Double-glazed windows provide insulation and help to reduce sound transmission as well as condensation. The use of traditional materials, construction techniques, and design show unity and coherence with the medieval setting of the old town. The exterior façade of the building is consistent with the climatic concerns and social statements that the building makes. At the same time, the addition of two new toilets and a kitchen gives a modern feel to the building interiors. Some old elements such as wooden doors and windows have been reused during the conservation of the building. Fully carpeted floors were introduced in the adjacent rooms in both houses to reduce different kinds of noise. Given the above discussion, the presence of the building becomes nuanced in the urban set up it is a part of (Fig. 6.2).

CH comes under real estate laws that state that the new owner has to pay a monthly amount to the government for the renovations and the general maintenance of the area. This amount is significant as CH is a part of a historical neighborhood that imparts a sense of communal ownership. In addition, CH attracts tourism to the town even though it is not used for tourist purposes. CH is a notable example of sustainable design and adaptive reuse as it has survived over 250 years. It also shows deeper adaptability to complex environmental constraints, patterns of functional utility, and retention of materials in the urban metabolic context of the Gävle city. The building promotes effective interdependence and the capacity to engage different groups and individuals such as Gävle County Administrative Board, municipality culture department, Gävle municipality, renovation companies, architectural firms and other private actors, community organizations, residents, and tourists. This engagement influences decision-making, healthy government interventions, and awareness of cultural heritage encouraging environmental and social management. CH is able to elevate the significance and continuity of history and culture through engaging various stakeholders further embedding it in its physical and social environment. Consequently, it lends identity to the place and helps the locals emotionally associate with and take care of their city.

FIG. A. Front elevation of Clergy House (CH)
FIG. B. Google top view of CH
FIG. C & E. Ground floor and first floor plan of CH
FIG. D & F. Front and side elevation drawing of CH.
© Images & drawings at Gävle municipality – Gävleborg County, Sweden (Report number Akt 2180K-P2020/15)

Fig. 6.2 A detailed illustrative view of Clergy House (CH), Sweden

6.4.3 Case Study 3: Esmail House (EH), Bahrain

This residential building, known as Esmail house, is selected due to its significance in the urban setting in the historic area of Muharraq, Kingdom of Bahrain. Bahrain is one of the Gulf Cooperation Council (GCC) countries located west and east of Qatar and Saudi Arabia, respectively. EH was originally built around the 1900s receiving several additions during the first half of the twentieth century. It underwent major alterations that affected the original structure in the late twentieth century. The conservation of EH was initiated by the Bahrain Authority of Culture and Antiquity (BACA). Originally, EH covered an approximate area of 470 m^2. However, BACA has conserved a part of the house spread over 170 m^2. The restored portion is used for joint development by an NGO, Ta'al Shabab, which promotes social and cultural activities for different age groups. The building is part of a historic urban setting, therefore, an introverted design that includes a courtyard with spaces looking toward it helps maintain the traditional values of its context such as privacy, family gatherings, and collaboration for carrying out various routine domestic tasks. This introverted design also assists in reducing heat gain during summers as well as ensuring good passive ventilation. The restoration of the portion is done with materials similar to the original construction with some addition of glass and steel that represent the structural traces of old and the new.

Following ecological wisdom, the building caters to many environmental considerations. In order to demonstrate energy efficiency, the building has been conserved with its original elements and materials such as *badger* and *roshanah* (small openings in the wall to provide ventilation). The skin (exterior facades) and other surfaces (interior walls, ceilings, and floors) have been restored with traditional and local materials such as coral stones, hydraulic lime mortar, and teak wood, which harmonize with the neighboring buildings. The construction comprises load-bearing walls and trabeated structure (post and lintel) system, which is a common style in the area. This system reduces the need for concrete and steel that have high embedded energy. Due to contemporary demands, some new additions including bathrooms and mechanical retrofits have been incorporated using 20–30% new materials such as glass, wood, and steel for vertical circulation between consecutive stories. Almond trees have been planted that thrive well in the local environment taking three to four years to grow fully. The ratio of covered to open area is 50%. This effectively supports passive ventilation helping to minimize the carbon footprint (Fig. 6.3).

The building's orientation is carefully set according to the sun path. The design respects the existing geography and climatic conditions through passive architecture having a courtyard, ventilation through different types of openings, and use of 0.5 m thick masonry walls as thermal mass. The new function of the building as the head office of Ta'al Shabab uplifts the awareness of traditional arts such as poetry, music, and calligraphy by organizing cultural events. The building is a model of moderate consumption that reuses the existing structure promoting circular economy and avoiding unnecessary demolition of existing infrastructure. The lifecycle of the building has been extended through careful and authentic conservation. The conservation process adopted minimum design intervention that retains the building's original character integrating it with the environment and sustaining cultural and heritage values. More than a hundred years old, the restoration of this residential building presents a good example of adaptive reuse bringing it back to habituation for an NGO as its head office.

6.5 Discussion: Adaptive Reuse of Historic Buildings to Promote Urban Sustainability

As discussed earlier, EW is considered as a framework for sustainable planning inputs for enhancing UM interaction with culture, place, and history. According to Yang et al. (2019), understanding the significance of natural resources is the pragmatic approach to the overall concept of UM and EW. Reusing heritage buildings presents unique opportunities to reduce stress on the environment and adapt to contemporary

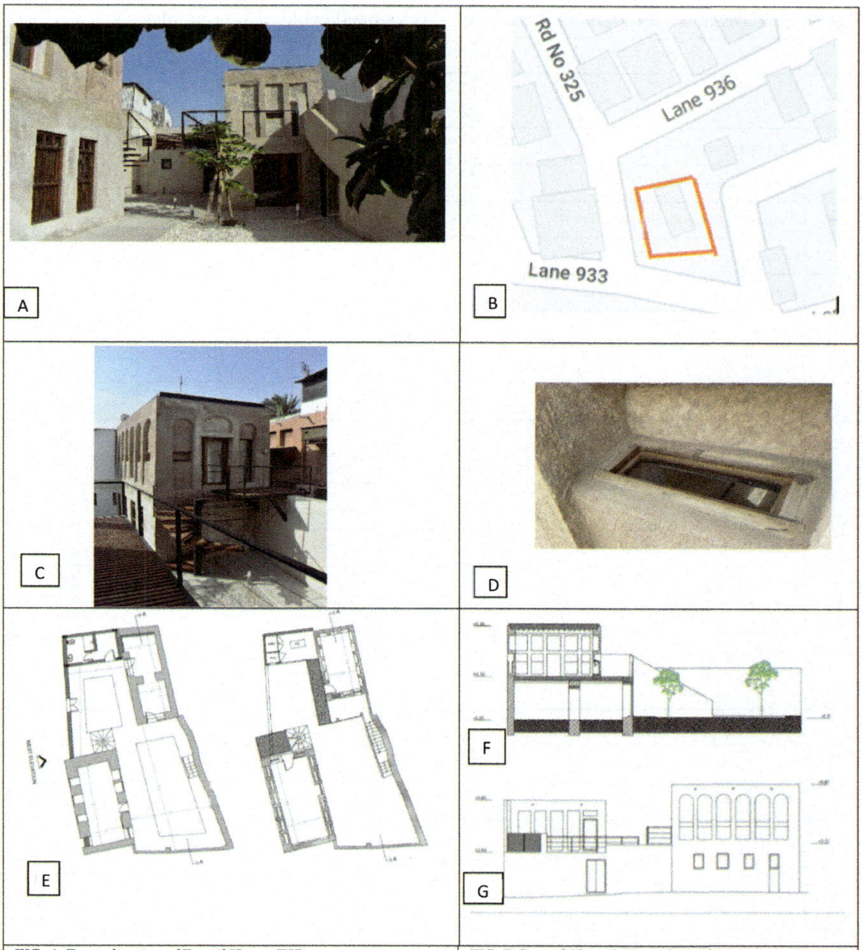

FIG. A. Front elevation of Esmail House (EH)
FIG. B. Google earth top view of EH
FIG. C. Perspective of EH
FIG. D Roshana of EH.

FIG. E Ground Floor & First Floor Plans
FIG F Section
FIG G ELEVATION
© Images & drawings by Bahrain Authority of Culture and Antiquities, Pearling Path Project

Fig. 6.3 A detailed illustrative view of Esmail House (EH), Bahrain

needs (Foster and Kreinin 2020). The literature review highlights the advantages of the reuse of heritage buildings by supporting UM in order to attain the outcomes of optimum material and energy use. The analysis from the case studies makes several contributions to the current literature by applying eight ecological indicators as identified through the literature review (Table 6.3). It shows to what degree each case study conforms to the principles.

Table 6.3 Case study comparison according to the defined ecological principles

Strong ● (blue) Medium ● (yellow) Low ● (red)

		Case study-1 Aman House	Case study-2 Clergy House	Case study-3 Esmail House
Interdependence	Community Engagement	Strong	Medium	Strong
	Cultural interdependence	Medium	Strong	Strong
	Authenticity (design, material, structure)	Strong	Strong	Medium
	Generate revenue	Strong	Medium	Low
Coherence	Unity and harmony with the environment	Strong	Strong	Strong
	Integration with the original design	Strong	Strong	Medium
	Relevance with locale	Medium	Strong	Strong
	Flexibility in design amendments	Strong	Medium	Medium
Embeddedness	Ecological embeddedness	Strong	Strong	Strong
	Building coincides with the existing geography	Strong	Strong	Strong
	Building material has a natural ability to decompose if demolish	Strong	Strong	Medium
	Embedded with social and cultural values	Medium	Strong	Strong
Functioning at Multiple Scales	The building is used for multiple functions	Medium	Strong	Strong
	Cultural activities	Medium	Low	Strong
	Tourist attraction	Medium	Medium	Medium
	Revenue generation	Medium	Low	Low
Nuanced approach	Building refers to universal values	Medium	Medium	Medium
	Elevates the sensibility of people	Strong	Strong	Strong
	Nuanced design	Strong	Strong	Strong
Circularity	The life span of the building is extendable	Strong	Strong	Strong
	Recyclable building materials	Strong	Strong	Strong
	Innovative use of traditional techniques	Medium	Strong	Strong
Conservation	A good example of conservation	Medium	Strong	Strong
	Sustains cultural value	Strong	Strong	Strong
	Appropriate building material & structure	Strong	Strong	Medium
	Requires minimum intervention	Strong	Strong	Medium
Adaptability	Easily adaptable	Medium	Medium	Medium
	Compatible with the contemporary use	Medium	Strong	Medium
	Suitable accessibility within the urban area	Strong	Strong	Strong
	Allows efficient functioning	Strong	Strong	Strong

Findings from the case studies suggest that maintenance, protection, and long-term use of the historic building share conceptual background with environmental sustainability (Chowdhury et al. 2021). This is so because adaptive reuse of these buildings not only preserves the "finite" cultural objects but the "infinite" cultural/social attributes that are equally significant and are also retained. Both aspects stress reflection on being more consciously sustainable for the current and future times. Similarly, conservation also enables one to account for and focus on existing resources instead of finding new ones. It has been argued that history transmits social meanings through time regardless of its nature. In this respect, heritage buildings are the symbolic embodiment of culture and association with a place lending it. The British guidance also supports this fact on the historic environment. Regarding this, PPG 15, states, "The presence of the physical survivals of our past adds to the quality of our lives by enhancing the familiar and cherished local scene and sustaining the sense of local distinctiveness, which is so important an aspect of the character and appearance of our towns, villages, and countryside".

Regarding embeddedness, all three case study buildings are anchored in their ecological settings and exhibit a strong character amongst the community, reflecting the physical environment and being a part of the residents' social practices and lived experiences. Durable materials used in the case studies showed coherence that has enabled the buildings to survive climate as well as periodic and remedial design interventions. Aman House, Pakistan, for instance, remained vacant for a long time but is now adopted as a drama/movie set regularly. The durability of materials, assemblies, and coherence among various building parts has made its survival to date. By reducing metabolic energy flows, the conservation of these buildings has improved the environment and added a variety to the urban area for diverse human liveability. Based on the application of various ecological principles, adaptive reuse of the selected case studies has engendered a complex set of considerations for a healthy and liveable urban area such as location, heritage, and current social and economic context (Bullen and Love 2011). These considerations are also in line with the Royal Australian Institute of Architects report (Reuse 2004) which states that a successful modification should reflect a building's historical context while also incorporating a contemporary layer in the building. Pointing out nuances, the case studies have become meaningful mainly because they offer numerous benefits by addressing social, spatial, and cultural issues. Adaptive reuse of the case studies has enabled the transition from heritage buildings into accessible and usable places that would have otherwise been left vacant, underused, or demolished. Interdependence is another key to the vitality of heritage buildings because the conservation process is not carried out in isolation. Instead, it is a collaborative and engaging practice concerning various institutions and stakeholders.

In essence, adaptive reuse of case study buildings has furnished benefits including significantly lowering environmental impacts, utilizing existing buildings as a resource for building material, reducing energy expenditure for demolishing old and constructing new structures, promoting traditional energy-efficient design, and encouraging an attitude shift toward reusing and recycling. Besides this, the cultural endowments in the form of these cases have strongly been acknowledged as valuable economic resource. Literature on adaptive reuse confirms that the economic value associated with the heritage buildings increases employment either directly or indirectly from the tourism activities (Bowitz and Ibenholt 2009; Alexandrakis et al. 2019). With reference to economic growth, the case studies enable sustainable development aligned with ecological wisdom by increasing local jobs, particularly in the case of Esmail house and Aman house. In the context of Gävle house, the rental value adds to the economic source of the owner. Each of the cases is generating welfare for society in its own capacity, often referred to as the economic value of cultural heritage (Ruijgrok 2006). Heritage is considered as a community as well as local authority asset that enhances productive economy (Greffe 1998). Given this, the conservation of the selected case studies has made sure to preserve these vital assets in the best possible condition to propel economic as well as cultural growth. It has also allowed the community frequenting and using the case studies to contribute to the present and future positively becoming a part of heritage and history.

6.6 Conclusion

By considering a city as an open ecosystem that demands a holistic and multidimensional view, this study advocates that the reuse and conservation of historic buildings are ecological principles and that adopting existing historical buildings and increasing their life cycles contributes toward desirable urban metabolism. EW is considered as a framework for sustainable planning inputs for enhancing UM's interaction with culture, place, and history. Findings from the three case studies from Pakistan, Sweden, and Bahrain and analysis of the application of ecological indicators of urban metabolism suggest that all three buildings in different regions strongly fulfill most of the criteria for assessing the relationship of urban metabolism from eight ecological perspectives. However, some exceptions exist, such as the functioning of heritage buildings at multiple scales and revenue generation. Multidisciplinarity is a core ecological principle; however, in our case study buildings, most of them fulfill one function (residence) at this moment. Similarly, interdependence in the conservation process is not strongly present yet in our case study buildings. On inspection, it can be discerned that the historic building's maintenance, protection, and long-term use have common ground with environmental sustainability. This conclusion is drawn based on the fact that we are not only preserving cultural objects that are "finite" but also concerned with how the movement can help us be more sustainable and conscious about our future. On a similar note, conservation also enables us to account for and focus on existing resources instead of finding new ones.

In contemporary heritage protection, adaptive reuse focuses on employing strategies that allow economic, cultural, and social preservation while nudging them toward a more sustainable route. If we look back at this field, we observe that architectural conservation has transitioned from simple preservation to regeneration and sustainability. This is a move in the right direction, given the global climate crisis. This is because adaptive reuse actively endorses alternatives to demolition and replacement, which were often employed in the past. This results in less energy expenditure and less impact on carbon emissions. With our society focusing more on ecological issues, demolition and similar activities are now frowned upon. This reinforcement by media and more societal awareness has facilitated architectural conservation to actively endorse adaptive reuse. Simply put, giving new life to a building allows for more societal benefits for the environment and the community. The methodology piloted in our study can help advance the theoretical explanations of EW and UM concerning adaptive reuse. Furthermore, a quantitative approach should be developed for future studies to operationalize these concepts. Finally, our study methodology and results can also provide important insights and guidelines to researchers and practitioners in conservation, ecology, and city planning.

References

Abubakar IR, Bununu YA (2020) Low carbon city: strategies and case studies. In: Leal Filho W, Azul A, Brandli L, Özuyar P, Wall T (eds) Sustainable Cities and Communities. Encyclopedia of the UN Sustainable Development Goals. Springer, Cham. pp 366–376. https://doi.org/10.1007/978-3-319-71061-7_24-2

Akbar N, Abubakar IR, Shah AA, Al-Madani W (2021) Ecological embeddedness in the Maya built environment: Inspiration for contemporary cities. Land 10(12):1360

Akbar N (2021) Surveying the ecologically embedded built environments of the preindustrial era with a focus on the classic Maya. Master's thesis, Environmental Design).

Aksamija A (2016) Regenerative design and adaptive reuse of existing commercial buildings for net-zero energy use. Sustain Cities Soc 27:185–195

Alexandrakis G, Manasakis C, Kampanis NA (2019) Economic and societal impacts on cultural heritage sites, resulting from natural effects and climate change. Heritage 2(1):279–305. https://doi.org/10.3390/heritage2010019

Almulhim AI, Abubakar IR (2021) Understanding public environmental awareness and attitudes toward circular economy transition in Saudi Arabia. Sustainability 13(18):10157

Bagheri N (2015) The built environment in the critical zone: from pre- to postindustrial cities. In: Developments in earth surface processes, vol 19. Elsevier B.V. https://doi.org/10.1016/B978-0-444-63369-9.00018-5

Bhadouria R, Upadhyay S, Tripathi S, Singh P (2022) Urban ecology and global climate change. Wiley-Blackwell

Bowitz E, Ibenholt K (2009) Economic impacts of cultural heritage—Research and perspectives. J Cult Herit 10(1):1–8 https://doi.org/10.1016/j.culher.2008.09.002

Burgess EW (1925) The growth of the city. In: Park RE, Burgess E, McKenzie RD (eds) The city, pp 47–62

Bullen P, Love P (2011) Factors influencing the adaptive reuse of buildings. J Eng Des Technol 9(1):32–46

Cárdenas-Mamani Ú, Perrotti D (2022) Understanding the contribution of ecosystem services to urban metabolism assessments: an integrated framework. Ecol Indic 136:1–18

Chang B, Hasanah A, Caesarina HM (2021) Cultural ecology of heritage building adaptation in tropical cities. In: IOP Conference Series: Earth and Environmental Science 780(1):012058. https://doi.org/10.1088/1755-1315/780/1/012058

Charruadas P (2012) The cradle of the city: the environmental imprint of Brussels and its hinterland in the High Middle Ages. Reg Environ Change 12(2):255–262. https://doi.org/10.1007/s10113-011-0212-2

Chowdhury M, Islam M, Rahman SM, Abubakar IR, Aina YA, Hasan M, Khondaker AN (2021) A review of policies and initiatives for climate change mitigation and environmental sustainability in Bangladesh. Environ Dev Sustain 23(2):1133–1161

D'Amico G, Taddeo R, Shi L, Yigitcanlar T, Ioppolo G (2020) Ecological indicators of smart urban metabolism: a review of the literature on international standards. Ecol Indic 118:106808

Dhingra M, Chattopadhyay S (2016) Advancing smartness of traditional settlements-case analysis of Indian and Arab old cities. Int J Sustain Built Environ 5(2):549–563. https://doi.org/10.1016/j.ijsbe.2016.08.004

Dimitrijevic B, Langford D, MacLeod I, Maver TW (2000) The sustainability of architectural heritage. In: Proceedings of the BIAT conference technological innovation in design and construction, Dublin

EBB 1977–1984 relates to an energy-saving plan for buildings in Sweden

Elefante C (2007) The greenest building is … one that is already built. Forum j: J Natl Trust Historic Preserv 21(4):26–36

Feldhoff T (2013) Shrinking communities in Japan: community ownership of assets as a development potential for rural Japan? Urban Des Int 18(1):99–109. https://doi.org/10.1057/udi.2012.26

Foster G (2019) Circular economy strategies for adaptive reuse of cultural heritage buildings to reduce environmental impacts. Resour Conserv Recycl. Vol 152:104507. https://doi.org/10.1016/j.resconrec.2019.104507

Foster G, Kreinin H (2020) A review of environmental impact indicators of cultural heritage buildings: a circular economy perspective. Environ Res Lett 15:043003

González RC (2020) Some insights into the theory and practice of heritage ecology. Doc d'anàlisi geogràfica, 66(1):133–158

Grammenos F, Russell PK (1997) Building adaptability: a view from the future. In: Proceedings from the second international conference: buildings and the environment, 9–12 June 1997, Paris, vol 2, pp 19–26

Girardet H (1996) The Gaia Atlas of cities: new directions for sustainable urban living. UN-HABITAT

Greffe X (1998) The economic value of heritage. University of Paris I (Pantheon-Sorbonne)

Gültekin A, Alparslan B (2011) Ecological building design criteria: a case study in Ankara. Gazi University J Sci 24:605–616

Hartel T, Dorresteijn I, Klein C, Máthé O, Moga CI, Öllerer K, Roellig M, von Wehrden H, Fischer J (2013) Wood-pastures in a traditional rural region of Eastern Europe: characteristics, management and status. Biol Cons 166:267–275. https://doi.org/10.1016/j.biocon.2013.06.020

Hornborg A (1998) Ecological embeddedness and the personhood: have we always been capitalists? Anthropol Today 14(2):3–5. https://doi.org/10.2307/2783277

Heinonen J, Säynäjoki A, Junnila S (2011) A longitudinal study on the carbon emissions of a new residential development. Sustainability 3:1170–1189. http://ec.europa.eu/environment/integration/research/newsalert/pdf/263na3.pdf

Kakoty S (2018) Ecology, sustainability and traditional wisdom. J Clean Prod 172:3215–3224. https://doi.org/10.1016/j.jclepro.2017.11.036

Kapoor V, Tripathi S, Devi RS, Srivastava P, Bhadouria R (2020) Ecological economics of an urban settlement: an overview. In: Verma P, Singh P, Singh R, Raghubanshi AS (eds.) Urban Ecology: Emerging Patterns and Social-Ecological Systems. Elsevier, pp 91–110

Langston C, Shen LY (2007) Application of the adaptive reuse application model in Hong Kong: a case study of Lui Seng Chun. Int J Strateg Prop Manag 11:193–207

Lucero LJ, Gonzalez Cruz J (2020) Reconceptualizing urbanism: insights from Maya cosmology. Front Sustain Cities 2:1–15. https://doi.org/10.3389/frsc.2020.00001

Mazraeh HM, Pazhouhanfar M (2018) Effects of vernacular architecture structure on urban sustainability case study: Qeshm Island, Iran. Front Architect Res 7(1):11–24. https://doi.org/10.1016/j.foar.2017.06.006
Michael H (2020) What is an ecological approach and how can it assist in understanding ICT take-up? Br J Edu Technol 51(3):853–866
Misirlisoy D, Günçe K (2016) Adaptive reuse strategies for heritage buildings: a holistic approach. Sustain Cities Soc 26:91–98
Mundaca L, Neij L, Worrell E, McNeil M (2010) Evaluating energy effciency policies with energy-economy models. Annu Rev Environ Resour 35:305–344
Municipality of Gävle 2020. Detailed plan för Väster 33:1, Lilla Vattugränd, proof of legal evidence Diary number: 18SB87 Livsmiljö Gävle
Newman PWG (1999) Sustainability and cities: extending the metabolism model. Landsc Urban Plan 44(25):219–226
Okubo T (2016) Traditional wisdom for disaster mitigation in history of Japanese Architectures and historic cities. J Cult Herit 20:715–724. https://doi.org/10.1016/j.culher.2016.03.014
Patten DT (2016) The role of ecological wisdom in managing for sustainable interdependent urban and natural ecosystems. Landsc Urban Plan 155:3–10. https://doi.org/10.1016/j.landurbplan.2016.01.013
Powter A, Ross S (2005) Integrating environmental and cultural sustainability for heritage properties. APT Bull: J Preserv Technol 36(4):5–11
Raffestin C, Lawrence R (1990) An ecological perspective on housing, health and well-being. J Sociol Soc Welfare 17(1):9. https://doi.org/10.15453/0191-5096.1932
Restrepo JDC, Morales-Pinzón T (2018) Urban metabolism and sustainability: precedents, genesis and research perspectives. Resour Conserv Recycl 131:216–224
Reuse DA (2004) Preserving our past, building our future. Department of Environment and Heritage. ACT: Department of Environment and Heritage, Commonwealth of Australia, p 17
Ruijgrok EC (2006) The three economic values of cultural heritage: a case study in the Netherlands. J Cult Herit 7(3):206–213
Shaowen H, Hanqi H (2017) Ecological cultures of traditional Hani villages. J Landscape Res 9(6):87–93
Shaw DJB (1977) Urbanism and economic development in a pre-industrial context: the case of southern Russia. J Hist Geogr 3(2):107–122. https://doi.org/10.1016/0305-7488(77)90208-0
Sallis JF, Cervero RB, Ascher W, Henderson KA, Kraft MK, Kerr J (2006) An ecological approach to creating active living communities. In: Annual review of public health, vol 27, pp 297–322
Tan LM, Arbabi H, Brockway PE, Densley Tingley D, Mayfield M (2019) An ecological-thermodynamic approach to urban metabolism: Measuring resource utilization with open system network effectiveness analysis. Appl Energy 254:113618
The British guidance on the Historic Environment, PPG 15. https://unece.org/DAM/env/pp/compliance/C2011-61/Communication/PlanningPolicyGuidance_PPG15_.pdf
Tomasetti DC (2016) Ecological approach. In: Encyclopedias in social sciences. The Wiley Blackwell Encyclopedia of family studies. Wiley, pp 1–6
Thomson G, Newman P (2018) Urban fabrics and urban metabolism—From sustainable to regenerative cities. Resour Conserv Recycl 132:218–229
Wilson L (2007) Lose or reuse—Managing heritage sustainability. Ulster Architectural Heritage Society, Belfast, UK
Wang X, Palazzo D, Carper M (2016) Ecological wisdom as an emerging field of scholarly inquiry in urban planning and design. Landsc Urban Plan 155:100–107. https://doi.org/10.1016/j.landurbplan.2016.05.019
Washsmuth D (2012) Three ecologies: urban metabolism and the society-nature opposition. Sociol Q 53(4):506–523
Xiang W-N (2019) Ecological wisdom: genesis, conceptualization, and defining characteristics. In: Ecological wisdom theory and practice

Yang B, Li S, Xiang W-N, Bishop I, Liao K-H, Liu J (2019) Where does ecological wisdom come from? Historical and contemporary perspectives. In: Ecological wisdom theory and practice, pp 33–56

Yin RK (2014) Case study research: design and methods, 5th edn. Sage Publication References, London

Young R (2019) Introduction: ecological wisdom as discourse. In: Ecological wisdom. Springer, Singapore, pp 1–9. https://doi.org/10.1007/978-981-13-0571-9

Zacharias J, Lei Y (2016) Villages at the urban fringe—The social dynamics of Xiaozhou. J Rural Stud 47(2016):650–656. https://doi.org/10.1016/j.jrurstud.2016.05.014

Zhang Y, Yang Z, Yu X (2015) Urban metabolism: a review of current knowledge and directions for future study. Environ Sci Technol 49(19):11247–11263

Zheng S, Han B, Wang D, Ouyang Z (2018) EcologicalWisdom and inspiration underlying the planning and construction of ancient human settlements: Case study of Hongcun UNESCO world heritage site in China. Sustainability (switzerland). https://doi.org/10.3390/su10051345

Chapter 7
Integrating Ecological and Social Concepts for Urban Metabolism Studies

Mangalasseril Mohammad Anees and Bhartendu Pandey

Abstract Urban sustainability efforts are increasingly related to concepts from self-sufficient natural ecosystems. Focusing on the metabolic characteristics and pathways of urban areas from an ecological lens can help advance our understanding of how they can be sustainable. The ecological lens allows the simultaneous notion of an urban area as a *system* and as a *natural entity* and offers insights into fundamental mechanisms that help identify natural solutions over purely technological ones. This chapter develops a systematic understanding of how urban metabolism aligns with ecosystem concepts and how social dimensions, and associated multi-dimensional and multi-scalar inequalities, add novel characteristics, tradeoffs, and synergies. In doing so, we develop an analytical framework that provides an outlook for sustainability goals relevant to urban areas including developing ecologically informed solutions and improving resource efficiency in urban areas.

Keywords Analytical frameworks · Ecosystem concepts · Social dimensions · Synergies · Tradeoffs · Urban ecology

7.1 Introduction

The concept of urban metabolism is loosely based on the analogy of the metabolism of organisms. It has also been compared to naturally existing ecosystems (Kennedy et al. 2011). Natural ecosystems are known to be self-sufficient and sustainable as they conserve mass, recycle, and are in sync with other ecosystems to maintain a balance. Urban ecosystems, or cities, on the other hand, have a high inflow of energy and mass from other ecosystems (artificial or natural) and an outflow of waste and pollutants (Verma et al. 2020). Unlike natural ecosystems, the over-dependence of urban ecosystems on other ecosystems, poor recycling of waste products and high

---

M. M. Anees (✉)
School of the Environment, Yale University, New Haven, CT, USA
e-mail: m.m.anees@yale.edu

B. Pandey
Department of Civil and Environmental Engineering, Princeton University, Princeton, NJ, USA

R. Bhadouria et al. (eds.), *Urban Metabolism and Climate Change*,
https://doi.org/10.1007/978-3-031-29422-8_7

concentration of by-product pollutants has led to an unsustainable future. The urban metabolism concept draws from the biological concept of metabolism (Rapoport 2011) where the consumption of products and release of waste and by-products is observed similar to how cities work on the whole (Bettencourt et al. 2007). Some even term cities as 'superorganisms' (Zhang et al. 2009, 2013). The concept is effectively a system-based approach to trace energy and matter usage, waste production, and flow among system components. An analogous notion of material flow, circulation, and stocks with that of ecological and social processes in cities has generated interdisciplinary interest in the subject of urban metabolism.

Formally, urban metabolism emerged as a concept when Wolman simulated a hypothetical American city and quantified its inflow and outflow rates (Wolman 1965). This led to a trending shift in research toward 'ecology of cities' from 'ecology in cities' (Pickett et al. 2016). In the recent decade, it has progressed from simple input–output processes (Wolman 1965) to the black box model (Girardet 1996) to network models (Zhang et al. 2009). Two distinct approaches to studying urban metabolism have evolved over time (Kennedy et al. 2011)—(i) 'emergy' based analysis, and (ii) material and energy-based analysis. Emergy-based analysis, the less popular of the two, employs a thermodynamic approach to compute solar equivalent energy (emjoules) used by human systems in an urban area (Odum 1983). This school of thought was developed by Odum using a systems ecologist perspective in the 1970s. The second approach follows the mass flux of cities' metabolism in terms of material and energy used, and the amount of waste produced (Li and Kwan 2017). This approach is the application of material flow analysis (MFA) developed in the 1990s. In the past two decades, this method has evolved into different simulation models and assessment methods like input–output analysis, ecological network analysis, ecosystem footprint assessment, and lifecycle assessment among others (Zhang 2013; Li and Kwan 2017). In Sect. 7.2, we explore the state of the art in the approaches development of main approaches.

Since the nascent stage, studies on urban metabolism were mostly formulated to address some form of environmental concern. For example, Wolman (1965) investigated air pollution and waste generated in US cities. Beginning in 2000, urban metabolism studies have developed to include aspects of a city's growth and its sustainability, such as, per capita food, water, material consumption (Stewart et al. 2014), nutrients (Burström et al. 1997), greenhouse gas emissions (Kennedy et al. 2009; Chen et al. 2020), material stocks and flows (Cui et al. 2019), among many others. Beyond providing an accounting and quantitative status of cities, the urban metabolism concept has been recognized as a potential tool for developing sustainable urban design and policy analysis.

The ecosystem is the basic unit of analysis in ecology. The discipline of urban ecology, and specifically the concept of 'ecology of cities', developed simultaneously with the concept of 'city as an ecosystem' to analyze their metabolism (Wolman 1965). Both these concepts benefit from the systems approach to delineate the human–environment interaction and help in understanding the interaction between subsystems of the urban area (Grimm et al. 2008). One of the key advantages cited for using the ecosystem concept in urban metabolism studies is the provision of using 'natural ecosystems as an archetype' (Broto et al. 2012). In Sect. 7.3, we discuss the relevance of the ecosystem concept in urban metabolism.

The decision-making and composition of the social system are one of the main drivers of all the energy and material flows within a city, and thus an integral part of its metabolism. This is mainly due to changes in demand for services/materials and supply of these services/materials. For example, low-density urban settlements lead to higher per capita carbon emissions due to transportation costs (Blecic et al. 2014). As social systems themselves depend on material flow, they have been incorporated into urban metabolism studies drawing from disciplines like political ecology and ecological economics. Such cross-discipline perspectives allow the incorporation of issues like urban inequality in the ambit of urban metabolism. In Sect. 7.4, we explore the development of social dimension and the importance of integrating it into mainstream urban metabolism studies. Overall, we aim to address two main themes of research in urban metabolism–sustainability of cities using the ecosystems concept and the integration of social dimension into this concept.

7.2 State of the Art

The roots of urban metabolism studies are set in accounting for the unsustainable metabolic flows of resources and energy within urban areas. Urban metabolism studies are mostly designed based on the understanding of metabolism processes and accounting. Over time, research on metabolism processes has evolved from a simple input–output linear process (Wolman 1965) to a cyclic process to a more inclusive and complex network process (Zhang 2013). Due to its ease of understanding, many of the studies still utilize the input–output linear process to understand the flow of materials and energy through the city. In terms of accounting, the two main approaches that developed over time are—MFA and emergy (Song et al. 2019). The main difference between the two approaches is the unit of measurement and ease of applicability of the former.

The MFA approach is more recognizable and understandable (Kennedy et al. 2011) by non-experts. As a result, the MFA approach is the choice of the majority of studies as compared to others (Voukkali and Zorpas 2022). Starting with Wolman's accounting of a hypothetical American city in 1965, the material flow was prioritized and applied in many cities around the world. The accounting of energy flow was a later addition (Haberl 2001) and has over time received equal consideration along with material flow (Zhang 2013). The familiarity of MFA is due to its accounting approach with the main source of the account being statistical data of various elements flowing into and out of the city. The approach has gained popularity due to its continued use in industrial ecology which focuses on the accounting of physical units of chemicals and/or bulk materials used in industries (Bringezu and Moriguchi 2002). But scaling up this approach at the city scale adds to the difficulty of deficiency of parcel-level data such as energy and water use, transportation, and economic activity data, especially in developing countries facing data constraints (Pincetl et al. 2012, 2014). Another difficulty in carrying out MFA at the city scale is the lack of a standard framework of methodology (Cui et al. 2019). Some standardized accounting frameworks at the national scale, such as *Eurostat*, focus on the physical interaction of the economy with the natural environment and provide the dataset for specific regions such as Europe (Eurostat 2018). Many studies have utilized such frameworks to attempt regional and urban-scale MFA (Cui et al. 2019; Wang et al. 2020). Currently, there are no standardized MFA frameworks at the city scale and thus comparisons among different studies are difficult. However, some studies provide indicators set to standardize the collection of urban metabolism data for cities (Kennedy and Hoornweg 2012) and specifically megacities (Kennedy et al. 2014).

The emergy approach differs from the popular MFA approach in terms of accounting for the metabolic flows using a standardized unit of measurement—available solar energy. Postulated by systems ecologist Odum in the 1970s, this approach overcomes a specific gap in other approaches, i.e., qualitative differences of mass or energy flows (Pincetl et al. 2012). A key concept in this approach is that of *transformity*, which is the energy input per unit of the energy output. Using a set of conversion factors for energy flows, material flows, and monetary flows, this approach can study both ecological and economic systems (Zhang et al. 2011). However, its application was limited due to challenges in comparing and converting a wide range of materials that flows through the city into a common unit of solar energy (seJ). In principle, the steps involved in this approach are similar to other approaches, such as defining systems boundaries, categorizing data, establishing links to identify flows, measuring cross-boundary flows (Yang et al. 2014).

As per the developed needs and to synthesize deeper understanding, metabolism studies have also focused on a specific metabolic flow in the city, for example—water (Lv et al. 2020), transport, electronic waste (Liu et al. 2006), solid waste (Liang and Zhang 2012), etc. Some studies have attempted to study the relationship between urban metabolism and land use. For example, Chen et al. (2011) estimated the relationship between different urban forms and energy consumption. To maximize the utility of MFA, some studies have combined it with Life Cycle Assessment (LCA) studies that also quantify the environmental impacts from the life cycle of in-use materials (Lopes Silva et al. 2015). Such addition of evaluating potential environmental impacts has broadened the scope of urban metabolism studies.

7.3 Ecosystem Concept in Urban Metabolism

Urban ecosystems can be viewed with a similar lens as used for other ecosystems in terms of their operating rules. The presence of natural features (such as forest patches, lakes, canals, and rivers) in urban ecosystems forms a hybrid system where such features, along with human developed features for habitation, transport, recreation, etc., and systems for energy and material exchange lead to a mosaic of artificial and natural subsystems. Being multifunctional, natural features are the source of valuable ecosystem services in urban areas. These features are either inherently part of the now-modified urban area or added to them later to gain ecosystem services in an unsuitable land. Many natural ecosystem studies have proven the benefits of natural designs in maintaining them and ensuring their sustainability. However, economic priorities, lack of advanced planning, and competition for land as a resource within cities have restricted or replaced natural features. Urban metabolism research, to an extent, recognizes the benefits of studying urban ecosystems through the operating rules of natural ecosystems that are governed by ecological principles (Zhang 2013). The idea of viewing the city as a system and natural entity—for accounting purpose and as an approach to simulate natural designs as a planning tool are two areas of research that needs to be curated and blended for a sustainable future. In this and the following section, we discuss these ideas and how social dimensions can fit into these.

While borrowing the fundamentals of the ecosystem concept, urban ecosystems represent a quasi-ecosystem that performs inefficiently in all three main aspects of metabolism—sources, consumption patterns, and recycling of resources. Urban ecosystems, by design, lack the production capacity to sustain their human population using various resources and thus are reliant on other artificially managed or natural ecosystems for resources. While many advocates for developing within-city resource generation capacities such as urban agriculture for food production, many of the other resources (such as minerals) and energy cannot be generated in-house to support the whole city. Due to the extensive dependence of cities on these vital inflows from other ecosystems, urban ecosystems are burdened with environmental problems. The problems are two-fold, one is off to maintain a sustainable supply of materials and energy, and the second of eliminating by-products that are generated as a result of their processing within the city for desirable use. These problems usually have a cascading effect within the city area due to commonly employed artificial (structural) solutions that are not in sync with natural features around it. Thus, focusing on reducing consumption (which decides the requirement of the resources from other sources) and improving the recycling/management of by-products (which will decide the extent of environmental problems caused within) is key to utilizing the urban metabolism concept (Zhang 2013). Due to the efficient cycling of materials and energy by biological systems or natural ecosystems, analogous comparisons can be derived for urban ecosystems to improve their efficiency. Although a system as large as a city can be far from replicating natural ecosystems in their efficiency, any improvement in urban systems would have many fold benefits as they have multi-scalar interconnections, and thus improvement in one may lead to improvement in many others. Also, by-products generated from one process can be used as an input in another, a key concept adopted by industrial ecology.

The expectation of urban ecosystems to be able to behave as any other ecosystem and achieve sustainability, given the consumption patterns of current urban populations, is often challenged (Broto et al. 2012). Urban ecologists have rather provided a more realistic model of understanding urban ecosystems by acknowledging their complexity, adaptivity, and interlinking between ecological and social systems (Alberti 2004, 2008). This branch of study focuses on the resilience of urban systems and challenges the notion that ecological principles and expectations can be the solution to developing an ideal city.

Despite the understanding of the natural ecosystem and borrowing their concepts to streamline accounting, urban metabolism studies lack integration of ecological principles that can further strengthen the planning and management of urban ecosystems and grow beyond an accounting technique. Urban metabolism studies can adopt the best practices of the urban ecosystems approach by focusing on the *Ecology of the city* paradigm of urban ecology, which, as system science, allows greater integration between disciplines. Here, the focus is on simplifying urban areas by studying their form, structure, and functioning (McPhearson et al. 2016). The advantage of focusing on this paradigm is its capability to include social patterns and processes. For example, inequality and its relation to vulnerability. Strong interconnections between the structure and function of urban ecosystem components, such as urban density and vulnerability to heat waves, can allow urban metabolism studies to derive pre-emptive measures that reduce the formation of structures with deleterious functions. The study of interrelationships between pattern and process in urban ecology showcases the best use of integrating multidisciplinary approaches as it allows a simplification of urban complexity and allows systems thinking. Urban ecosystems or specifically the field of urban ecology interacts with fields like sociology, public health, and urban planning to relate the users who benefit from ecosystem services to natural features which provide them. Urban metabolism research can also adopt vision from such interconnections and move closer to humans and their needs from an ecological viewpoint, rather than focusing only on materialistic and energy consumption, and their related environmental problems. Thus, the adoption of a transdisciplinary approach, following in the footsteps of urban ecology research, is expected to orient urban metabolism research into a more human-centric approach. Some have also pitched for a more comprehensive framework that integrates urban ecology and urban metabolism (McPhearson et al. 2016). A unifying framework, that integrates these overlapping subjects, and further includes the social dimension in a way that looks beyond the human-centric problems and focuses on human-centric benefits is necessary.

Urban metabolism, or the parent field of industrial ecology, has evolved since its inception by adopting various ecological principles. Its biological analogy of resource recycling and efficient cycling of materials and energy is well-recognized (Ayres Robert and Ayres Leslie 2002). Other, more specific ecological analogies are superficial and often debated for their true adoption, for example, the concept of ecological resilience. But beyond these concepts, there are many other concepts and principles that define how an ecosystem function.

7.4 Social Dimensions in Urban Metabolism

In the context of urban ecosystems, the concept of metabolism has interfaced with different disciplines used to study urban areas, such as societal metabolism, cultural anthropology, social geography, biology, ecology, and urban ecology (Ayres Robert and Ayres Leslie 2002). In the field of urban ecology, this is associated with the evolution of the '*ecology of the cities*' that utilizes systems theory principles while '*ecology in the cities*' continues as a separate line. The term metabolism has its origins in material exchanges within the human body and over time been applied to other levels of the biological hierarchy, most importantly at the ecosystem level. The concept of metabolism, due to its origins in the basic biological domain, about the metabolism of cells and overall, a single person, is restrictive in explaining the collective metabolism of cities where the human population lives together. Thus, the expansion of the idea and analogy of the metabolism of a 'member of the population' to the 'metabolism of a system composed of all the members' and their associated material and energy flow is challenging, yet promising. This is due to the advantage of principles of metabolism that are used to achieve complex self-organizing processes required to maintain the body to achieve sustainability. The social organization of humans as a species affects its individual metabolism, i.e., materials required for some parts of individuals' sustenance are not just through direct exchanges with the environment, but through other materials and energy found somewhere else (Ayres Robert and Ayres Leslie 2002).

Integration of social dimension to urban metabolism is strongly supported by scholars of political ecology, who view the city as a system where humans shape the metabolism through desires and needs, and there exists a politics of distribution of energy and materials (Broto et al. 2012). The production and reproduction of social and infrastructural inequalities closely intertwine with urban metabolism. Energy and material resource exchanges as well as flows of people and information are foundational to urban systems. These interactions manifest within and between urban areas (Bettencourt 2013) and extend between rural and urban areas (Seto et al. 2012). The socioenvironmental characteristics of these interactions tend to be unequal, with emergent socio-spatial inequalities that cut across geographical scales (Dinarès 2014). For instance, inequalities in per capita or per-household energy and material resource consumption endure within (Tong et al. 2021) and across cities (Kennedy et al. 2015), often with patterns that highlight socio-economic disparities (Goldstein et al. 2022). While a multitude of inter- and intra-disciplinary theoretical constructs exist outlining the significance of social inequalities vis-à-vis urban metabolism, an empirically based understanding is relatively underdeveloped (Weisz and Steinberger 2010). Advancing an empirically based understanding of the complexity of urban resource consumption and flows in terms of socio-spatial heterogeneities and inequalities necessitates fine-scale datasets, i.e., at the individual, household, or neighborhood scales (Clark et al. 2022). Such datasets are seldom gathered and analyzed. These datasets can offer a critical source of information in unpacking patterns of social inequalities pertaining to urban metabolic inputs and provide input toward advancing social equity and bringing to the fore underserved population groups and spaces.

Beyond consumption, infrastructure, a critical enabler of urban metabolism, remains intrinsically subject to unequal distributions (Pandey et al. 2022). Inequalities in urban infrastructure availability and access are often produced and reinforced through differentials predicated on economic and political hierarchies (Broto et al.

2012). However, some inequalities in infrastructure distributions are expected due to biophysical and economic constraints. For instance, designing a road network that serves everyone equitably often has foundational design constraints, which can be theoretically resolved (Brelsford et al. 2018). Still, implementing a fully connected spatial network with relatively equal link lengths is often challenging. Furthermore, changing urban infrastructure and population growth dynamics can create lags in infrastructure provisioning (Pandey et al. 2022). Overall, both urban consumption and infrastructure access are subject to social inequalities, but their characteristics can vary across regional contexts.

7.5 Analytical Frameworks

For urban metabolism to truly adopt an ecological perspective and strengthen social dimension, concepts from these fields must be relooked at and integrated. This is expected to comprehend a better understanding of processes and flows that can be used in metabolism studies. But this requires integration of isolated notions of currently used and those concepts and principles that are yet to be studied from a metabolism perspective. As flows, systems, and nested systems are already utilized in the urban metabolism concept, incorporating core principles of ecosystem functioning and urban inequality such as—networks, cycles, ecosystem services, resilience, and infrastructural and social inequality can be easily integrated into studies (Fig. 7.1).

Networks are an integral part of urban metabolism studies and are well-represented by links between different sources, consumers, and sinks. However, from an ecological perspective, the network of biotic interactions is not connected to urban metabolic flows. The biotic network and interdependence which forms a community can be disturbed by a particular metabolic flow. For example, over-extraction of a particular plant species for timber or medicinal value leads to a cascade of impact in the community and is unaccounted for in metabolic flows. Thus, the impact of a metabolic flow on several interrelated species remains unaccounted for and as a result can leads to the deterioration of a particular species over time and ultimately hamper the metabolic flow.

Cycles are also an integral part of urban metabolism studies and represent the main principle behind envisioning a circular metabolism for cities. However, metabolism studies are focused mostly on the cycles of material and its consumption through industries/human usage. The impact of human-led metabolic flows on the cycle of resources that keeps within and surroundings of the urban ecosystem in balance is unaccounted for.

Ecosystem services are the benefits obtained from natural features of the landscape but are often neglected in accounting. However, ecosystem services on their own have been developed as an accounting framework that monetizes benefits humans obtain from nature, and thus their incorporation into urban metabolism studies can be simplified. A monetary value attached to the supply and usage of services from nature

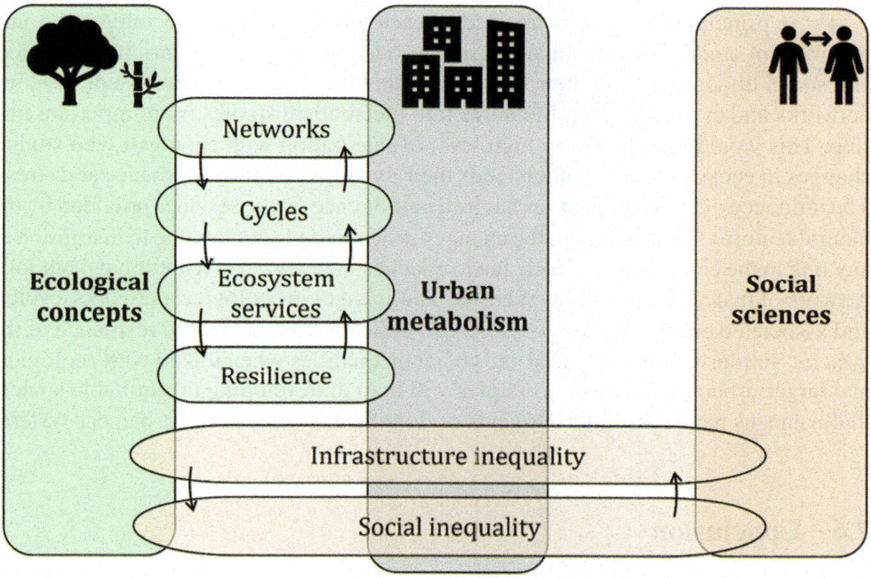

Fig. 7.1 Analytical framework showing the integration of key concepts from ecology and social sciences into urban metabolism

can lead to the development of metabolism flows that acknowledge the necessity of maintaining a healthy status of natural features. Recent literature has highlighted the value of such ecosystem service-urban metabolism integration.

Resilience is the property of ecosystems to cope with disturbances or stress and return to a stable state. Urban ecosystems, due to their complexity and dynamic interaction within the system, are always in one form of stability rather than a single state. The shifting between these stable states is through the resilience property. Identification of the resilience state, when different components of the ecosystem adapt to disturbances or stress and accounting for the flows in different states, is important to capture the true metabolic status.

Infrastructural inequalities are a characteristic of urban areas and can manifest as the amount of variation in infrastructure distributions considering availability, provisioning, or access, within and across urban areas and regions. Infrastructure inequalities with combined interactions with social inequalities can influence the bi-directional interactions between urban areas and natural ecosystems (Hamann et al. 2018).

Social inequalities, one of the core topics of study in social sciences, are multi-dimensional and more experienced in urban areas than in rural areas (Nijman and Wei 2020). Like infrastructure inequalities, social inequalities shape and are shaped by urban metabolism. A better understanding of social inequalities in the urban metabolism context can benefit from investigations that draw upon ecological as well as social science concepts (Groffman et al. 2017).

These principles are interconnected to each other and thus their integration into metabolism studies requires a wholesome understanding of the various ecological and social dimensions and their functioning that form an urban ecosystem. Linking networks and cycles of natural features into metabolism studies is an important first step. This would require a very high level of integration with ecologist who studies changes in ecosystems and understands their dynamics under disturbance and stress. The influence of network changes has a clear influence on the services provided by the natural features. Thus, gauging the changes in accounted services due to disruption of any link in the ecological network due to changes in the usual conventional metabolic network of materials and energy is easier. The changes in ecological networks, cycles, and associated ecosystem services lead to a collective impact on the resilience of the system. Similarly, infrastructural and social inequalities when linked with ecological and larger urban metabolism principles will help in developing an equitable society that balances nature with the demands of dominant species in the urban ecosystem.

7.6 Conclusion

A key advantage of metabolism studies is the ease of reorienting indicators to answer the sustainability quotient of any city. This is due to the inherent link between energy use, material consumption, waste recycling, and infrastructure development and sustainability of an area. To strengthen the adaptation of ecological concepts in urban metabolism studies, it is necessary to discuss the critiques of our assumptions and analogy with an organism and other ecosystems. Golubiewski (2012) puts forward an ecological critique of urban metabolism concept in urban settings and questions the organismic comparison of cities. An important note is that '*Ecology is not synonymous with biology but rather developed into a distinct discipline*'. This is based on the established understanding that ecology is the study of interactions between an organism and its surrounding while biology is the study of the organism. Thus, rather than comparing cities to organisms and their metabolism, it is best to accept cities as ecosystems, like other ecosystems. A key problem observed with organismal comparison is the artificial boundary that is assumed for cities. Not limiting to organismic comparisons, the analogy drawn with other ecosystems is also problematic to a certain extent, and thus researchers should be cautious of borrowing concepts. A key disadvantage of analogy is the inability to truly integrate socio-economic components into metabolism studies as noted in the previous section. The integration of social components into urban metabolism is even more important given that human's relationship with the environment, and thus influenced metabolism, is influenced by societal norms and human behavior (McDonald and Patterson 2007). Going forward, urban metabolism research should integrate key concepts of ecology, but by considering cities as an ecosystem on its own and attempt to answer—*How can we incorporate ecosystem concepts in urban settings?* The other area of focus can be on integrating the social component into metabolism studies by avoiding the pitfalls of deriving analogy with the organism and/or ecosystem, i.e., *How can we incorporate social component as equally as other material and energy flow in the cities?* For both these questions, it is important to acquire and strengthen methodological skills in the areas of—systems thinking and data science and develop multidisciplinary synthesis. This synthesis needs to be one that relates urban metabolism to different domains of urban sustainability (Song et al. 2019).

References

Alberti M (2008) Modeling the urban ecosystem: a conceptual framework. In: Marzluff JM, Shulenberger E, Endlicher W et al. (eds) Urban Ecology. Springer, Boston, MA, pp 623–646. https://doi.org/10.1007/978-0-387-73412-5_41

Alberti M (2004) Ecological resilience in urban ecosystems: linking urban patterns to human and ecological functions Urban Ecosyst 7:241–265

Ayres RU, Ayres LW (2002) A handbook of industrial ecology. Edward Elgar Publishing, p 704

Bettencourt LMA (2013) The origins of scaling in cities. Science 340(June):1438–1442

Bettencourt LMA, Lobo J, Helbing D, Kühnert C, West GB (2007) Growth, innovation, scaling, and the pace of life in cities. Proc Nat Acad Sci USA 104(17):7301–7306. https://doi.org/10.1073/pnas.0610172104

Blecic I, Cecchini A, Falk M, Marras S, Pyles DR, Spano D, Trunfio GA (2014) Urban metabolism and climate change: a planning support system. Int J Appl Earth Obs Geoinf 26:447–457. https://doi.org/10.1016/j.jag.2013.08.006

Brelsford C, Martin T, Hand J, Bettencourt LMA (2018) Toward cities without slums: topology and the spatial evolution of neighborhoods. Sci Adv 4(8):1–9. https://doi.org/10.1126/sciadv.aar4644

Bringezu S, Moriguchi Y (2002) Material flow analysis. A handbook of industrial ecology, pp 79–90. https://doi.org/10.4337/9781843765479.00017

Broto VC, Allen A, Rapoport E (2012) Interdisciplinary perspectives on urban metabolism. J Ind Ecol 16(6):851–861. https://doi.org/10.1111/j.1530-9290.2012.00556.x

Burström F, Brandt N, Frostell B, Mohlander U (1997) Parallel session C: MFA for regional and local materials management material flow accounting and information for environmental policies in the city of Stockholm

Chen Y, Li X, Zheng Y, Guan Y, Liu X (2011) Estimating the relationship between urban forms and energy consumption: a case study in the Pearl River Delta, 2005–2008. Landsc Urban Plan 102(1):33–42. https://doi.org/10.1016/j.landurbplan.2011.03.007

Chen S, Chen B, Feng K, Liu Z, Fromer N, Tan X, Alsaedi A, Hayat T, Weisz H, Schellnhuber HJ, Hubacek K (2020) Physical and virtual carbon metabolism of global cities. Nat Commun 11(1):182. https://doi.org/10.1038/s41467-019-13757-3

Clark LP, Tabory S, Tong K, Servadio JL, Kappler K, Xu CK, Lawal AS, Wiringa P, Kne L, Feiock R, Marshall JD, Russell A, Ramaswami A (2022) A data framework for assessing social inequality and equity in multi-sector social, ecological, infrastructural urban systems: focus on fine-spatial scales. J Ind Ecol 26(1):145–163. https://doi.org/10.1111/jiec.13222

Cui X, Wang X, Feng Y (2019) Examining urban metabolism: a material flow perspective on cities and their sustainability. J Clean Prod 214:767–781. https://doi.org/10.1016/j.jclepro.2019.01.021

Dinarès M (2014) Urban metabolism: a review of recent literature on the subject. Universitat Autonoma de Barcelona 60:551–571. https://doi.org/10.5565/rev/dag.134

Eurostat (2018) Economy-wide material flow accounts HANDBOOK 2018 edition. https://doi.org/10.2785/158567

Girardet H (1996) The Gaia Atlas of cities: new directions for sustainable urban living. Gaia, UK

Goldstein B, Reames TG, Newell JP (2022) Racial inequity in household energy efficiency and carbon emissions in the United States: an emissions paradox. Energy Res Soc Sci 84:102365. https://doi.org/10.1016/j.erss.2021.102365

Golubiewski N (2012) Is there a metabolism of an urban ecosystem? An ecological critique, pp 751–764. https://doi.org/10.1007/s13280-011-0232-7

Grimm NB, Faeth SH, Golubiewski NE, Redman CL, Wu J, Bai X, Briggs JM (2008) Global change and the ecology of cities. Science 319(5864):756–760. https://doi.org/10.1126/science.1150195

Groffman PM, Avolio M, Cavender-Bares J, Bettez ND, Grove JM, Hall SJ, Hobbie SE, Larson KL, Lerman SB, Locke DH, Heffernan JB, Morse JL, Neill C, Nelson KC, O'Neil-Dunne J, Pataki Di E, Polsky C, Chowdhury RR, Trammell TLE (2017) Ecological homogenization of residential macrosystems. Nat Ecol Evol 1(7):0191. https://doi.org/10.1038/s41559-017-0191

Haberl H (2001) The energetic metabolism of societies part I: accounting concepts. http://mitpress.mit.edu/JIE. http://www.univie.ac.at/iffsocec

Hamann M, Berry K, Chaigneau T, Curry T, Heilmayr R, Henriksson PJG, Hentati-Sundberg J, Jina A, Lindkvist E, Lopez-Maldonado Y, Nieminen E, Piaggio M, Qiu J, Rocha JC, Schill C, Shepon A, Tilman AR, Van Den Bijgaart I, Wu T (2018) Inequality and the biosphere. Annu Rev Environ Resour 43:61–83. https://doi.org/10.1146/annurev-environ-102017-025949

Kennedy C, Hoornweg D (2012) Mainstreaming urban metabolism. J Ind Ecol 16(6):780–782. https://doi.org/10.1111/j.1530-9290.2012.00548.x

Kennedy C, Steinberger J, Gasson B, Hansen Y, Hillman T, Havránek M, Pataki D, Phdungsilp A, Ramaswami A, Mendez GV (2009) Greenhouse gas emissions from global cities. Environ Sci Technol 43(19):7297–7302. https://doi.org/10.1021/es900213p

Kennedy C, Pincetl S, Bunje P (2011) The study of urban metabolism and its applications to urban planning and design. Environ Pollut 159(8–9):1965–1973. https://doi.org/10.1016/j.envpol.2010.10.022

Kennedy C, Stewart ID, Ibrahim N, Facchini A, Mele R (2014) Developing a multi-layered indicator set for urban metabolism studies in megacities. Ecol Indic 47:7–15. https://doi.org/10.1016/j.ecolind.2014.07.039

Kennedy CA, Stewart I, Facchini A, Cersosimo I, Mele R, Chen B, Uda M, Kansal A, Chiu A, Kim KG, Dubeux C, La Rovere EL, Cunha B, Pincetl S, Keirstead J, Barles S, Pusaka S, Gunawan J, Adegbile M et al (2015) Energy and material flows of megacities. Proc Natl Acad Sci USA 112(19):5985–5990. https://doi.org/10.1073/pnas.1504315112

Li H, Kwan M-P (2017) Advancing analytical methods for urban metabolism studies. Resour Conserv Recycl, July, 0–1. https://doi.org/10.1016/j.resconrec.2017.07.005

Liang S, Zhang T (2012) Comparing urban solid waste recycling from the viewpoint of urban metabolism based on physical input-output model: a case of Suzhou in China. Waste Manage 32(1):220–225. https://doi.org/10.1016/j.wasman.2011.08.018

Liu X, Tanaka M, Matsui Y (2006) Generation amount prediction and material flow analysis of electronic waste: a case study in Beijing, China. Waste Manag Res 24(5):434–445. https://doi.org/10.1177/0734242X06067449

Lopes Silva DA, De Oliveira JA, Saavedra YMB, Ometto AR, Rieradevall I Pons J, Gabarrell Durany X (2015) Combined MFA and LCA approach to evaluate the metabolism of service polygons: a case study on a university campus. Resour Conserv Recycl 94:157–168. https://doi.org/10.1016/j.resconrec.2014.11.001

Lv H, Yang L, Zhou J, Zhang X, Wu W, Li Y, Jiang D (2020) Water resource synergy management in response to climate change in China: from the perspective of urban metabolism. In: Resources, conservation and recycling, vol 163. Elsevier B.V. https://doi.org/10.1016/j.resconrec.2020.105095

McDonald GW, Patterson MG (2007) Bridging the divide in urban sustainability: from human exemptionalism to the new ecological paradigm. Urban Ecosyst 10(2):169–192. https://doi.org/10.1007/s11252-006-0017-0

McPhearson T, Pickett STA, Grimm NB, Niemelä J, Alberti M, Elmqvist T, Weber C, Haase D, Breuste J, Qureshi S (2016) Advancing urban ecology toward a science of cities. Bioscience 66(3):198–212. https://doi.org/10.1093/biosci/biw002

Nijman J, Wei YD (2020) Urban inequalities in the 21st century economy. Appl Geogr 117(March):102188. https://doi.org/10.1016/j.apgeog.2020.102188

Odum HT (1983) Systems ecology: an introduction. Wiley, Singapore

Pandey B, Brelsford C, Seto KC (2022) Infrastructure inequality is a characteristic of urbanization. Proc Natl Acad Sci USA 119(15):4–11. https://doi.org/10.1073/pnas.2119890119

Pickett STA, Cadenasso ML, Childers DL, Mcdonnell MJ, Zhou W (2016) Evolution and future of urban ecological science: ecology in, of, and for the city. Ecosyst Health Sustain 2(7):e01229. https://doi.org/10.1002/ehs2.1229

Pincetl S, Bunje P, Holmes T (2012) An expanded urban metabolism method: toward a systems approach for assessing urban energy processes and causes. Landsc Urban Plan 107(3):193–202. https://doi.org/10.1016/j.landurbplan.2012.06.006

Pincetl S, Chester M, Circella G, Fraser A, Mini C, Murphy S, Reyna J, Sivaraman D (2014) Enabling future sustainability transitions: an urban metabolism approach to Los Angeles Pincetl et al. Enabling future sustainability transitions. J Ind Ecol 18(6):871–882. https://doi.org/10.1111/jiec.12144

Rapoport E (2011) Interdisciplinary perspectives on urban metabolism. In: University College London, Development Planning Unit. https://doi.org/10.1111/j.1365-2699.2010.02377.x

Seto KC, Reenberg A, Boone CG, Fragkias M, Haase D, Langanke T, Marcotullio P, Munroe DK, Olah B, Simon D (2012) Urban land teleconnections and sustainability. Proc Natl Acad Sci USA 109(20):7687–7692. https://doi.org/10.1073/pnas.1117622109

Song Y, van Timmeren A, Wandl A (2019) A literature review and categorisation of sustainability aimed urban metabolism indicators: a context, indicator, mechanism, outcome analysis. Reg Stat 9(1):54–71. https://doi.org/10.15196/RS090103

Stewart I, Kennedy C, Facchini A (2014) Metabolism of megacities: a review and synthesis of the literature

Tong K, Ramaswami A, Xu C, Feiock R, Schmitz P, Ohlsen M (2021) Measuring social equity in urban energy use and interventions using fine-scale data. Proc Natl Acad Sci USA 118(24):e2023554118. https://doi.org/10.1073/pnas.2023554118

Verma P, Singh P, Singh R, Raghubanshi AS (2020) Urban ecology: emerging patterns and social-ecological systems. Elsevier, ISBN: 9780128207307

Voukkali I, Zorpas AA (2022) Evaluation of urban metabolism assessment methods through SWOT analysis and analytical hierocracy process. Sci Total Environ 807:150700. https://doi.org/10.1016/j.scitotenv.2021.150700

Wang X, Li Y, Liu N, Zhang Y (2020) An urban material flow analysis framework and measurement method from the perspective of urban metabolism. J Cleaner Prod 257:120564. https://doi.org/10.1016/j.jclepro.2020.120564

Weisz H, Steinberger JK (2010) Reducing energy and material flows in cities. Curr Opin Environ Sustain 2(3):185–192. https://doi.org/10.1016/j.cosust.2010.05.010

Wolman A (1965) The metabolism of cities. Sci Am 213(3):179–190. https://doi.org/10.1038/scientificamerican0965-178

Yang D, Kao WTM, Zhang G, Zhang N (2014) Evaluating spatiotemporal differences and sustainability of Xiamen urban metabolism using emergy synthesis. Ecol Model 272:40–48. https://doi.org/10.1016/j.ecolmodel.2013.09.014

Zhang Y (2013) Urban metabolism: a review of research methodologies. Environ Pollut 178:463–473. https://doi.org/10.1016/j.envpol.2013.03.052

Zhang Y, Yang Z, Yu X (2009) Ecological network and emergy analysis of urban metabolic systems: Model development, and a case study of four Chinese cities. Ecol Model 220(11):1431–1442. https://doi.org/10.1016/j.ecolmodel.2009.02.001

Zhang Y, Yang Z, Liu G, Yu X (2011) Emergy analysis of the urban metabolism of Beijing. Ecol Model 222(14):2377–2384. https://doi.org/10.1016/j.ecolmodel.2010.09.017

Part III
Sustainable Urban Metabolism and Urban Planning

Chapter 8
Role of Sustainable Urban Metabolism in Urban Planning

Mushtaq Ahmad Dar, Riya Raina, Rishikesh Singh, Shalinder Kaur, Harminder Pal Singh, and Daizy R. Batish

Abstract The term "metabolism" refers to the fundamental process whereby living organisms maintain a constant exchange of products and materials to their surroundings in order to sustain their growth and development. Likewise, urban metabolism is considered as a multidisciplinary field for understanding various aspects of cities. In-depth understanding of various dimensions of urban metabolism shows strong interlinkages with the sustainability of a city. Urban metabolism based on the judicious resource utilisation, self-sufficiency, and resilience of a city provide a road map for the sustainable urban planning. It is proposed that urban metabolism is measured as part of the planning practises, where various metabolic elements like resources (inputs), processes (metabolism of resources), wastes/emissions (outputs), and their fate and transport (linear and/or circular) be evaluated at different scales. The estimation of these components needs to be specific and balance the complexity and practicability of the planning practises. Land, a major component of urban planning, plays a crucial role in supporting the metabolic structure of a city. This chapter investigates academic scope of urban metabolism, which includes its relationship with city sustainability and urban planning. Scopus-based literature survey is performed to observe the global research scenario on urban metabolism and urban planning. The findings revealed that the topic is well-explored and has been given ample attention by the scientific community in the developed nations, however, there is a paucity of information from the developing and other nations. The limitations are also examined in light of urban sustainability knowledge in order to identify research gaps and requirements within the field of urban metabolism. Overall, appropriate urban planning, particularly in the developing economies, based on the principles of urban metabolism can be a sustainable approach for reducing the carbon footprints of cities and mitigating climate change in the coming decades.

Mushtaq Ahmad Dar and Riya Raina contributed equally.

M. A. Dar · R. Singh · S. Kaur · D. R. Batish (✉)
Department of Botany, Panjab University, Chandigarh 160014, India
e-mail: daizybatish@pu.ac.in; daizybatish@yahoo.com

R. Raina · H. P. Singh
Department of Environment Studies, Panjab University, Chandigarh 160014, India

R. Bhadouria et al. (eds.), *Urban Metabolism and Climate Change*,
https://doi.org/10.1007/978-3-031-29422-8_8

Keywords Climate change · Developmental processes · Resource recycling · Sustainability indicators · Urban ecology

8.1 Introduction

Cities are the complex systems characterised by pervious and impervious surfaces, built structures, and are inhabited by a wide range of living organisms, including humans, animal life, and vegetative cover (Verma et al. 2020). Cities convert crude ingredients like fuel and water to build the environment as well as human organic matter and effluents (Decker et al. 2000; Piña and Martínez 2014). As an outcome of the energy usage, the system produces goods and generates effluents or wastes (Verma et al. 2021). Thus, the idea that urban areas (cities) or municipalities are like ecological systems is also appropriate. Assessing the link among resource usage, goods, and waste generation is essential to understand how urban ecosystems survive. The concept of urban metabolism is essential for understanding the emerging municipalities and societies with respect to their sustainability (Wachsmuth 2012; Carréon and Worrell 2018). Urban metabolism represents the sum of total amount of analytical and socio-economic methods or processes that take place in the municipalities leading to power production, development, and reducing wastes (Maranghi et al. 2020). The idea of urban metabolism is partially built on a similarity with organism's metabolism, though similarities among city centres and ecological systems can also be drawn in other ways (Thomson and Newman 2018). Wolman (1965) defined "urban metabolism" by equating the city with an ecological system and outlining how resources and materials flowed into the system in a manner similar to how living creatures in an environment consume resources like light from the sun and produce food (Tan et al. 2019). Urban metabolism was described by Kennedy et al. (2007) as a combination of energy production, waste reduction, and sustainable living techniques. Technically, learning of urban metabolism of a city requires a detailed assessment of city's inputs and outputs, as well as the quantity of energy, water, nutrients, resources, and effluents/wastes utilised and generated in a given time frame (Walker et al. 2014).

Nowadays, urban metabolism research has become quite popular in the scientific community (Perrotti 2020; Padovan et al. 2022; Singh et al. 2022). Modern cities have a large linear metabolic activity with increased transfers of materials and energy. Land as a commodity has been identified as a major element in bridging the theoretical and practical divides in the metabolic structure (Oliveira and Vaz 2021). This necessitates a detailed appreciation of the physical form of cities, such as the most recent advances in urban morphology, as well as the identification of city centres as ecosystems via improved relationships among green and established areas and development of ecological services (Oliveira and Vaz 2021). Urban centres, like organisms, intake raw materials from their environment and secrete waste materials. Healthy ecosystems are usually self-reliant, and are subsidised by sustainable inputs, and they frequently preserve mass via composting by decomposing microorganisms (e.g., detritivores) (Singh et al. 2020). The urban areas would be much more self-sustaining if they possessed such characteristics. In some ways, the concept of an ecological system is the goal for promoting sustainable urban areas (Golubiewski 2012; Zhang et al. 2015). Once a system is unable to gain the resources, it requires to remain alive from within (i.e., through primary production as plants do in an ecological system), and it must attain those resources from the surroundings that facilitate the structure. Likewise, if the existing system is unable to accumulate the components and effluents generated by its metabolic reactions, those components and garbage must be oxidised. Thus, study on urban metabolism needs to focus on the inputs and consumption of raw materials as well as their cycling within the system and sludge production, treatment, and reusing (Kennedy et al. 2007; Mohan et al. 2020).

Several researchers created understandings and modifications of Wolman's idea of urban metabolism after him. After seeing that a linear progression from a town's input of natural resources to its generation of products and effluents did not adequately replicate how actual life-forms affect the world's life-support system, Girardet (1990) proposed a cyclical urban metabolic design. Brunner (2007) later highlighted the significance of metabolic functions for a recycling city, which have been examined from both a planning and a metabolic activity point of views. Newman (1999) expanded the definition of urban metabolism and includes human aspects (e.g., citizen safety, prices, and schooling). Duan (2004) explored various processes and noticed that, in divergence to natural metabolism, pathways of urban metabolism are large, with ineffective and insufficient resource and energy exchange. To tackle these questions, he created a conceptual framework for urban metabolism based on current concepts that represents the major aspects involved in urban resource metabolism, as well as their interactions and related metabolic activities, in order to provide factual support for improving and regulating urban metabolism (Dijst et al. 2018).

Researchers studied both energy and resource flow in industries, households, as well as their ecological consequences. The disparities among urban economic and social development as well as the ecological integrity have become more visible as advanced industrial growth has speeded up (Bao and Fang 2012). Due to the intricacy of these observations, other researchers have concentrated on metabolism of a single system, enabling for a more detailed knowledge of that flow. Water and energy metabolism are common manifestations. Boyden et al. (1981) investigated Hong Kong's metabolic condition through the lens of the inhabitant's way of life. They explained the methods through which people respond to possibly stressful circumstances such as extremely growing populations and also the boundaries to human adaptability by taking into account factors such as accommodation, demography, air quality, death rates, happiness, violence, as well as the upcoming human habitation in urban areas. According to Newman (1999), the metabolism method to urban centres is biologically determined, but urban areas are more than just a method for manufacturing resources and generating wastes. As a result of these studies, our perception of urban metabolic activities has gradually increased (Van Broekhoven and Vernay 2018). The understanding of urban metabolism, thus, helps in proper urban planning and sustainable utilisation of resources. In this chapter, the role of urban metabolism in sustainable urban planning has been highlighted, followed by a brief insight on different policies governing urban planning and enlisting of a few limitations of urban metabolism. The chapter also outlines the information on the global distribution of research on urban metabolism and urban planning based on the search results of Scopus database.

8.2 Literature Survey and Analysis

For precise literature survey on the topic "urban metabolism and urban planning", Scopus database was explored in the last week of December, 2022. The initial search query was: (TITLE-ABS-KEY ("urban metabolism*" AND "urban planning*") AND PUBYEAR > 1990 AND PUBYEAR < 2023) which was further refined by (LIMIT-TO (DOCTYPE, "ar") OR LIMIT-TO (DOCTYPE, "cp") OR LIMIT-TO (DOCTYPE, "re") OR LIMIT-TO (DOCTYPE, "ch")) AND (LIMIT-TO (LANGUAGE, "English")), i.e., by selecting "articles, conference papers, reviews, and book chapters" published in English language as highlighted in Singh et al. (2022). The search query yielded 158 documents published till 2022. Though the literature search was performed for 1990–2022 time period, the first publication on this topic emerged in 1997 (Fig. 8.1). The topic gained less attention until 2011, with < 5 documents a year. From 2011 onwards a considerable attention has been given on the interlinkage of urban metabolism and urban planning. Since the year 2017, an increase (>12 documents a year) in publication was observed on this topic (Fig. 8.1). In 2015, United Nations released 17 Sustainable Development Goals (SDGs) with 169 targets and indicators covering a range of features for the sustainable development on the Earth system. A specific goal, i.e., SDG-11 with 10 targets, covers the "*Sustainable Cities and Communities*" with an agenda "*to make cities and human settlements inclusive, safe, resilient and sustainable*". This provides a platform to the research communities to focus and understand the potential of cities with a precise lens of view. Thus, the increase in number of publications after 2017 coincided with the emergence of SDGs. Of the cumulative number of publications on this topic, > 60% are published after 2015 (Fig. 8.1).

Further, country-wise, affiliation/institution-wise, author-wise, and source/journal-wise distribution of literature was performed to identify the global pattern of research, leading scientists/researchers, and journals publishing on the topic "urban metabolism and urban planning" (Tables 8.1, 8.2 and 8.3). Country-wise and affiliation-wise distribution are provided in Table 8.1. The country-wise data revealed that the concept is getting wider attention in the developed countries, particularly from the European and American continents, except for China from the Asia (Table 8.1). Italy, the United States, China, Australia, the Netherlands, and Portugal are identified as the leading countries supporting research on the topic "urban metabolism and urban planning" by publishing 28, 24, 18, 15, 15, and 15 documents, respectively. Amongst institutions, Wageningen University & Research (the Netherlands), Beijing Normal University (China), University of Louvain (Belgium), University of Aveiro (Portugal), and University of Lisbon (Portugal) are identified

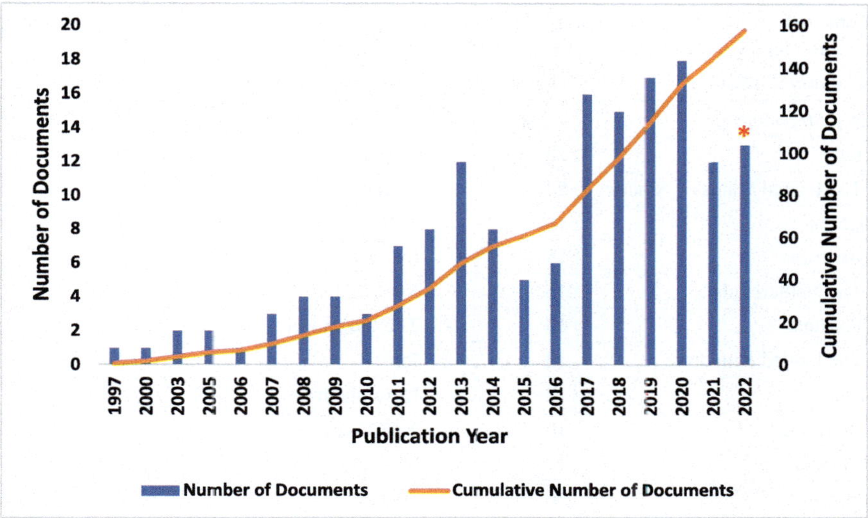

Fig. 8.1 Annual and cumulative number of documents published on the topic "urban metabolism and urban planning" (*Source* Scopus 2022). * represents that the data for 2022 is incomplete

as five leading institutions supporting research on this topic by publishing 10, 9, 7, 6 and 6 documents, respectively (Table 8.1). Among top 30 institutions actively publishing research on urban metabolism and planning, around 20 institutions have their origin in the European region. The government initiatives and policies, funding possibilities, improved infrastructures, and people's willingness to respond to the research questions might be the possible reasons for these observations.

Amongst the leading authors, Perrotti D, Chrysoulakis N, Zhang Y, Ioppolo G, Musango JK, and Niza S are identified as the top five/six researchers working and publishing research on different aspects of urban metabolism in relation to urban planning, and have published 8, 5, 5, 4, 4, and 4 documents, respectively (Table 8.2). The author-wise distribution results revealed that there are a number of researchers who have started working on the topic, and are having only 1 or 2 publications till date. Another viewpoint on these observations can be that we may witness an increased number of publications contributed by the researchers affiliated with the institutions of different developing nations, in the near future. Further, Journal of Industrial Ecology, Journal of Cleaner Production, Sustainability, Resources Conservation and Recycling, Sustainable Cities and Society, WIT Transactions on Ecology and the Environment, and Landscape and Urban Planning are recognised as the leading sources/journals publishing 15, 12, 9, 8, 6 and 6 documents, respectively, on this topic (Table 8.3). Amongst top 20 journals, most of the journals have multidisciplinary research publication scope. However, a few journals like Sustainable Cities and Society, Landscape and Urban Planning, Cities, Environment and Planning B: Urban Analytics and City Science, International Journal of Sustainable Building Technology and Urban Development, Frontiers in Sustainable Cities, and Journal

Table 8.1 A list of top 30 counties and affiliations/universities/institutions supporting/publishing research on the topic "urban metabolism and urban planning"

Country/Affiliation	No. of articles	Country/Affiliation	No. of articles
Top 30 countries			
Italy	28	Austria	4
The United States of America	24	Finland	4
China	18	India	4
Australia	15	Sweden	4
The Netherlands	15	Switzerland	4
Portugal	15	The Czech Republic	3
The United Kingdom	13	Japan	3
Spain	12	Luxembourg	3
Belgium	10	New Zealand	3
Canada	10	Norway	3
France	6	Algeria	2
Germany	6	Brazil	2
Greece	6	Ecuador	2
South Africa	6	Israel	2
Ireland	5	Poland	2
Top 30 affiliations/universities/institutions			
Wageningen University & Research, the Netherlands	10	University of California, Davis, United States	4
Beijing Normal University, China	9	Technical University of Madrid, Spain	4
University of Louvain, Belgium	7	Commonwealth Scientific and Industrial Research Organisation, Australia	4
University of Aveiro, Portugal	6	Yale University, the United States of America	3
University of Lisbon, Portugal	6	University of Porto, Portugal	3
Stellenbosch University, South Africa	5	French National Centre for Scientific Research (CNRS), France	3
Trinity College Dublin, Ireland	5	Ca' Foscari University of Venice, Italy	3
University of Toronto, Canada	5	Monash University, Australia	3
Foundation for Research and Technology-Hellas, Greece	5	Autonomous University of Barcelona (UAB), Spain	3
Polytechnic University of Milan, Italy	5	Tsinghua University, China	3
The University of Queensland, Australia	5	Griffith University, Australia	3

(continued)

Table 8.1 (continued)

Country/Affiliation	No. of articles	Country/Affiliation	No. of articles
Euro-Mediterranean Center on Climate Change (CMCC), Italy	5	Institute of Environmental Science and Technology (ICTA-UAB), Spain	3
Instituto Superior Técnico (Higher Technical Institute), Spain	4	Luxembourg Institute of Science and Technology, Luxembourg	3
University of Sassari, Italy	4	Free University of Brussels (ULB), Belgium	2
University of Messina, Italy	4	Nagoya University, Japan	2

Source Scopus (2022)

Table 8.2 A list of top 30 authors working/publishing research on the topic "urban metabolism and urban planning"

Authors	No. of articles	Authors	No. of articles
Perrotti D	8	Rijnaarts HHM	3
Chrysoulakis N	5	Rugani B	3
Zhang Y	5	Spano D	3
Ioppolo G	4	Stremke S	3
Musango JK	4	Voskamp IM	3
Niza S	4	Yang Z	3
Blecic I	3	Agudelo-Vera CM	2
Cristiano S	3	Almenar JB	2
Elliot T	3	Athanassiadis A	2
Gonella F	3	Bahers JB	2
González A	3	Berezowska-Azzag E	2
Kenway SJ	3	Caputo P	2
Liu G	3	Cecchini A	2
Lopes M	3	Chen B	2
Mitraka Z	3	Cucurachi S	2

Source Scopus (2022)

of Environmental Planning and Management are exclusively having focus on urban planning as their core objectives (Table 8.3). The source-wise distribution of publication further revealed the multidisciplinary nature of the topic and its reach to a wider group of readers. Overall, the literature survey on the topic "urban metabolism and urban planning" gives an idea that the topic is getting wider attention recently, and a further increase in the number of publications covering different aspects of urban ecology can be seen in the near future.

Table 8.3 A list of top 20 sources/journals publishing research on the topic "urban metabolism and urban planning"

Source/Journal	No. of articles	Source/Journal	No. of articles
Journal of Industrial Ecology	15	Environment and Planning B: Urban Analytics and City Science	3
Journal of Cleaner Production	12	International Journal of Sustainable Building Technology and Urban Development	3
Sustainability (Switzerland)	9	Science of the Total Environment	3
Resources Conservation and Recycling	8	Environmental Pollution	2
Sustainable Cities and Society	6	Frontiers in Sustainable Cities	2
WIT Transactions on Ecology and the Environment	6	International Journal of Environmental Research and Public Health	2
Landscape and Urban Planning	5	International Journal of Life Cycle Assessment	2
Ecological Modelling	4	Journal of Environmental Planning and Management	2
Cities	3	Water Research	2
Ecological Indicators	3	Ambio	1

Source Scopus (2022)

8.3 Applications of Urban Metabolism for Sustainable Urban Planning

According to Newman (1999), the definition of metabolism applicable to urban centres can illustrate the real meaning of self-sufficiency. However, one such application must go beyond the general measurement of inputs and outputs creating specific figure of methodological choices and not on various components but on cities (Oliveira and Vaz 2021). According to Pinho et al. (2013), the concept of urban metabolism allows for a more comprehensive evaluation of the effects of every action (especially large projects) on the urban system. This spatial dimension estimation entails understanding both flows and dynamics related to each plan in relation to the specific urban fabric into which it aspires to fit. As such, enhancing our understanding of different flows that occur in urban centres as well as their metabolism is critical to ensuring their assimilation into planning activities (Nijkamp and Kourtit 2013; Allam and Newman 2018). Because transportation, resources, as well as land use are all vital factors of the urban system, investigating their interdependence from the standpoint of metabolism will permit assessing the significant ramifications of planning policies and tactics as well as knowledge on how urban management is efficient (Pincetl et al. 2012; Nijkamp and Kourtit 2013; Mostafavi et al. 2014). In addition, urban planning as a medium- and long-term path needs relevant data, devices, and techniques for evaluating and making decisions (Lee et al. 2013).

Sanli and Townshend (2018) distinguished among ideology and practises. Contextual differences and social customs influence not only layouts and social aspects but also the legal and executive system as well as the numerous processes that frame the output of the physical environment. As a result, every culture develops its particular scope or structure (Sanli and Townshend 2018). Moreover, since planning tries to resolve overall pattern and qualities (Caruso et al. 2016), it is critical to comprehend the scope into which it would incorporate. However, research is essential for urban planning (Vogelij 2015), but if it is removed from its context, it risks to be becoming obsolete. As a result, for effective strategic planning, the conceptual and indicative perspectives of concept and the rational viewpoint of practise must coexist (Vogelij 2015). Understanding how the notion of urban metabolism is perceived in urban planning and how it needs to be articulated is crucial for promoting its integration without enlisting the help of ineffective local government specialists, and preventing the creation of cumbersome, inefficient methodologies and tools. Some features of urban metabolism having their role in urban planning are elaborated in the following sub-sections:

8.3.1 Sustainability Indicators

Since urban metabolism provides indicators of a city's self-sufficiency, its study is an essential element of State of the Environment (SOE) research (Kennedy et al. 2011). The urban metabolism contains information on energy accuracy, resource cycling, sewage disposal, and urban infrastructure (Voskamp et al. 2018). These dimensions typically adhere to Maclaren's (1996) principles of noble sustainability indicators: they are scientifically effective (based on mass and energy conservation values), representative, attentive, applicable to planners and residents based on statistics that are comparable over time, comprehensible, and clear. The primary goals of SOE coverage are to examine and define significant environmental situations and patterns, as well as to aid as a pioneer to the policymaking (Maclaren 1996). Thus, sustainability indicators can act as the effective tools for sustainable urban planning.

8.3.2 Contributions to the Central Greenhouse Gas Estimation

With cities attempting to cut down their greenhouse gases (GHGs) emission, estimating urban GHGs emission is an extremely useful application of urban metabolism dimensions (Pincetl et al. 2014). Carbon dioxide, methane, and other air pollutants emitted by cities are valid constituents of urban metabolism. Emissions from a city, on the other hand, are typically broader in scope, including a few gases generated outside of urban boundaries, such as from electricity generation by the power plants or the removal of exported waste (McGranahan and Satterthwaite 2003). Whether GHG emissions occur within or outside the city limits, they should be measured using data on energy consumption, flow of goods, and residual wastes by urban metabolism (Verma et al. 2021). GHG emissions are widely calculated in a variety of areas by multiplying an activity level by an emission level factor in accordance with Intergovernmental Panel on Climate Change (IPCC) guidelines (Verma et al. 2021). GHG emissions from a community's power generation, for example, are calculated by multiplying usage by the regional or national generation capacity's GHG strength (Hultman et al. 2011). Emission factors for energy sources used in shipping or manufacturing are well generated as a result of national GHG inventory reporting; they are grounded on the combustion properties of fuel. Estimates for such segments, such as wastes, are, however, more complex for urban GHG inventories (Verma et al. 2021).

8.3.3 Mathematical Models for Policy Examination

Although most studies have relied heavily on urban metabolism as a foundation for financial statements, others have started implementing numerical simulations of urban metabolism mechanisms (Zhang 2013; Currie et al. 2017). Such numerical methods, which are typically used to investigate substances such as metallic and non-metallic materials in urban and regional metabolism, were largely designed by the material flow analysis (MFA) (Lotfi and Daneshpour 2016; Cui et al. 2019). Model platforms are based on the software SIMBOX and STAN (Lotfi and Daneshpour 2016). Sub-processes, equities, and flows inside the metabolism are often linked with economic inputs or outputs in such models. Although the models can be used to estimate existing material equities and flows, they can moreover be utilised to forecast upcoming variations in urban metabolism as a consequence of technological initiatives or regulations. The models are especially valuable in recognising long-term remedies that extend beyond "end of life" approaches (Elliot et al. 2019).

8.3.4 Plan Apparatus

The application of urban metabolism concept in urban planning and design is comparatively a recent concept. Oswald and Baccini (2003) start to determine how a combination of morpho-physiological methods could be employed in Netzstadt's "*long process of recreating the city*". Their journey begins with the realisation that the centre-periphery city model is no longer viable; moreover, the novel urbanity is not self-sustaining. The four pillars of city redesign are then presented as: shapability, sustainable development, rebuilding, and obligation. In a design method that contains an assessment of urban metabolism, four parameters of urban reliability are, thus, attempted to seek, viz., acknowledgment, diversity, level of resource efficiency, and self-sufficiency (Kuznecova and Romagnoli 2014; Currie et al. 2017). Baccini and Brunner (1991) defined four urban activities, viz., nurturing and restoring, washing, transporting, and interacting. These operations are established on four aspects of urban metabolism, viz., water, biomass (food), building materials, and power (Bristow and Kennedy 2013). Numerous examples demonstrate how morpho-physiological perspectives can be combined. For example, urban metabolism was employed to speed up the re-establishment of New Orleans after Hurricane Katrina. MFA was used by John Fernandez and his MIT students to assist them in developing eco-friendlier city models (Quinn and Fernandez 2007).

The urban metabolism is also used to teach economic viability to students at the University of Toronto. The students are confronted with design issues that are generally at the local community level, as well as the implementation of multiple facilities using the concept of neighbourhood metabolism (Kennedy et al. 2007; Codoban and Kennedy 2008). The students use guiding principles in green building construction, sustainable shipping, and renewable energy devices in their task. By identifying the water flows, power, substances, and nutrients via an urban system, it is possible to create circuits that limit resource input and waste output. One institution's urban metabolism study demonstrates how near it came to a basically sustainable construction, with treated wastewater being utilised for proper sanitation as well as open air use, and sewage sludge being used to grow food in public gardens. The energy produced by the imported solid waste from municipalities not only powered the structures, but also powered the light-rail system and revived numerous grid surpluses (Premalatha et al. 2013). Furthermore, fly-ash released from the waste gasification plant was recovered for use as a building material. Electricity, water, equipment, and essential minerals have all been drastically reduced as a result of partially closing such loops. Experts also oversee material and energy flow in urban design to minimise environmental impacts. Thus, various dimensions of urban metabolism can be useful in designing urban systems. Some of the policies governing the urban planning are outlined in the next section.

8.4 Policies Governing Urban Planning

A sustainable urban structure is defined by three key features: higher density development, the existence of advancement across public transit corridors, and the establishment of a mixed land use arrangement (McCormick et al. 2013). Many experts see those as indicators of a decrease in the need, and therefore, the tendency to travel through personal transport (Newman 1999), and thus, reduction in energy consumption. As explained previously, many urban planning approaches have openly sought to put such principles into practice. Whereas the goals, devices, situations, and focused performers of such initiatives vary greatly. This is obviously only one feasible clustering, although it is useful in attempting to classify what urban guidelines are attempting to accomplish, as the goals of these initiatives can be ambiguous in some instances, beyond perhaps attaining "sustainability". Transition to low-carbon emission city is an essential component of sustainable urban metabolism which is governed by the urban form and density, directly regulating the GHG emissions (Davoudi and Sturzaker 2017). However, none of this can be considered as an absolute success as various unsustainable practises such as lengthy transportation and urban growth unabatedly continue in almost all of the urban centres of the cities (Davoudi and Sturzaker 2017). It is, thus, beneficial to conduct a more in-depth analysis of the approach used in order to recognise any common cause of success or failure, in order to add to the creation of enhanced approaches that encourage high-performance urban metabolism. Jabareen (2006) categorises sustainable urban forms into four major designs, viz., neo-traditional development, urban confinement, compact city, and eco-city models, which are as follows:

8.4.1 Neo-Traditional Development Model

According to Jabareen's (2006), "*the best neo-traditional city might be self-contained, closely aggregated, pedestrian friendly, and arranged on American tiny city of pre–World War II*". A strategy centred on learning and information initiatives, and behavioural modification promotion is suggested by such an initiative statement (Villena 2019). The key indicators used were fresh density targets and the percentage of houses being built on earlier industrial lands (DCLG 2006). As a result, instead of drawing people away to cities as policy rhetoric proposed, initiatives are designed to drive people out of other places. Jabareen (2006) defines transit-oriented development (TOD) as just another type of neo-traditional advancement with three basic elements: combined growth, improvement near to and well offered by transit, and growth favourable to transportation riding. It has been implemented in a number of urban centres all over the North America such as Calgary, Portland, and San Francisco. The focus in TOD has frequently been on planning and regulatory policies. Despite criticism for its narrow capability (Goodwill and Hendricks 2002), the TOD idea remains popular in the United States.

8.4.2 Compacted City Model

Compact city model, as the name suggests, is based on small size, and thus, higher concentration of inhabitants in an area (Bibri et al. 2020). Those in favour of this system argue that in denser urban areas, journey intervals are reduced, reducing GHG emissions, regional amenities are aided, and regional communities become significantly more self-sufficient (Jabareen 2006). Decision makers of the Netherlands have also advocated for the compacted city model (Dieleman et al. 1999), although initially under the highly specific de-concentration ideology, which was later referred to as poly-centricity (Davoudi 2003). The primary strategy employed has also been the development of restrictions intended at limited urban growth. This, along with higher levels of regeneration financing, has also been mentioned as "*instrumental in remaking urbanisation in urban centres as well as in recreating a vibrant pedestrian setting in city centres*" (van der Burg and Dieleman 2004). Furthermore, the initiative is not fully successful, as van der Burg and Dieleman (2004) state that there is a housing supply scarcity due to the lack of flexibility in the planning activities, and thus, the patterns have not altered to the level that was expected. As, in spite of their own reputation for transportation use, Dutch transportation systems are frequently subjected to huge overcrowding (van der Burg and Dieleman 2004).

8.4.3 City Confinement Model

At its core, urban containment restricts the urban domain from expanding outward and compels the improvement segment to look inward (Jabareen 2006). As a consequence of the large percentage of expansion in rural areas in the 1970 and 1980s, decision makers in the United States attempted to implement urban confinement initiatives according to Jabareen (2006). Moreover, urban control has a much strong history in other nations. For instance, one of the post-war preparing orthodoxies in England has become a nearly exclusively strong dependence on planning rules initiatives, with the Green Belt being the most prominent of these (Sturzaker and Mell 2016). One example is the negative economic rewards implied to the property operators by imposing a greater tax rate on redeveloping as well as retrofitting structures compared to the residential development on Greenfield land (Adams et al. 2000). Likewise, a common rule of making the price of vehicle ownership low and more recently going to ease limitations on parking facilities in urban areas all in the name of deciding to end the "battle on drivers" (DCLG 2011). These contradictory rules demonstrate the deeply political nature of making plans as was already noticed somewhere else (Randolph 2004). They immensely impact the efficiency of urban policies directed at thermally saving urban forms by operating occasionally ideologically opposed to one another. Another example is the ABC plan in the Netherlands, which can be largely defined as urban containment. This was challenged less by competing policies from the similar sphere of government and more by disagreements amid central government policy and the incorporation of such a strategy by local governments. Thus, such models need attention of the policymakers to make it more applicable and acceptable.

8.4.4 Eco-City Model

The term "eco-city" refers to a broad spectrum of ecological approaches that intent to attain urban sustainability. Moreover, Jabareen (2006) continues to state that the concept focuses on urban planning rather than a specific urban form. The 'eco-city' strategy implemented in the United Kingdom amid 2007 and 2010 is an intriguing topic in this regard. It encouraged small new settlements of at least 520,000 apartment's (DCLG 2007), following the footsteps of Ebenezer Howard's Garden Cities and new towns, which have been constructed in stages throughout the United Kingdom from the beginning of the twentieth century (TCPA 2011). The eco-city plan aimed to encourage regional basis driven settlement expansion, and the Government's advice appears to be intentionally documented to avoid clarifying a suitable urban form, aside from highlighting "the best new architectural design" (DCLG 2007). The eco-city policy was guided by a socially progressive approach rather than a regulatory planning approach, as well as the usage of encouraging rhetoric.

All the urban planning models have certain advantages as well as limitations leading to their limited adaptation figures. A few limitations of urban metabolism features leading to limited application in urban planning are presented in the next section.

8.5 Limitations of Urban Metabolism

The concept of urban metabolism can signal the need for assessment of planning (Oliveira and Pinho 2010), especially setting up a framework of ideas for that analysis. Modern urban centres face a wide range of major challenges. Proper scientific planning and research theory presumptions, ideas, and performance are not always appropriate to planning practises. Planning practises help to examine such difficulties in order to capitalise on all of the possibilities that towns can provide to their citizens in terms of cultural, financial and environmental well-being. The relationship among scientific strategy and development practice is not simple. On the other hand, the former's impact on the others is typically limited and slow. There are numerous reasons for this. Despite the fact that some research findings attempt to address such goals (Chrysoulakis et al. 2013; Pinho et al. 2013), there is still a long way to go to make the concept beneficial for urban planning (Kennedy et al. 2011; Pistoni and Bonin 2017; Oliveira and Vaz 2021), provided the broad range of concerns and dimensions at which it must function. As a result, it is crucial to take into account the urban landscape's spatial and qualitative aspects as well as to comprehend how urban metabolism might be incorporated into regular urban planning (Pistoni and Bonin 2017). Since a number of researchers believe that incorporating the idea will indeed support planning, the difficulty of several scholarly outcomes as well as the overreliance on huge amounts of information for quantification limit their possible application in day-to-day practice. One more factor that complicates the application is the absence of feedback on that practice. Some of the major limitations are enlisted as:

8.5.1 Data Scarcity at the City Level

It is broadly accepted that there is a scarcity of information on urban/city materials and energy flow (Kennedy et al. 2007). This information scarcity is a problem both for production- and consumption-based methods.

8.5.2 Urban Metabolism Studies Have Extensive Data and Resource Needs

Despite the absence of information at the urban level, conducting urban metabolism studies necessitates a large amount of data due to the system's approach of the concept (Shahrokni et al. 2015). Owing to the globalisation of consumption and production chains, there is a necessity for accuracy in the explanation of metabolic flows required to identify environmental problem shifting related with policies, as well as a requirement for global system boundaries and consumption-based accounting (Broto et al. 2012; Shahrokni et al. 2015). When conducting urban metabolism studies, such an undertaking imposes an important resource and time-burden.

8.5.3 Difficulties in Recognising "Cause and Effect" Associations

The long interval among urban metabolism analyses within the same city suggests a lack of data analysis of the causes that alter a city's energy and resource flows. As such, slight attention has given to the political and socio-economic forces that drive material flows, as well as the function of an urban system and its ecological practises (Chrysoulakis et al. 2009; Shahrokni et al. 2015). Furthermore, it is important to understand how urban metabolism changes over time as a consequence of changing policy devices and decision-making.

8.6 Conclusions and Future Prospects

The general public's understanding of urban metabolism is growing nowadays. The concept of urban metabolism can help with proper planning by explaining how cities function in terms of basic supplies and flows. It can offer a broader picture of the multifaceted operation of urban systems, exploring essential parts of the interactions among facilities and infrastructure, residence designs and behaviour patterns, and natural supporting systems, in addition to the standard management analysis of urban mechanisms. The research findings of urban metabolism provide useful insights into how specific locations operate at specific periods; however, there is always a lot more to learn. By incorporating activity analysis in a more combined and repetitive perception, it can add unique techniques to monitor urban sustainability of the overall ecosystems. There are, however, a few major challenges which must be incorporated into policy formulation, acknowledging different practices and discussing different metabolic components at various scales. For example, by performing a thorough discussion on all metabolic aspects at the national level whereas relatively brief discussion on various components at comparatively small scales. To enable this, the meaning of each component should be empirically and theoretically sound, and the determination of such metabolic elements must be specific, trying to balance research ambiguity and training rationality.

Although the concept of urban metabolism has advanced considerably over the past few years, there are numerous areas where more study is needed, such as:

1. There have been several studies of various cities, but time series experiments are rare.
2. There is a need to increase our understanding of carbon-related metabolic procedures in order to offset the influence of urban metabolism on temperature rise and climate change.
3. It is needed to regulate the mechanisms of urban metabolic networks to improve their performance.
4. More studies are required to investigate the impacts of social indicators on the metabolic flows of an urban system.

Acknowledgements MAD is thankful to Council of Scientific and Industrial Research (CSIR), New Delhi, India for providing financial assistance as Senior Research Fellowship. RS gratefully acknowledges the financial support from DST-SERB, New Delhi, India as National Post-Doctoral Fellowship Scheme (Grant number: PDF/2020/001607).

References

Adams CD, Disberry A, Hutchison N, Munjoma T (2000) Taxes, subsidies and the behaviour of brownfield owners. Land Use Policy 17:135–145

Allam Z, Newman P (2018) Redefining the smart city: culture, metabolism and governance. Smart Cities 1:4–25
Baccini P, Brunner PH (1991) Metabolism of the anthroposphere: analysis, evaluation, design. Springer, Berlin, Heidelberg
Bao C, Fang CL (2012) Water resources flows related to urbanization in China: challenges and perspectives for water management and urban development. Water Res Manag 26:531–552
Bibri SE, Krogstie J, Kärrholm M (2020) Compact city planning and development: emerging practices and strategies for achieving the goals of sustainability. Develop Built Environ 4:100021
Boyden S, Millar S, Newcombe K, O'Neill-Canberra B (1981) The ecology of a city and its people: the case of Hong Kong. Australian National University Press, Canberra
Bristow DN, Kennedy CA (2013) Urban metabolism and the energy stored in cities: implications for resilience. J Ind Ecol 17:656–667
Broto VC, Allen A, Rapoport E (2012) Interdisciplinary perspectives on urban metabolism. J Ind Ecol 16:851–861
Brunner PH (2007) Reshaping urban metabolism. J Ind Ecol 11:11–13
Carréon JR, Worrell E (2018) Urban energy systems within the transition to sustainable development. a research agenda for urban metabolism. Resour Conserv Recycl 132:258–266
Caruso N, Hammami F, Peker E, Tulumello S, Ugur L (2016) Differences and connections beyond universal theories in planning, urban, and heritage studies. Urb Res Pract 9:219–224
Chrysoulakis N, Vogt R, Young D, Grimmond SCB, Spano D, Marras S (2009) ICT for urban metabolism: the case of BRIDGE. In: Environmental informatics and industrial environmental protection: concepts, methods and tools. EnviroInfo, Berlin, pp 175–185
Chrysoulakis N, Lopes M, San José R, Grimmond CSB, Jones MB, Magliulo V, Cartalis C (2013) Sustainable urban metabolism as a link between bio-physical sciences and urban planning: the bridge project. Landscap Urb Plan 112:100–117
Codoban N, Kennedy CA (2008) Metabolism of neighbourhoods. J Urban Plan Develop 134:21–31
Cui X, Wang X, Feng Y (2019) Examining urban metabolism: a material flow perspective on cities and their sustainability. J Clean Prod 214:767–781
Currie PK, Musango JK, May ND (2017) Urban metabolism: a review with reference to Cape Town. Cities 70:91–110
Davoudi S, Sturzaker J (2017) Urban form, policy packaging and sustainable urban metabolism. Resourc Conserv Recycl 120:55–64
Davoudi S (2003) European briefing: Polycentricity in European spatial planning: from an analytical tool to a normative agenda. Euro Plan Stud 11:979–999
DCLG (2006) Planning policy statement 3: Housing. Stationery Office, London [Rev.] edn
DCLG (2007) Eco-towns prospectus. Department of Communities and Local Government London
DCLG (2011) Pickles and Hammond to end the war on motorists [Online], Available from http://www.communities.gov.uk/news/newsroom/1809347 Accessed 08 Aug 2011
Decker EH, Elliott S, Smith FA, Blake DR, Rowland FS (2000) Energy and material flow through the urban ecosystem. Ann Rev Energ Environ 25:685–740
Dieleman FM, Dijst MJ, Spit T (1999) Planning the compact city: the Randstad Holland experience. Euro Plan Stud 7:605–621
Dijst M, Worrell E, Böcker L, Brunner P, Davoudi S, Geertman S, Zeyringer M (2018) Exploring urban metabolism—towards an interdisciplinary perspective. Resourc Conserv Recycl 132:190–203
Duan N (2004) Urban material metabolism and its control. Res Environ Sci 17:75–77
Elliot T, Babí Almenar J, Niza S, Proença V, Rugani B (2019) Pathways to modelling ecosystem services within an urban metabolism framework. Sustainability 11:2766
Girardet H (1990) The metabolism of cities. In: Cadman D, Payne G (eds) The living city towards a sustainable future. Routledge, London, pp 170–180
Golubiewski N (2012) Is there a metabolism of an urban ecosystem? An ecological critique. Ambio 41:751–764

Goodwill J, Hendricks SJ (2002) Building transit oriented development in established communities. Center For Urban Transportation Research University of South Florida, Tampa
Hultman N, Rebois D, Scholten M, Ramig C (2011) The greenhouse impact of unconventional gas for electricity generation. Environ Res Lett 6:044008
Jabareen YR (2006) Sustainable urban forms: their typologies, models, and concepts. J Plan Edu Res 26:38–52
Kennedy C, Cuddihy J, Engel-Yan J (2007) The changing metabolism of cities. J Ind Ecol 11:43–59
Kennedy C, Pincetl S, Bunje P (2011) The study of urban metabolism and its applications to urban planning and design. Environ Pollut 159:1965–1973
Kuznecova T, Romagnoli F (2014) System dynamics approach within the context of city resilience and urban metabolism. In: 9th international conference "Environmental Engineering" at Vilnius, Lithuania. https://doi.org/10.3846/enviro.2014.125
Lee JH, Phaal R, Lee SH (2013) An integrated service-device-technology roadmap for smart city development. Technol Forecast Social Chang 80:286–306
Lotfi A, Daneshpour SA (2016) Analysis and evaluation of the ecological restoration of the city with an emphasis on urban metabolism. Open J Ecol 6:377–386
Maclaren VW (1996) Developing indicators of urban sustainability. ICURR Press
Maranghi S, Parisi ML, Facchini A, Rubino A, Kordas O, Basosi R (2020) Integrating urban metabolism and life cycle assessment to analyse urban sustainability. Ecol Indic 112:106074
McCormick K, Anderberg S, Coenen L, Neij L (2013) Advancing sustainable urban transformation. J Clean Prod 50:1–11
McGranahan G, Satterthwaite D (2003) Urban centers: an assessment of sustainability. Ann Rev Environ Resourc 28:243–274
Mohan SV, Amulya K, Modestra JA (2020) Urban biocycles–closing metabolic loops for resilient and regenerative ecosystem: a perspective. Biores Technol 306:123098
Mostafavi N, Farzinmoghadam M, Hoque S (2014) A framework for integrated urban metabolism analysis tool (IUMAT). Build Environ 82:702–712
Newman PW (1999) Sustainability and cities: extending the metabolism model. Landscap Urban Plan 44:219–226
Nijkamp P, Kourtit K (2013) The "new urban Europe": global challenges and local responses in the urban century. Euro Plan Stud 21:291–315
Oliveira V, Pinho P (2010) Evaluation in urban planning: advances and prospects. J Plan Lit 24:343–361
Oliveira V, Vaz R (2021) Is the concept of urban metabolism useful for planning practice? Euro Plan Stud 29:411–424. https://doi.org/10.1080/09654313.2020.1740173
Oswald F, Baccini P (2003) Netzstadt-designing the urban. Birkha¨user, Basel, Boston
Padovan D, Cristiano S, Gonella F (2022) Strategies of socio-ecological transition for a sustainable urban metabolism. Front Sustain Cities 4:875912. https://doi.org/10.3389/frsc.2022.875912
Perrotti D (2020) Urban metabolism: old challenges, new frontiers, and the research agenda ahead. In: Verma P, Singh P, Singh R, Raghbanshi AS (eds) Urban ecology: emerging patterns and social-ecological systems. Elsevier, pp 17–32
Piña WHA, Martínez CIP (2014) Urban material flow analysis: an approach for Bogotá, Colombia. Ecol Ind 42:32–42
Pincetl S, Bunje P, Holmes T (2012) An expanded urban metabolism method: toward a systems approach for assessing urban energy processes and causes. Landscap Urban Plan 107:193–202
Pincetl S, Chester M, Circella G, Fraser A, Mini C, Murphy S, Sivaraman D (2014) Enabling future sustainability transitions: an urban metabolism approach to Los Angeles. J Ind Ecol 18:871–882
Pinho P, Oliveira V, Cruz SS, Barbosa M (2013) Metabolic impact assessment for urban planning. J Environ Plan Manag 56:178–193
Pistoni R, Bonin S (2017) Urban metabolism planning and designing approaches between quantitative analysis and urban landscape. City Territ Arch 4:1–11
Premalatha M, Tauseef SM, Abbasi T, Abbasi SA (2013) The promise and the performance of the world's first two zero carbon eco-cities. Renew Sustain Energ Rev 25:660–669

Quinn D, Fernandez J (2007) Urban metabolism: ecologically sensitive construction for a sustainable New Orleans. School of Architecture and Planning MIT Cambridge, pp 40–41
Randolph J (2004) Environmental land use planning and management. Island Press, Washington DC
Sanli T, Townshend T (2018) Hegemonic power relations in real practices of spatial planning: the case of Turkey. Euro Plan Stud 26:1242–1268
Scopus (2022) https://www.scopus.com/results/ Accessed 28 Dec 2022
Shahrokni H, Lazarevic D, Brandt N (2015) Smart urban metabolism: towards a real-time understanding of the energy and material flows of a city and its citizens. J Urban Technol 22:65–86
Singh VK, Solanki P, Ghosh A, Pal A (2020) Solid waste management and policies toward sustainable agriculture. In: Handbook of solid waste management: sustainability through circular economy, pp 1–22
Singh R, Verma P, Singh VK, Srivastava P, Kumar A (2022) Urban ecology and climate change: challenges and mitigation strategies. In: Bhadouria R, Upadhyay S, Tripathi S, Singh P (eds) Urban ecology and global climate change. Wiley Ltd, pp 1–29. https://doi.org/10.1002/9781119807216.ch1
Sturzaker J, Mell I (2016) Green belts: past; present; future? Routledge
Tan LM, Arbabi H, Brockway PE, Tingley DD, Mayfield M (2019) An ecological-thermodynamic approach to urban metabolism: measuring resource utilization with open system network effectiveness analysis. Appl Energ 254:113618
TCPA (2011) Re-imagining garden cities for the 21st Century. Town and Country Planning Association London
Thomson G, Newman P (2018) Urban fabrics and urban metabolism–from sustainable to regenerative cities. Resourc Conserv Recycl 132:218–229
Van Broekhoven S, Vernay AL (2018) Integrating functions for a sustainable urban system: a review of multifunctional land use and circular urban metabolism. Sustainability 10:1875
van der Burg AJ, Dieleman F (2004) Dutch urbanisation policies: from 'compact city' to 'urban network.' J Econ Human Geograph 95:108–116
Verma P, Singh P, Singh R, Raghbanshi AS (2020) Urban ecology: emerging patterns and social-ecological systems. Elsevier. https://doi.org/10.1016/C2018-0-05562-0
Verma P, Kumari T, Raghubanshi AS (2021) Energy emissions, consumption and impact of urban households: a review. Renew Sust Energ 147:111210
Villena VH (2019) The missing link the strategic role of procurement in building sustainable supply networks. Prod Operat Manag 28(1149):1172
Vogelij J (2015) Is planning theory really open for planning practice? Plan Theory Pract 16:128–132
Voskamp IM, Spiller M, Stremke S, Bregt AK, Vreugdenhil C, Rijnaarts HH (2018) Space-time information analysis for resource-conscious urban planning and design: a stakeholder-based identification of urban metabolism data gaps. Resourc Conserv Recycl 128:516–525
Wachsmuth D (2012) Three ecologies: urban metabolism and the society-nature opposition. Sociol Quart 53:506–523
Walker RV, Beck MB, Hall JW, Dawson RJ, Heidrich O (2014) The energy-water-food nexus: strategic analysis of technologies for transforming the urban metabolism. J Environ Manage 141:104–115
Wolman A (1965) The metabolism of cities. Sci Am 213:179–190
Zhang Y (2013) Urban metabolism: a review of research methodologies. Environ Poll 178:463–473
Zhang Y, Yang Z, Yu X (2015) Urban metabolism: a review of current knowledge and directions for future study. Environ Sci Technol 49:11247–11263
Zhang Y, Yang Z, Liu G, Yu X (2011) Emergy analysis of the urban metabolism of Beijing. Ecol Model 14:2377–2384

Chapter 9
Urban Metabolism in the Circular Bio-economy of Tomorrow

G. Venkatesh

Abstract A bio-economy is grounded on the use of renewable, biotic resources. A circular economy, on the other hand, emphasises resource conservation in general. When one refers to a circular bio-economy, one gets the best of both these worlds, so to say. It is a set of 'many-to-many' relationships, which are perfectly symbiotic at best. In the longer run, in a circular bio-economy, one can expect economies of scale and scope. It follows that one would then have accelerated the pace of sustainable development—that never-ending journey towards the elusive goal of 'sustainability'. It goes without saying that urban metabolism in a circular bio-economy will metamorphose into something extremely transformative—for combating climate change and its repercussions, as well as a host of other challenges, attaining several Sustainable Development Goals in the process. Well-begun is half done, and here is where cities ought to learn from each other. All new ventures can be inspired by the paradigm of a circular economy (bio-economy wherever that is possible), while simple symbiotic relationships amongst diverse entities within a city can be uncovered with some out-of-the-box thinking. Needless to state, as all know, challenges lurk where opportunities abound to supplant the current 'take-make-use-dispose' culture of a linear economy with a beneficial 'grow-make-use-share-partake-restore' paradigm of a circular bio-economy. Smart cities will be the ones that avail of these opportunities and strive towards the 'more from less' goal.

Keywords Circular bio-economy · Resource valorisation · Sustainable development · Urban metabolism

9.1 Introduction

Homo sapiens take in oxygen (which may also be supplied by plants and trees around them) and release carbon dioxide (CO_2), which is fodder for the flora in the biosphere, generating energy for themselves in the process by the combustion of the

G. Venkatesh (✉)
Department of Engineering and Chemical Sciences, Karlstad University, 65188 Karlstad, Sweden
e-mail: Venkatesh.govindarajan@kau.se

R. Bhadouria et al. (eds.), *Urban Metabolism and Climate Change*,
https://doi.org/10.1007/978-3-031-29422-8_9

'carbonaceous energy sources' they consume. The action of oxygen on these energy sources also results in coordination with the excretory system downstream of the digestive system, and urine and excreta are discharged to the sewer network. Just as the CO_2 they exhale comes back to them as oxygen and also as the 'carbonaceous energy sources', the urine and the excreta may also be made to come back to them as food in a circular bio-economy if their fertiliser value is harnessed (Venkatesh 2021; also see Fig. 9.1). After all, cities are built, designed, and managed, and therefore, need to be sustained by *Homo sapiens* whose lives are steered by the metabolism which happens involuntarily within their mortal frames (Venkatesh and Govindarajan 2014).

In a bio-economy, the focus is on resources which are renewable—so called fund and flow resources. In a circular economy, the emphasis is on resource conservation. One gets to the best of both these worlds, so to say, in a circular bio-economy (Venkatesh 2021). As the concepts of Circular Economy and Circular Bio-economy (the latter being a subset of the former) entrench themselves slowly but surely and steadily, it will be interesting to understand urban metabolism or rather the transformation in urban metabolism in a budding circular bio-economy, which bases itself on a grow-make-use-share-partake-restore paradigm (Barros et al. 2020). Talking of urban settings, the two main sectors of interest are municipal and industrial solid waste and sewage management (MISWSM) and energy. The inflows into cities are both biotic (organic) and abiotic (inorganic) in nature, and this applies to the outflows too—as solid wastes and sewage. While the management of both these is within the scope of a circular economy, the focus of this chapter is only on the organic/biotic flows and their circularisation by resorting to innovative technologies and creative approaches to valorise what would otherwise be looked upon as wastes sans economic value.

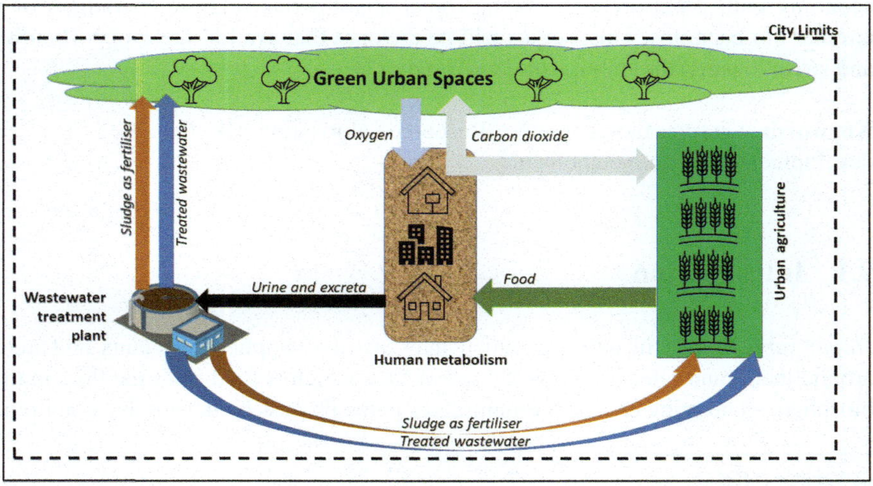

Fig. 9.1 Human metabolism inspires urban metabolism (illustration done by the author)

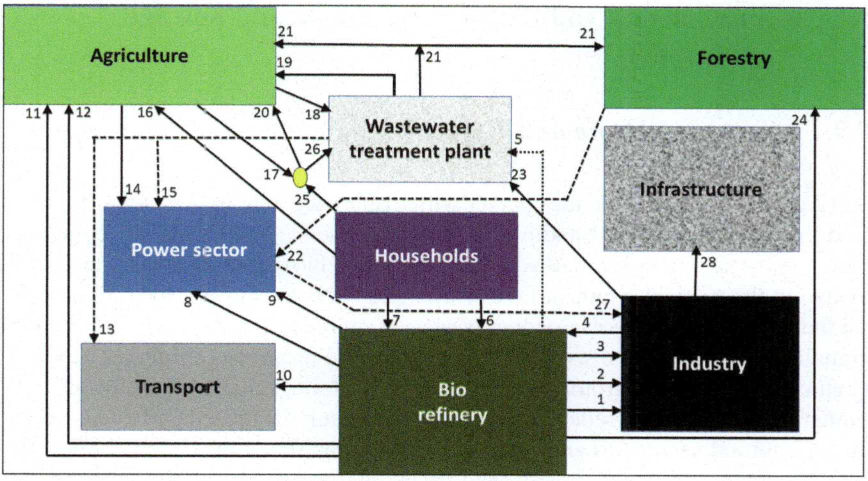

Fig. 9.2 Graphical representation of the transformative metabolic flows in an urban setting of a future circular bioeconomy, discussed in this chapter. *The numbered flows have been identified in the text in the first paragraph of section 3*

While holistic sustainability assessments—social, economic, environmental—are called for before any new approach can be commercialised, in favour of a circular (urban) bio-economy, the synergies and the conflicts notwithstanding, such efforts will take cities down the path of sustainable development, towards the attainment of several Sustainable Development Goals, charted out by the United Nations in year-2015, for the year-2030 (United Nations 2015; also refer Fig. 9.2).

The discussion which follows is based on and bolstered by selected recent research in the area of 'circular bio-economy', applicable to urban metabolism. Food, water, and energy (and thereby the water-energy-food nexus) are fundamental to socio-economic welfare in urban settings. The author has focused on MISWSM and bio-energy. The former has been segmented into the metabolism associated with sewage handling and food wastes. Needless to state, clear synergies exist between MISWSM and bio-energy (heat, electricity, and transport fuel). The material and energy flows of this 'futuristic urban metabolism in a circular bio-economy', illustrated in Fig. 9.2, encapsulate the content of this chapter, for the readers.

9.2 Municipal and Industrial Solid Waste and Sewage Management

9.2.1 Sewage—Urine and Excreta

Since we have commenced the chapter with a reference to human metabolism, let us start this section with the handling of wastewater/sewage or rather, its 'component parts', so to say. Coconut palms do belong to the biosphere of coastal cities (and towns) in the tropical regions of the world, and thus, constitute a local food source for the populations thereof. Coconut wastes, at source or point of end-use, can yield granular activated carbon with excellent adsorptive properties (Simha et al. 2018). Urea can be extracted from urine diverted from toilets (for which obviously, the sanitation system has to redesigned) using this granular activated carbon and sold to the agricultural sector (urban or otherwise). This, as stated by Simha et al. (2018), can effectively and economically (and in an environment-friendly manner) supply about one-fifth of the nitrogen fertiliser requirements for global food production. When one recalls that the production of synthetic nitrogen fertilisers is an energy-intensive and thereby often a climate-change-inducing process, the entrenchment of this circularity in the metabolism is undeniably much recommended.

When you decide to do something different with one purpose in mind, serendipity uncovers synergies. While recovery of nitrogen can be maximised if done upstream (vis-à-vis its recovery as ammonium sulphate downstream at the wastewater treatment plant—Szymańska et al. 2019), it also renders the wastewater downstream ammonia-deficient and less inhibitory to anaerobic bacterial activity. It follows thereby that the biomethane-generation potential can also be enhanced as a consequence of augmenting nitrogen recovery upstream (Simha et al. 2020). Theory inspires practise and supports experimentation. Theoretically, as calculated by Duan et al. (2020), if one could collect and anaerobically digest all the human wastes generated in China, after separating urea upstream as advocated by Simha et al. (2020), and if the resulting biomethane replaces coal as a source of electricity, 142 kt of CO_2-eq per day could disappear from China's GHG-footprint. The truncation would, however, be smaller (55 kt of CO_2-eq daily) if biomethane wends its way into the transportation sector and eats into gasoline's market share. This knowledge can serve as a guideline for decision-makers who may erroneously mandate that all biomethane be used for electricity generation.

Bio-energy for heating and/or transportation and/or electricity, and nitrogen fertiliser apart, urban sewage will be targeted more and more in the years to come for its phosphorus content, as observed by Szymańska et al. (2019), Werle and Sobek (2019), and Jarvie et al. (2019). Phosphorus recovery will become indispensable in the wake of dwindling resources of phosphates, which is sure to imperil food security around the globe. Let us not forget that the agricultural sector will anyway bear the brunt of climate change in many parts of the world. And, more importantly, the population of the world is bound to keep increasing. Indeed, sludge is often directly added to arable soil with the intention of utilising all the 'desirable' nutrients therein, and neglecting the effect some non-desirable organics and toxic heavy metals may have on crop yield and subsequent bio-amplification in the food chain, as noted by Rigueiro-Rodríguez et al. (2018). Werle and Sobek (2019) have uncovered a synergy between the avoidance of bio-amplification and the maximisation of phosphorus recovery from sewage sludge. Gasification, according to those authors, yields a by-product that is on par with naturally occurring phosphate rock when it comes to its phosphorus pentoxide (P_2O_5) content. Two birds with one stone, as the popular adage goes! However, it is not *'fait accompli'* necessarily if one just decides to use organic phosphorus fertilisers in lieu of the synthetic counterparts. The plant phosphorus uptake–which depends on the availability of the phosphorus in the fertiliser added–has to be acceptable, reminds Huygens and Saveyn (2018).

There was a mention of 'desirable nutrients' earlier, and these include, in addition to nitrogen and phosphorus, the three macronutrients (magnesium, potassium, and calcium) and the six micronutrients (iron, sodium, copper, manganese, cobalt, and zinc). All these are found in struvite, a valuable by-product of fertiliser value (recovered especially from agro-industrial wastewaters), as stated in Taddeo et al. (2018). Indeed, here, it is agriculture practised within city limits which is being considered, as our focus is only on the 'urban' metabolism. However, since we are focusing on a circular bio-economy to which a city belongs, the effluents from within the city—like activated sludge from wastewater treatment, for example—can be carbonised hydrothermally, and the resultant aqueous stream can be utilised as a source of nutrients as well as carbon for microalgae-cultivation. While the algal biomass can be supplied to the bioenergy sector (within or outside the city limits) which has the potential to slam the brakes on global warming caused by the dependence on fossil fuels, the treated effluent can be pumped to fields located beyond city limits, as has been happening in Israel for instance (Belete et al. 2019).

9.2.2 Food Wastes

Moving upstream, we can now focus on food waste generated in urban areas. Cities, needless to state, are the centres of consumption, and thereby waste generation, of humungous amounts of food—from processing units within city limits and households (Tsai 2020; Albizzati et al. 2021). Organic food waste accounts for a sizable fraction of municipal and industrial waste. If collected and valorised, it can yield huge quantities of biomethane (gas), biochar (solid), and bio-oil (liquid). While the former is a source of energy and a most welcome substitute for fossil fuels, the second-named finds use in agriculture (urban or otherwise), and the latter, as gathered from Borole et al. (2018), when subjected to a range of valorisation technologies, yields electricity, hydrogen (a bio-fuel), bio-chemicals like esters, diols, alcohols, and medium-chain fatty acids. These bio-based alternatives, needless to state, have climate change mitigation potential. While the possibility of economically diversifying the spectrum of valuable outputs by centralising the bio-refinery and scaling up capacity exists (Cristóbal et al. 2018), often simpler non-technical approaches like vermicomposting can be practised by restaurants, eateries, and households. This, as Sharma and Garg (2019) reiterate, is perhaps the best way to produce rich organic fertilisers (claimed to be better than compost and digestate by some researchers) which may be sold within the city or to farmers on its outskirts.

In the previous sub-section, we dwelt on the well-known anaerobic digestion of sewage sludge. If livestock manure, fruit-and-vegetable wastes, spent coffee grounds, and waste from the fish-canning industry are added to it, the biomethane yield can be enhanced substantially (Duarte et al. 2020). However, the fact that food waste can be valorised into bio-energy, bio-fertiliser, and bio-chemicals does not mean that the urban resident can pass on the responsibility of sustaining a circular bio-economy to the bio-refineries managed by the public and/or private sectors of the economy. Oldfield et al. (2016), in an Irish case study, hark back to 'Reduce', which readers recognise as the first 'R' in the waste management hierarchy, and stress its indispensability for urban metabolism in a circular bio-economy in the future. The fact that minimising food waste provides immense environmental and economic benefits, is common knowledge, even though it is not often incorporated by urban residents in their daily lives. But there are inspiring examples to learn from if inquisitive entrepreneurs wish to modify the metabolism in their cities. One such is the social reverse logistics system, InnovOleum, in Cyprus (Loizides et al. 2019). Something like used cooking oil which not many may even think about as a 'raw material' with the potential to yield valuable products, is collected from households and transesterified to biodiesel and/or bio-lubricants. While this is a clear trendsetter that can be emulated globally, it also emphasizes that the consumers need to cooperate and collaborate with the private/public sector enterprises, which are striving to transform the urban metabolism. Social acceptability and top-down regulatory measures must meet halfway, as recommended in a paper from Taiwan by Tsai (2020), in which the researcher reports that a 500% increase was noticed over a 9-year period, in the quantities of food wastes collected for digesting (anaerobically) and composting (aerobically)—a clear indication of the willingness on the part of the urban residents to separate wastes at source. If one looks for value, one finds it…as pointed out by Saadoun et al. (2020), who recommend harnessing the leachate from the food waste fraction of municipal solid wastes to produce medium-chain fatty acids.

Glucose and fructose ideally would be produced from sugarcrops on agricultural land. How about substituting sugarcrops with suitable non-contaminated agricultural and industrial wastes from the food and beverage sector? While techno-economic feasibility studies will show how profitable the saccharification of such wastes can be, Kwan et al. (2018) and Zhang et al. (2020) have demonstrated the possibility. Farms and pastures may not necessarily be parts of urban settings, but urban metabolism includes inflows from farms, and thereby circularity can motivate an outflow as well, in the form of organic municipal solid waste being availed of as animal feed. This has been shown to be environmentally a better alternative than valorising the said waste fraction in urban bio-refineries into higher-value bio-products, by Khounani et al. (2021) and Albizzatti et al. (2021). There is however an opportunity cost involved here—the sacrifice of the economic benefits which one could avail of, by opting for the bio-refinery route. A similar conundrum is uncovered in a UK study (Schmidt-Rivera et al. 2020), in which biodiesel production from spent coffee grounds (SCG), while being economically superior, is environmentally inferior to incineration, anaerobic digestion, landfilling (all three with energy recovery), direct application, and composting for a host of impact categories. While decision-making is indeed complicated by the multiplicity of criteria involved and one cannot have the cake and eat it too if the economic and techno-functional dimensions are prioritised (for some justifiable reason), the density and cold flow properties of a blend of SCG oil methyl ester, Euro diesel, pentanol, octanol, and butanol, have been shown to be comparable to those of pure Euro diesel, by Atabani and Al-Rubaye (2020). Upstream of the coffee supply chain, coffee silverskin which is a by-product of coffee roasting, when pyrolyzed has been shown to yield energy-rich biochar, oils, and antioxidants (Del Pozo et al. 2019, 2020).

Hildebrandt and Bezama (2019) and Briassoulis et al. (2019) have alluded to the 'design for recycling' concept (also refer Fig. 9.2) while discussing the gradual proliferation of bioplastics in the urban technosphere. They concur that these need not be looked upon as sources of energy necessarily at the end of their useful functional lives, but with appropriate recycling infrastructure in place, their material value can be 'circularly' harnessed. Panagiotou et al. (2018) have proven that waste eggshells (from anywhere along the egg supply chain, right down to the consumers and MSW management) adsorb phosphorus in effluent wastewater from anaerobic digesters. Thereafter, they can serve as bio-fertilisers in the form of brushite—hydrated calcium biphosphate. Thanks to their adsorptive capacity, they are excellent bio-flocculating media which can be used to harvest microalgal cells, just as granular activated carbon from coconut shells helps to extract urea from urine (Simha et al. 2018).

9.2.3 Bioenergy

In an interesting Swedish case (Mikhelkis and Venkatesh 2020), the authors have performed a techno-economic analysis of a state-of-the-art technology for the so-called carbon capture, utilisation, and storage or CCU/S, in a circular bio-economy. This smartly bridges agriculture (urban or otherwise), municipal solid waste (MSW) management, and forestry on the one hand as sources of wastes, the bio-energy sector as a potential user of the same, and the construction sector as the secondary user of the biogenic CO_2 emitted by the combined heat and power (CHP) plants. Agriculture and forestry here are the primary sources of bio-wastes (inflows of the urban metabolism). MSW management is a secondary source, and the bio-energy sector (which is one of the destinations of the bio-waste flows) is a tertiary sector of 'recyclables', which in this case is biogenic CO_2. The biogenic CO_2 is finally sintered into high-quality concrete blocks, achieving carbon negativity in the process. Carbon negativity is what needs to be targeted, moving forward, if climate change has to be effectively tackled. Talking of concrete blocks as entrapments for biogenic CO_2, designing concrete constructions from a circular bioeconomy point of view is the focus of Chiaia et al. (2017). One may think of 'anthropospheric jungles' taking over part of the responsibility of carbon sequestration from the forests in the biosphere.

Biodiesel production yields glycerol (a tertiary source output in this case), which Giacomobono et al. (2019) look upon as a valuable raw material for citric acid production. Citric acid ultimately finds use in the food and pharmaceutical sectors. Incidentally, glycerol also finds use as a carbon source in the denitrification process in wastewater treatment in some cities of the world [Oslo being one of them, as gathered from Venkatesh (2011)], and bio-based glycerol replacing the conventional petroleum-based ethanol or methanol in this regard will enable wastewater treatment plants to truncate their greenhouse-gas footprints further.

9.3 Summary and Conclusions

Figure 9.2 summarises the transformative metabolic flows discussed in the previous section. The bio-refinery supplies glycerol (1), bio-chemicals/lubricants (2), and bio-electricity (3) to the industrial sector within the city, while receiving food wastes (4) from the food and beverage industries for valorising. Glycerol (5) from the bio-refinery can be used as a carbon source in the denitrification process in biological wastewater treatment. Households supply the refinery with food wastes (6) and used cooking oil (7), and the agricultural sector with vermicompost (16). Coconut shells (17) are used to separate urine (yellow water) from mixed domestic wastewater (25), producing urea (20) in the process, and forwarding the urine-free wastewater (26) to the treatment plant which also receives wastewater from the industries (23). Hydrated calcium brushite (19)—obtained with the aid of waste eggshells from the poultry sector (18), and struvite (21) from the wastewater treatment plant, bio-fertilisers (12) and biochar (11, 24) from the bio-refinery—find use as fertilisers and soil-stabilisers in both forests and urban agriculture. The power sector receives bio-oil (8) and hydrogen (9) from the bio-refinery, wood (22) from the forests, algal biomass (14) from agriculture, and biogas (15) from sewage sludge digesters in the wastewater treatment plant. Public transport within the city benefits from biomethane (10) from the refinery and biogas (13) from the wastewater treatment plant. The difference between biomethane and biogas is simply that the former is upgraded biogas with almost 100% methane, while the latter is a blend of methane and carbon dioxide. Biogenic carbon dioxide from the power sector (27) is sequestered in 'carbon-sink' concrete (28) which finds use in urban infrastructure.

It is, after all, humans who build, design, and manage cities, and thereby the urban metabolism associated with it. Focusing on the word, 'design' here, urban planners have a crucial part to play in the transformation of the prevalent urban metabolism to one that would be more conducive and apt for a circular bio-economy. Some salient points to be borned in mind are listed hereunder:

- While socio-economic factors and environmental attitudes shape urban metabolism and thereby the status of a city in a circular bio-economy (smart or otherwise), urban design and planning—be that for new cities or for existing cities which need to be redesigned—ought to factor in these three dimensions of sustainable development, as illustrated in Fig. 9.3. As Kapoor et al. (2020) advocate, the urban ecosystem services have different kinds of value—monetary, socio-cultural, and ecological.

- A city which is a component of a circular bio-economy must not simply focus on combating climate change. Environmental impact categories other than climate change (global warming in other words), must not be swept under the carpet (D'Amato et al. 2020). The social and economic dimensions of sustainability and thereby the related sustainable development goals (which have been identified in Fig. 9.3) must not be overly compromised.
- The possibility of partnering with neighbouring cities and towns (as both demand centres and supply sources of valorised products) must be explored. This will improve the economic feasibility of bio-refineries. After all, a city is a part of a larger provincial/national circular bio-economy. That necessitates good governance.
- Numerous barriers will have to be overcome (Gottinger et al. 2020), and stakeholders—the different sectors in an urban setting which need to cooperate and collaborate to make 'circular bio-economic urban metabolism' a reality—have to be convinced by the urban planners that they are on the right path (Biber-Freudenberger et al. 2020).
- It needs to be appreciated that wherever challenges lurk, there will always be opportunities to be availed of.

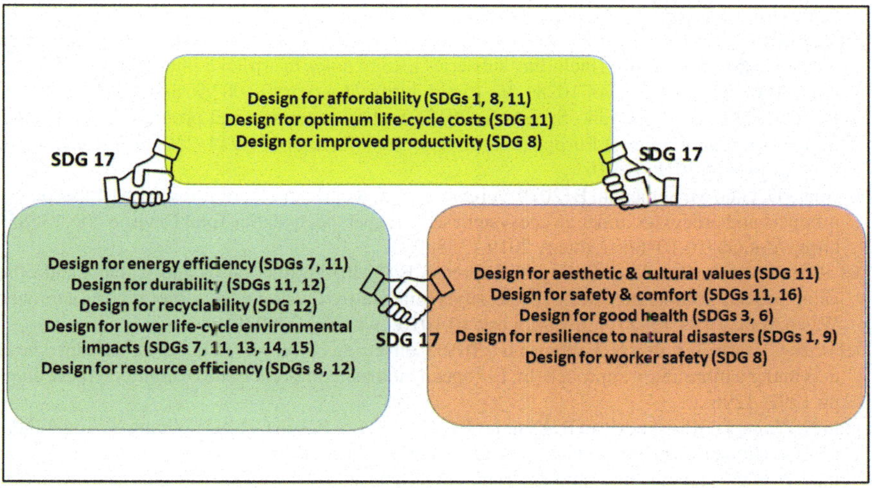

Fig. 9.3 Designing sustainable cities holistically for circular bio-economies, to adapt to climate change

Acknowledgements This chapter is dedicated to the fond memory of my wife Varshita.

References

Albizzati PF, Tonini D, Astrup TF (2021) High-value products from food waste: an environmental and socio-economic assessment. Sci Total Environ 755:142466. https://doi.org/10.1016/j.scitotenv.2020.142466

Atabani AE, Al-Rubaye OK (2020) Valorization of spent coffee grounds for biodiesel production: blending with higher alcohols, FT-IR, TGA, DSC, and NMR characterizations. Biomass Convers Biorefine 12:577–596. https://doi.org/10.1007/s13399-020-00866-z

Barros MV, Salvador R, de Francisco AC, Piekarski CM (2020) Mapping of research lines on circular economy practices in agriculture: from waste to energy. Renew Sustain Energy Rev 131:109958. https://doi.org/10.1016/j.rser.2020.109958

Belete YZ, Leu S, Boussiba S, Zorin B, Posten C, Thomsen L, Wang S, Gross A, Bernstein R (2019) Characterization and utilization of hydrothermal carbonization aqueous phase as nutrient source for microalgal growth. Bioresour Technol 290:121758. https://doi.org/10.1016/j.biortech.2019.121758

Biber-Freudenberger L, Ergeneman C, Förster JJ, Dietz T, Börner J (2020) Bioeconomy futures: expectation patterns of scientists and practitioners on the sustainability of bio-based transformation. Sustain Dev 28(5):1220–1235. https://doi.org/10.1002/sd.2072

Borole AP, Tsouris C, Pavlostathis SG, Yiacoumi S, Lewis AJ, Zeng X, Park L (2018) Efficient conversion of aqueous-waste-carbon compounds into electrons, hydrogen, and chemicals via separations and microbial electrocatalysis. Front Energy Res 6:94. https://doi.org/10.3389/fenrg.2018.00094

Briassoulis D, Pikasi A, Hiskakis M (2019) End-of-waste life: inventory of alternative end-of-use recirculation routes of bio-based plastics in the European Union context. Crit Rev Environ Sci Technol 49(20):1835–1892. https://doi.org/10.1080/10643389.2019.1591867

Chiaia B, Fantilli A, Peruccio PP (2017) A systemic approach to concrete constructions. In: Lecture notes in computer science (including subseries lecture notes in artificial intelligence and lecture notes in bioinformatics), vol 10281, pp 15–24. https://doi.org/10.1007/978-3-319-57931-3_2

Cristóbal J, Caldeira C, Corrado S, Sala S (2018) Techno-economic and profitability analysis of food waste biorefineries at European level. Bioresour Technol 259:244–252. https://doi.org/10.1016/j.biortech.2018.03.016

D'Amato D, Gaio M, Semenzin E (2020) A review of LCA assessments of forest-based bioeconomy products and processes under an ecosystem services perspective. Sci Total Environ 706:135859. https://doi.org/10.1016/j.scitotenv.2019.135859

Del Pozo C, Bartrolí J, Alier S, Puy N, Fàbregas E (2020) Production of antioxidants and other value- added compounds from coffee silverskin via pyrolysis under a biorefinery approach. Waste Manag 109:19–27. https://doi.org/10.1016/j.wasman.2020.04.044

Del Pozo C, Bartrolí J, Puy N, Fàbregas E (2019) Converting coffee silverskin to value-added products under a biorefinery approach. In: European biomass conference and exhibition proceedings, pp 1292–1296

Duan N, Zhang D, Khoshnevisan B, Kougias PG, Treu L, Liu Z, Lin C, Liu H, Zhang Y, Angelidaki I (2020). Human waste anaerobic digestion as a promising low-carbon strategy: operating performance, microbial dynamics and environmental footprint. J Clean Prod 256:120414. https://doi.org/10.1016/j.jclepro.2020.120414

Duarte E, Fragoso R, Smozinski N, Tavares J (2020) Enhancing bioenergy recovery from agro-food biowastes as a strategy to promote circular bioeconomy. J Sustain Dev Energy Water Environ Syst 9(1):1–13. https://doi.org/10.13044/j.sdewes.d8.0320

Giacomobono R, Albergo R, Valerio V, Bari ID (2019) Conversion of crude glycerol to citric acid by yarrowia lipolytica. European Biomass Conference and Exhibition Proceedings, pp 1471–1474

Gottinger A, Ladu L, Quitzow R (2020) Studying the transition towards a circular bioeconomy—A systematic literature review on transition studies and existing barriers. Sustainability (switzerland) 12(21):1–27. https://doi.org/10.3390/su12218990

Hildebrandt J, Bezama A (2019) Cross-fertilisation of ideas for a more sustainable fertiliser market: the need to incubate business concepts for harnessing organic residues and fertilisers on biotechnological conversion platforms in a circular bioeconomy. Waste Manag Res 36(12):1125–1126. https://doi.org/10.1177/0734242X18815988

Huygens D, Saveyn HGM (2018) Agronomic efficiency of selected phosphorus fertilisers derived from secondary raw materials for European agriculture. a meta-analysis. Agron Sustain Dev 38(5):52. https://doi.org/10.1007/s13593-018-0527-1

Jarvie HP, Flaten D, Sharpley AN, Kleinman PJA, Healy MG, King SM (2019) Future phosphorus: advancing new 2D phosphorus allotropes and growing a sustainable bioeconomy. J Environ Qual 48(5):1145–1155. https://doi.org/10.2134/jeq2019.03.0135

Kapoor V, Tripathi S, Devi RS, Srivastava P, Bhadouria R (2020) Ecological economics of an urban settlement: an overview. In: Verma P, Singh P, Singh R, Raghubanshi AS (eds) Urban Ecology: Emerging Patterns and Social-Ecological Systems. Elsevier, pp 91–110. https://doi.org/10.1016/B978-0-12-820730-7.00006-9

Khounani Z, Hosseinzadeh-Bandbafha H, Moustakas K, Talebi AF, Goli SAH, Rajaeifar MA, Khoshnevisan B, Salehi Jouzani G, Peng W, Kim K-H, Aghbashlo M, Tabatabaei M, Lam SS (2021) Environmental life cycle assessment of different biorefinery platforms valorizing olive wastes to biofuel, phosphate salts, natural antioxidant, and an oxygenated fuel additive (triacetin). J Clean Prod 278:123916. https://doi.org/10.1016/j.jclepro.2020.123916

Kwan TH, Ong KL, Haque MA, Kwan WH, Kulkarni S, Lin CSK (2018) Valorisation of food and beverage waste via saccharification for sugars recovery. Bioresour Technol 255:67–75. https://doi.org/10.1016/j.biortech.2018.01.077

Loizides MI, Loizidou XI, Orthodoxou DL, Petsa D (2019) Circular bioeconomy in action: collection and recycling of domestic used cooking oil through a social, reverse logistics system. Recycling 4(2):16. https://doi.org/10.3390/recycling4020016

Mikhelkis L, Venkatesh G (2020) Techno-economic and partial environmental analysis of carbon capture and storage (CCS) and carbon capture, utilization, and storage (CCU/S): case study from proposed waste- fed district-heating incinerator in Sweden. Sustainability (Switzerland) 12(15):5922. https://doi.org/10.3390/SU12155922

Oldfield TL, White E, Holden NM (2016) An environmental analysis of options for utilising wasted food and food residue. J Environ Manage 183:826–835. https://doi.org/10.1016/j.jenvman.2016.09.035

Panagiotou E, Kafa N, Koutsokeras L, Kouis P, Nikolaou P, Constantinides G, Vyrides I (2018) Turning calcined waste egg shells and wastewater to Brushite: phosphorus adsorption from aqua media and anaerobic sludge leach water. J Clean Prod 178:419–428. https://doi.org/10.1016/j.jclepro.2018.01.014

Rigueiro-Rodríguez A, Amador-García A, Ferreiro-Domínguez N, Muñoz-Ferreiro N, Santiago-Freijanes JJ, Mosquera-Losada MR (2018) Proposing policy changes for sewage sludge applications based on zinc within a circular economy perspective. Land Use Policy 76:839–846. https://doi.org/10.1016/j.landusepol.2018.03.025

Saadoun L, Campitelli A, Kannengiesser J, Stanojkovski D, Fels EAE, A, Mandi L, Ouazzani N, (2020) Potential of medium chain fatty acids production from municipal solid waste. Waste Manage 120:503–512. https://doi.org/10.1016/j.wasman.2020.10.013

Schmidt Rivera XC, Gallego-Schmid A, Najdanovic-Visak V, Azapagic A (2020) Life cycle environmental sustainability of valorisation routes for spent coffee grounds: from waste to resources. Resour Conserv Recycl 157:104751. https://doi.org/10.1016/j.resconrec.2020.104751

Sharma K, Garg VK (2019) Vermicomposting of waste: a zero-waste approach for waste management. In: Sustainable resource recovery and zero waste approaches, pp 133–164. https://doi.org/10.1016/B978-0-444-64200-4.00010-4

Simha P, Karlsson C, Viskari E-L, Malila R, Vinnerås B (2020) Field testing a pilot-scale system for alkaline dehydration of source-separated human urine: a case study in Finland. Front Environ Sci 8:570637. https://doi.org/10.3389/fenvs.2020.570637

Simha P, Zabaniotou A, Ganesapillai M (2018) Continuous urea–nitrogen recycling from human urine: a step towards creating a human-excreta-based bio–economy. J Clean Prod 172:4152–4161. https://doi.org/10.1016/j.jclepro.2017.01.062

Szymańska M, Szara E, Sosulski T, Wąs A, Van Pruissen GWP, Cornelissen RL, Borowik M, Konkol M (2019) A bio-refinery concept for n and p recovery—A chance for biogas plant development. Energies 12(1):en12010155. https://doi.org/10.3390/en12010155

Taddeo R, Honkanen M, Kolppo K, Lepistö R (2018) Nutrient management via struvite precipitation and recovery from various agro-industrial wastewaters: process feasibility and struvite quality. J Environ Manage 212:433–439. https://doi.org/10.1016/j.jenvman.2018.02.027

Tsai W-T (2020) Turning food waste into value-added resources: current status and regulatory promotion in Taiwan. Resources 9(5):53. https://doi.org/10.3390/RESOURCES9050053

United Nations—Department of Economic and Social Affairs: Sustainable Development (2015) The 17 goals. https://sdgs.un.org/goals. Accessed 11 Nov 2020

Venkatesh G (2011) Ph.D. thesis—Systems performance analysis of Oslo's water and wastewater system. Norwegian University of Science and Technology, Trondheim, Norway-7491. ISBN 978-82-471-2623-3, http://urn.kb.se/resolve?urn=urn:nbn:no:ntnu:diva-12664

Venkatesh G, Govindarajan R (2014) Metabolism—Not the preserve of the biological sciences. J Am Water Works Assoc 106(7):44–48

Venkatesh G (2021) Circular bio-economy—Paradigm for the future: systematic review of scientific journal publications from 2015–2021. Circular Econ Sustain 2:231–279. https://doi.org/10.1007/s43615-021-00084-3

Werle S, Sobek S (2019) Gasification of sewage sludge within a circular economy perspective: a Polish case study. Environ Sci Pollut Res 26(35):35422–35432. https://doi.org/10.1007/s11356-019-05897-2

Zhang C, Kang X, Wang F, Tian Y, Liu T, Su Y, Qian T, Zhang Y (2020) Valorization of food waste for cost-effective reducing sugar recovery in a two-stage enzymatic hydrolysis platform. Energy 208:118379. https://doi.org/10.1016/j.energy.2020.118379

Chapter 10
Closing the Urban Waste Loop-Delivering Environmental and Financial Sustainability

Bill Butterworth

Abstract The current level of global population and urbanisation has now reached an evidenced level of threat to the well-being of biodiversity, life generally, and, potentially, to the survival of the human species. The link between urban wastes, soil mycorrhizae, photosynthesis, and climate change is examined as an opportunity to close the urban-waste loop. A range of biological processes which are able to treat urban wastes as resources is reviewed. Composting is examined as a way to make wastes safe and to recycle them as useful fertiliser products, to give synergies in reducing pollution, increasing biodiversity, and locking up Carbon, so helping to limit or reduce atmospheric Carbon dioxide. The place and safe mechanisms of the soil mycorrhizae are outlined and discussed. Experience of recycling over 5 million tonnes of a range of urban wastes containing a wide spectrum of organic Carbon molecules, and many non-organic chemicals, including a laboratory study of recycling a microplastic, which can be safely recycled at local level and at low cost, is detailed with the cash drivers to make urban waste recycling both environmentally and financially sustainable. The possibilities to reduce irrigation need, and reclaim desert are qualitatively and digitally outlined.

Keywords Climate change · Composting · Irrigation need · Microplastics · Mycorrhizae · Organic matter · Pharmaceuticals · Soil · Sustainable · Wastes

Abbreviations

IPCC	The Intergovernmental Panel on Climate Change
GHG's	Greenhouse Gases
HDD	Horizontal Directional Drilling
MSW	Municipal Solid Wastes
OECD	Organisation for Economic Co-operation and Development

B. Butterworth (✉)
Land Research Ltd., Devizes, Wiltshire SN10 2ES, UK
e-mail: bill@landresearchonline.com

R. Bhadouria et al. (eds.), *Urban Metabolism and Climate Change*,
https://doi.org/10.1007/978-3-031-29422-8_10

10.1 Introduction

10.1.1 Urban Wastes and Global Warming

The reality of global warming, climate change, and environmental stress is widely discussed elsewhere, particularly the Intergovernmental Panel on Climate Change (IPCC) reports (IPCC 2017, 2022; Bhadouria et al. 2022). The evidence heavily supports the conclusion that most, if not all, is man-made and that the biggest single pollutant is the volume of Carbon dioxide from burning hydrocarbon fuels. There is a classic way of solving a problem which is to find a mirror image and put the two together so as each cancels out the other and, when done right, also generates synergies of advantage to the wider society. "Wastes" can and do become resources in the generation of fertilisers for food production, for locking up Carbon dioxide, reducing pollution of water and air, combating drought stress in crop production, and in stabilising climate change (Butterworth 2009a). If wastes from urban activity can be used to fertilise arable crops, reclaim marginal and desert land, and grow more crops and trees, with less irrigation, then there is an opportunity to make a significant contribution to balancing the two problems of climate change and urban wastes. Further, the use of proximity farm and forestry land could reduce municipal costs and bolster the local rural economy.

All cities produce "waste". Sustainability necessarily demands that society, led by municipal authorities, must organise the recycling of wastes as resources, involving the wider environment including the air and areas of land and water around urban areas, to deliver not just re-use but the reduction of energy and resource use in developing sustainability in the ecosystem of society as a whole. The quantities and composition of urban wastes produced globally are directly related to population size and wealth. Globally, current production of municipal solid waste (MSW) is in the region of 2.5 billion tonnes. As populations get wealthier, that figure rises. World Bank figures (Hoonveg and Blada-Tata 2012) indicate very wide variation, from 0.09 kg per person per day in parts of North Africa to 2.0 kg in countries in the Organisation for Economic Co-operation and Development (OECD). Much of this waste is put to landfill or landraise. Material put to landfill will often emit significant methane which, in many cases, will be released to atmosphere (and so have a further effect on climate change). For landraise material, there will be a mixture of methane and carbon dioxide emissions (Butterworth 2021a, b). Composting separated MSW material and putting to farm and forestry land will dramatically slow down the emissions, potentially to sustainable levels, and reduce or eliminate the energy and emissions cost of producing mineral fertilisers (Butterworth 2009b, 2021a).

There are many other issues with MSW being put to landfill. Just one example is that Mishra et al. (2020) observed that such a disposal route is a major potential pollution threat to the nearest aquifer system. This, then, is yet another example of not closing the loop causing greater environmental stress. Source-separated MSWs, i.e., the householder puts materials into separate bins, does work but not very well. Land Research Ltd.'s experience of several million tonnes of source separated MSW is that such material is dramatically more polluted, and unpredictable so, than factory wastes. Factory wastes are generally formed from a controlled process (Singh et al. 2008; Butterworth 2009c) which is relatively constant and predictable. MSW from households is disciplined by individuals with little or no external supervision. As an alternative, it may well be better to collect garbage (MSW) unseparated and separate in the relatively controlled environment of an MRF—Materials Recycling Facility. Nevertheless, while this material looks untidy, but the level of separation may be seen as acceptable and not having unacceptable risks attached to it.

10.1.2 Wider Issues of Pollution, Especially by Pharmaceutical Drug Residues

We have known for over 30 years that hormone residues from women taking the "contraceptive pill" passed through the body, into the sewage systems, through treatment, into the rivers, and then up the food chain into male stickleback fish which produced eggs (the eggs were infertile, of course). In 2018, some Australian researchers (Klein 2018) identified 69 named pharmaceutical drugs in river mud and up the food chain into fish and found doses close to those given to humans in legitimate treatment prescribed by a doctor. The effects of such doses on wildlife were quoted in the research paper as "unknown" but it is not logical to assume that there would be little or none. The New Scientist quoted another research paper (Richmond et al. 2018) as saying, "*If we went to a doctor and said we were on 69 drugs, they would probably be concerned*". This research shows evidence that pharmaceutical drug residues which have passed through the human body are significantly worse than previously, commonly understood. Up to 90% of drugs administered to humans pass out in their urine (Butterworth 2021a, b). Drugs residues do get through the conventional, widespread, sewage collection and treatment systems into riverbed muds and back up the food chain. In the countries with higher population density and greater average age than Australia, the risks from pollution with pharmaceutical drug residues are likely to be higher and more complex. Indeed, there is reasonable evidence (WWQA 2021) that rivers round the world are heavily polluted with pharmaceutical residues and there are significant risks involved. Without doubt, this means that food and drinking water are widely polluted, globally. What this report shows is that composting can and does provide a processing route to reduce these environmental risks from urban wastes.

10.1.3 Formulating a Strategy

Land Research did, back in the early 1990s, ask if and how composting urban "wastes" could deliver safe recycling on a much larger, potentially global scale.

10.1.3.1 Energy in Fertilisers

Nitrogen fertiliser is an example of benefit covering folly. To make ammonium nitrate, air is pumped through a continuous, 2 m diameter, electric arc. It takes enormous amounts of electricity to do this. Bear in mind that one tonne of Nitrogen nutrient, almost exactly 3 tonnes of purchased ammonium nitrate fertiliser, made in a modern USA factory does, according to the United Nations-sponsored research (Gellings and Parmitter 2004) takes 21,000 (yes, no mistake, twenty-one thousand) kWh of electricity to manufacture and deliver. According to Land Research Ltd. (Butterworth 2021a, b), some factories in production elsewhere in the world were up to a factor of 26 times less efficient. Global nitrogen fertiliser production is approaching 150 million tonnes. Most of that electricity will have come from burning fossilised, hydrocarbon fuels (see www.landresearchonline.com 20 December 2014). As a guestimate, even at an efficient factory rate, that may be *more than* a global consumption of a thousand, trillion kWh of electricity per annum (That is 1000,000,000,000 kWh). Now, take coal as a common fuel and use that to indicate how much CO_2 would be produced if coal was used (as is the substantial case) to generate the electricity to make the Nitrogen fertiliser as in the above paragraph. The answer is 340,000,000,000 kg or 340 million tonnes of CO_2.

10.1.4 Biological Processes

The sustainable link between all wastes, urban, industrial, and agricultural, is embodied in photosynthesis. The author of this text has persistently taken the view that, in the time (IPCC 2022) we have available to significantly reduce environmental over-stress, there are two issues of fundamental importance: reducing the burning of hydrocarbon fuels and stepping up global photosynthesis coupled to raising the global soil carbon sink. All the other solutions, potentially important as they certainly are, will be too little and too late. Globally, the rich nations tend to be excited by developing new, and often expensive, industrial-based processes on a big scale. These processes often have high capital cost and therefore demanding volumes of wastes committed for long periods ahead, resulting in high transport costs and inhibiting change. What, so often, happens is that proximity processing to support green leaves gets forgotten.

10.1.4.1 Anaerobic Digestion (AD) and Thermophilic Aerobic Digestion (TAD)

How AD Works?

All living creatures "breathe" and feed in some way, and produce "waste" products. All micro-organisms are capable of digestion and either do need oxygen (i.e., they are "aerobic") or work in the absence of Oxygen and produce gases (for example, some produce methane and, because they work in the absence of Oxygen are "anaerobic"). Generally speaking, AD, Anaerobic Digestion, is a process which produces methane and Carbon dioxide, commonly around 50% of each. This "Biogas" will burn but does so much better if the Carbon dioxide is removed first (by "scrubbing" with water). The resultant and relatively pure methane can be burned as fuel in an internal combustion engine, to drive a generator, to produce electricity. In so burning the gas, it will produce a "clean" burn producing Carbon dioxide and water and very little else.

Digestion of Wastes in Mesophilic Processes

Historically, sewage products have caused very significant public nuisance when spread to land. Raw, untreated sewage can no longer be legally spread to land in the UK or most of Europe. Whenever and wherever it is so spread, it stinks. The first line of processing, which many sewage companies use, is mesophilic digestion in the absence of air (AD). This allows natural (at UK ambient temperatures) microorganism process to a temperature of up to around 38 °C. There is a very significant reduction in odour and, with sewage and many other organic wastes, there is a usable production of methane. Indeed, this is the basis of AD.

Generally, anaerobic digestion works well with liquids up to about 10% dry matter. As dry matter increases, the process becomes increasingly difficult to manage and costs rise (from grinding up solids in order to allow the process to proceed). Historically, AD for high dry matter operations has really not worked. However, mesophilic AD has been used for many decades in China and other Asian countries on a small scale in one-family digesters to produce gas for cooking. Feedstocks are often all family wastes (Ntostoglou et al. 2021).

Digestion of Wastes in Thermophilic AD Processes

These processes, by definition, involve going up to higher temperatures, usually around 60 °C to get pasteurisation. Gas production is also increased, and therefore, the organic Carbon residue is also yet more reduced. The world is expanding in population (World Bank 2019) and people need food and warmth for cooking and home comfort. That takes energy and energy security is directly related to national security in the wider, including military, sense. So, for this reason alone, AD to produce electricity is compelling. Also, where subsidies are available to support AD, an individual can make a profit. Further, the technology of AD is getting better. One word of caution, it is as well to remember that methane is potentially an explosive gas. The downside of AD is that the process turns the Carbon-based organic matter into methane which, when used as a fuel, produces Carbon dioxide.

Outputs from AD–Energy and Crop Nutrients

Soluble nutrients from AD liquor may enter the plant root in a crop. This happens in solution, or hydroponically. It works, very much as ammonium nitrate fertiliser works, i.e., the problem is that 30–50% of Nitrogen in solution (probably mainly as ammonium and nitrate ions) is lost to groundwater (Butterworth 2012, 2015; IPCC 2022).

10.1.4.2 Thermophilic Aerobic Digestion (TAD)

Composting is actually a thermophilic aerobic process. However, the label of TAD is usually used with reference to the processing liquids. Basically, this process involves blowing air through the mass of "waste" liquid. This can be done outside with materials such as pig manure in a lagoon in order to reduce odour. It can also be done inside which will cost very much more. The indoor route is useful for treatment of potentially high-risk materials such as animal bi-products or high odour materials. In both cases, total enclosure gives potentially total control. Blowing air will raise temperatures in suitable materials (those with enough nutrient concentration for feed the microorganisms) to above 70 °C which will give adequate pathogen control in most circumstances. With some feedstocks, up to 100 °C can be reached (although the energy cost of blowing the air may become very significant). If extracted air is put through a carbon filter, then a very high degree of odour control can be achieved.

Thermophilic Aerobic Digestion (TAD) will not produce methane, and therefore, is of no value in producing electricity. However, it has three very significant advantages over AD (Butterworth 2012). The equipment is very significantly of lower cost. The output liquor is of significantly higher value in agriculture and horticulture because it has maintained nearly all the organic carbon, the nutrients in low-leachability form and has low odour. Thirdly, there is no explosion risk. There is another potentially dramatically important possibility with TAD, it will break the molecules of pharmaceutical drug residues. There is reasonable evidence that pharmaceutic residues do generally get through conventional sewage treatment works (STW's) and into groundwater, and in some cases at least, into drinking water for human consumption (Klein 2018; Butterworth 2022). Fortunately, there is also reasonable evidence that TAD will satisfactorily digest these residues and make them safe. This leaves the question as to whether the volumes of sewage produced by high concentrations of population (such as cities) could be treated this way.

10.1.4.3 Composting

Square Static Windrow

There is a "hybrid" of ideas that has most of the advantages of the other composting methods and less of the disadvantages, it has been labelled Square Static Windrows.

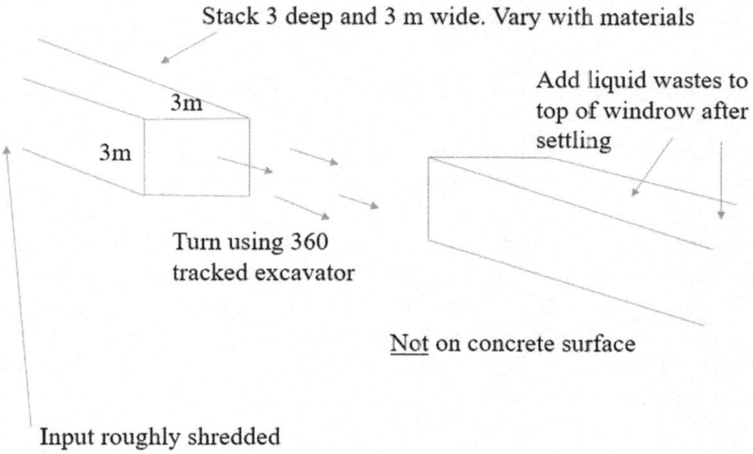

Fig. 10.1 Square static windrows—with and without turning, on or off concrete

This still uses a long, thin windrow (to let heat out, air in, and reduce fire risk), but it is square-topped (to be able to manage rain and liquid additions such as sewage, animal slurries, or industrial liquid wastes) and "static" because it is turned infrequently or, sometimes, not at all. The square windrow can be left without turning, and it will then take months or even years to mature to be a spreadable compost. In many farm situations, this length of process time really does not matter. To speed things up, it can be turned using a tracked 360 excavator as in Fig. 10.1. This way of composting is not likely to be as fast as USA-style windrowing but it is of much lower financial cost, much lower environmental cost and just as good in terms of fixing Carbon and plant nutrients to return to the soil. It has one major plus and that is to be able to safely operate off concrete (Concrete has a very high energy cost).

10.1.4.4 Economic Drivers

There are basically two drivers to make recycling happen; encouragement and deterrence. Encouragement usually involves cash, i.e., financial gain either by commercial trading or by government subsidy. Deterrence involves regulation by commercial, local municipal authority of central/regional government, supported by adequate policing. However, in the UK and EU, and indeed many other countries, disposal to landfill (or land raise) is progressively taxed. That only works if there is adequate policing to stop dumping illegally. If and when legal landfill is limited in volume, then the operators of a recycling site can charge a fee for entry (a "gate" fee) and that, coupled to tax, becomes an incentive to recycle. Farmers taking wastes can therefore charge a little less of a gate fee and develop a business which earns cash and manufactures organic fertiliser.

10.1.5 Problems in Implementation

10.1.5.1 Low-Volume Air Extraction and Biobeds

Blowing large volumes of air in at the bottom of a compost heap needs large amounts of power and produces an odour problem. If air is blown through a granular or mixed particle material, it tends to "track" through the pathway of least resistance. This leaves some areas over-ventilated and some under-ventilated. What happens here is that air in the voids travels at a rate which becomes turbulent. At what is known as the critical flow rate, that turbulence becomes so violent that the air just goes round and round in that cavity, rather than travel to the next void. Almost always, putting more power into the fan and increasing air flow, increases this turbulence effect, making the lack of uniform ventilation worse, not better. Eventually, the air will track through a limited pathway, leaving most unventilated, and therefore, of over-heating and a fire risk. Extracting air from the bottom of the heap, by suction at very low volumes, avoids this turbulence and results in a uniform air flow through all the voids. A fan of only 1–2 kW will ventilate several hundred tonnes of compost.

If the extracted air is pumped (possibly with the same fan) out through a bio-bed which will extract all the odour once it has become biologically active. The bio-bed is made from coarsely shredded wood, preferably root or stump timber which does not degrade (compost) so quickly. Alternatively, the bed can be made of plastic particles, but what is needed is a large surface area. Natural wood, especially root timber, will already be "dirty" and carry microorganisms, and it is these that will digest the odour and anything else in the airflow. The bed will be around 1 m deep and take a couple of weeks to become biologically active. It will take many months, but, eventually, the process will digest the wood chip and, as the particles become smaller, the resistance to the airflow will increase, and the chip will need replacing (Fig. 10.2).

10.1.5.2 Separating "Good" and "Bad" Wastes

Source separation works quite well in industry, but for mixed wastes in general there is a range of engineering solutions to separation including electro-magnets, air flotation, and of course hand labour. Figure 10.3 shows an option for low grade compost or compost contaminated with trash such as plastic bottles. This can be used as aerobic landfill for hedgerow boundaries or forestry plantings. In hot climates this can be used for planting energy crops such as Jatropha (Butterworth 2009b). The purpose of the polypropylene netting is to stop uncontrolled spreading of visible rubbish if there is excessive wind-blow.

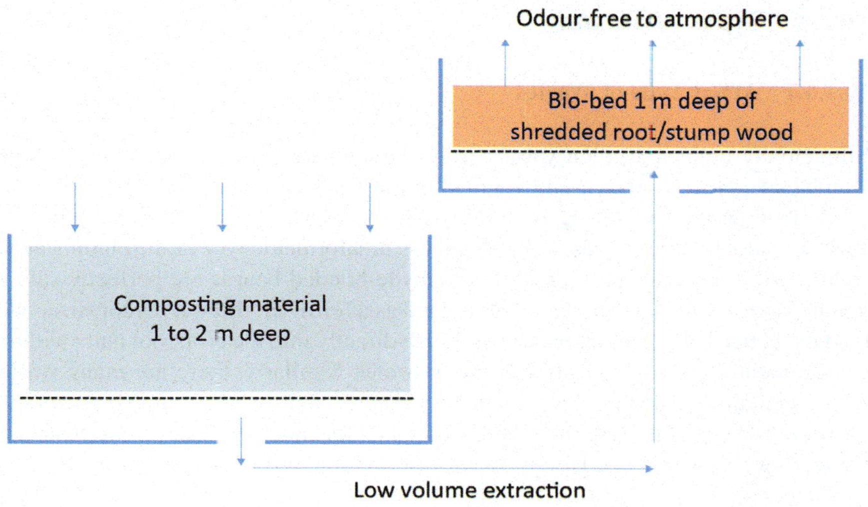

Fig. 10.2 Composting with low volume air extraction

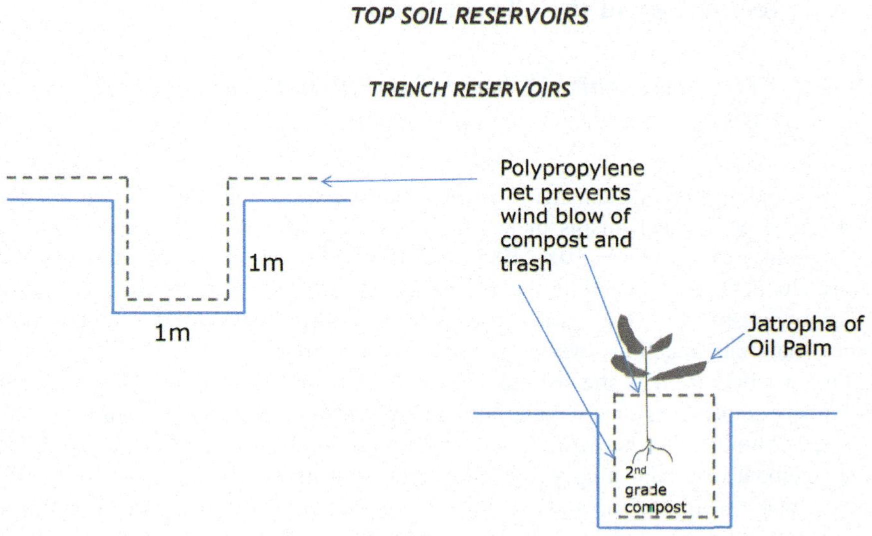

Fig. 10.3 Trench reservoirs—aerobic landfill

10.1.6 *Types of Wastes*

10.1.6.1 What Type of Wastes

Municipally collected green wastes from gardens are often a good starting point which may give a good base and volume for a composting operation. There are often other good sources of nitrogen, notably Medium Density Fibreboard (MDF) and similar manufactured boards, all of which use ureaformaldehyde as the binding agent or glue. After manufacture, ureaformaldehyde-bonded boards are perfectly safe to handle, and when composted and put to land as a fertiliser, the soil mycorrhizae will feed the N in the ureaformaldehyde digestate directly into the crop root hairs without environmentally significant loss to groundwater. Similarly, there are many wastes which contain Nitrogen as part of their structure, and therefore, make potentially useful fertilisers (Butterworth 1997, 1998a, b, 2000, 2008, 2009a, b; CORDIS—The Community Research and Development Information Service, EPA 1998).

10.2 Managing Environmental Safety: The Basic Technology of the Closed Loop

10.2.1 *The Mycorrhizal Closed Conduit and Why Natural Ecosystems Do Not Leak Nutrients*

Understanding the mechanisms in what is commonly called "the closed loop" and managing those mechanisms makes recycling to land dramatically safer in environmental terms. Figures 10.4, 10.5, and 10.6 below show the principles of the closed loop. Organic materials, and inorganic ones which have food value for the soil microorganisms, do *NOT* break down directly to form humus (Butterworth 1986). Such materials added as "waste" are digested by microorganisms and turned into their own bodies. It is the breakdown of these bodies which form the complex, relatively stable, black tarry material which gives soils their dark colour, generally termed "humus". So, knowing how to feed these organisms is the first step in the management of composting wastes and of the soil. It is also important to see that the compost heap and the soil are not separate operations. Mostly, everything that goes on in a compost heap would also happen in the soil, even pathogen destruction. The big advantages to farming of composting before spreading to land are to use the temperature to kill weed seeds and to fit the logistics of the farm cropping seasons (Butterworth 2009a, b). The micro-organisms feed, multiply, and die, then break down into humus. Humus is an extremely complex mixture of heavy molecules of hydrocarbons (the same process which makes crude oil), carbohydrates and proteins (which lock up the Nitrogen). These molecules are large and insoluble, and there is no limit to the quantities that can be put onto the soil safely. The evidence for this can be found in any natural ecosystem such as the Fens in the Eastern part of England in the UK. When the Dutch engineer Vermuyden drained them nearly 300 years ago, some were 10–15 m deep of black, alluvial, high-organic Carbon soil. Up until now, farmers could grow crops there every year (for 300 years!) exporting the harvested products with the nutrients they contained, including the Nitrogen, and *never* needed to add any fertiliser. Clearly, there had been an enormous reservoir of crop nutrients, but the local dikes and large areas of open water are not polluted with green slime and dead fish. Those large molecules of organic matter as humus will remain for centuries until long strands of soil fungi, called mycorrhiza, linked at one end to plants requiring food, start digesting them. These mycorrhizae either go upto or envelop the plant root hair, rather like the placenta in a baby mammal in the womb, or actually cross the root hair wall into the plant (Butterworth 1999, 2008). This is a closed conduit, and why natural ecosystems do not lose their nutrients (as do systems based on soluble fertilisers) and do not leak enough nutrients to pollute themselves or groundwater.

As common sense might indicate, the green-leaved plants and the soil fungi evolved together over millions of years, and they operate at the same temperatures, so the system is demand-led. This system is how all natural ecosystems not only eliminate nitrate pollution, they also eliminate all such out of balance pollution including phosphates, potash, etc. There is a further advantage, as the system locks up Carbon in the soil. Around 100 million tonnes of "waste" produced in the UK per

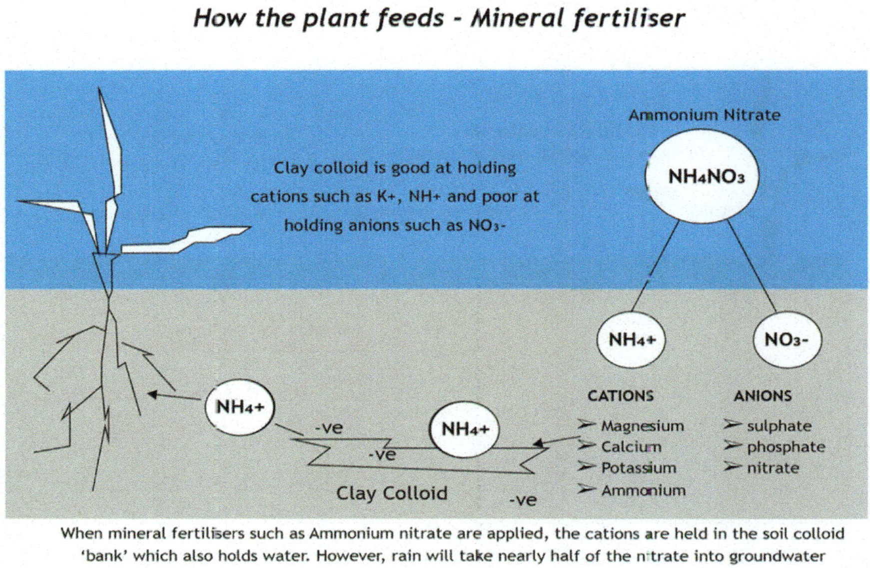

Fig. 10.4 How the plant feeds—mineral fertiliser. The conventional view of how plants feed with the assumption that nutrients get to the plant via solution in the groundwater. With mineral fertilisers, this is probably either partly or completely true

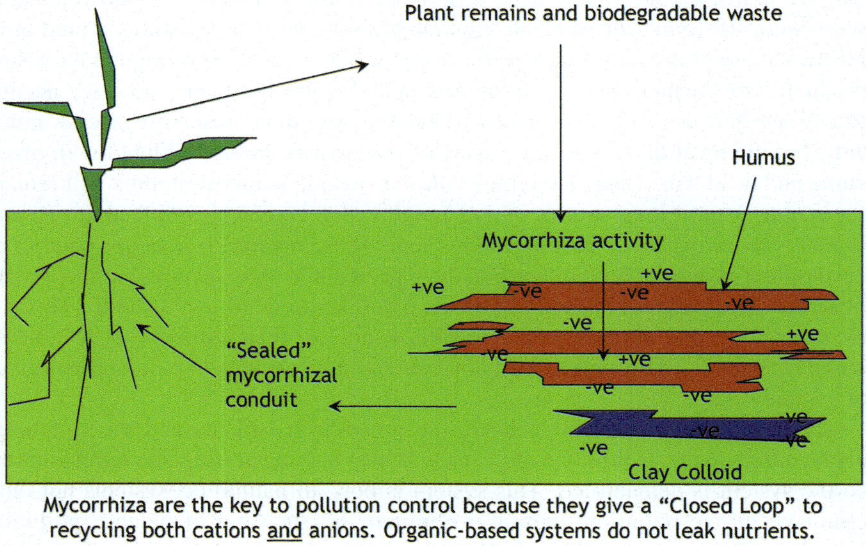

Fig. 10.5 How the plant feeds—natural ecosystem

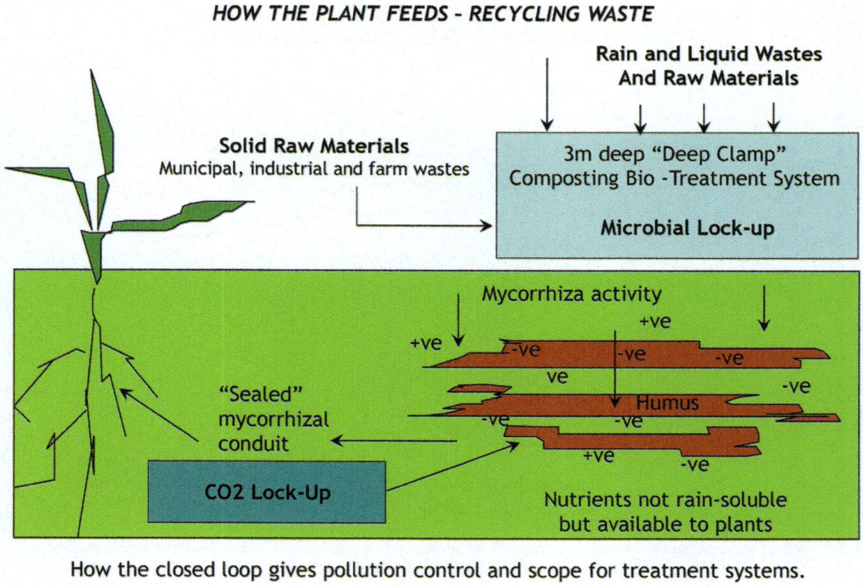

Fig. 10.6 How the plant feeds—recycling system

annum (DEFRA 2021) and which could be recycled to land would, if incinerated, produce around 75 million tonnes of Carbon dioxide per annum which is 10% of the Kyoto Protocol estimate of total UK emissions. Composting to land can lock that up. Nevertheless, regulation in the EU and UK in particular, is progressively limiting recycling to land and increasingly incinerating wastes under the guise of "Energy from Waste"—EfW. EfW is really little more than incineration, producing GHGs. Figures 10.4, 10.5, and 10.6 show how natural ecosystems manage to "leak" enough, and only enough, to keep the system working without pollution or starvation, i.e., in balance or "sustainably" (Butterworth 1998b, 2002, 2009a, b). Figure 10.4 shows a conventional view of how the system works.

In hydroponic crop production, nutrients are in solution. The same applies to soil-based farm production using mineral fertilisers. However, in soil-based systems which have a significant proportion of clay, or more so if there is significant organic matter, then at least some of the mineral ions will be held colloidally and less likely to leach out to groundwater and be lost. Nevertheless, there will still be losses of mineral ions at a level which may cause significant pollution. This, of course, does not occur in natural ecosystems. Figure 10.5 shows how such pollution is avoided and shows the mycorrhizal conduit which is the central mechanism in what is commonly referred to as the "Closed Loop". It is the mechanism which stops leakage at a level of pollution. It is this very same mechanism which feeds plants and protects them from disease (Butterworth 1998a, b, 1999, 2009c).

Natural ecosystems depend partly on the colloidal properties of clay but mainly on organic matter. Humus is a complex mixture of heavy molecules which are not soluble in water. Albrecht (1919–1997) and Kinsey (1999) pointed out that this humus has several times the colloidal capacity of clay and will hold onto anions as well as cations. That, however, still did not explain how the nutrients got into the plant without leakage. It was the American PGA (Professional Golfers Association), back in the 1980/90s, who pursued this investigation to show that the soil fungi, known as mycorrhiza, fed at one end of their hyphae on the humus and the other end went not up to somewhere near the plant root hair but actually *crossed the root hair wall into the plant*. This finding was added to by researchers at Aberystwyth in South Wales (Allen 1996) who showed that there was another type of mycorrhiza which went up to the root hair and wrapped around it much as the placenta in a mammal. This is a molecular-level relationship and a closed conduit. In natural ecosystems, plant nutrients do not enter into solution in the ground water in order to enter the plant. That is why the natural ecosystems do not leak enough to cause pollution (Butterworth 1999). What composting can do is provide a "buffer" between a controlled process and the soil. That buffer can isolate physical, chemical, and biological risks in order to allow processing, monitoring, and safety controls to operate.

10.2.2 Why Soil Universes Do Not Pollute Themselves?

It is, perhaps, useful in understanding how these mechanisms work to first understand what pollution actually is. Pollution is, incidentally, not just quite natural but fundamental to life itself. Part of the definition of a living organism (as distinct from not living—a chair, a tractor or a computer for example) is that the living organism is producing pollutants. These pollutants are products from the body of the organism which must be got rid of outside that body. The question then arises as to when that production of pollutants becomes "pollution". The answer, in the scientific sense, is *always*. In the practical or legal sense, the exact definition of pollution depends on the following. If the production of pollutants is at a level where the local environment cannot, in "reasonable" time, bring the system back to what the ecosystem previously was, if there is a shift in the ecological equilibrium, then, it may be said, pollution may have occurred. At the end of any discussion, "pollution" is in the eye of the regulator which tends to restrict the innovation we need.

The two ways a soil combats pollution are by providing a "buffer" to buy time and by digesting the pollutants and passing them into the ecological chain. Biological systems are rarely instant, and commonly not even fast. Certainly, digestion is slow, and the bigger the molecule, the slower the process. However, given time and possibly enough dilution, nature will digest almost anything, including such molecules as polychlorinated hydrocarbons (PCB's) and the explosive TNT (tri-nitro toluene) (USEPA 1998).

Soils which are substantially sands have little buffering capacity and little ability to hold crop nutrients. For example, by adding ammonium nitrate fertiliser to sand, the ammonium cations and the nitrate anions will leach out very easily with probably more than half going into the groundwater with rain or irrigation. That is a significant economic loss and potential pollution of groundwater. However, add the same material to a clay soil, and the colloidal capacity of the clay will retain much of the ammonium cations and possibly some of the nitrate anions, too. Add the same material to humus, and there will be a retention of most, may be all, of both ions (Butterworth 2009b). There will be no leaching with rain or irrigation of either the ammonium or the nitrate ion, and pollution of groundwater will be eliminated (Butterworth 1999, 2000). So, different soils will have different buffering effects, and we can alter that capacity by adding and managing the organic matter levels of soils, specifically the humus content.

10.2.3 Heavy Metal Tolerance

The soil mycorrhizae have hyphae which digest humus at one end and cross the root hair wall (or wrap round it like the placenta of a mammal) at the other, so forming a closed conduit to feed the crop or forest trees and plants. Observations suggest that this conduit is cleverly selective. If there is more than a "normal" and acceptable amount of a heavy metal present, this mechanism seems to be able to tolerate it. What may happen is that the heavy metal areas inhibit or kill the mycorrhizae activity, and so the heavy metal does not get fed to the plant. May be, eventually, there is some oxidation of the humus and the heavy metals get leached into the groundwater. Alternatively, the heavy metals remain organically bound and just stay in position. In this way, which appears likely (if not the whole story), organic-based systems are much more tolerant of heavy metal pollution than systems-based on soluble fertilisers such as ammonium nitrate.

10.2.4 Professional Enablement Versus Regulation

Regulation is necessary in all societies, even under anarchy. However, history shows that regulation inevitably grows and inhibits progress and development (Collins 2001). There is, however, an alternative. In the UK, there used to be a relatively small, separate, completely independent body called HMIP—Her Majesty's Inspectorate of Pollution. It acted if, and only if, there was actual pollution or danger of imminent and significant pollution. That was backed up by taking cases to Court for legal proceedings. Suppose, then, there is no further regulation. None at all. Instead, a local authority can be empowered to appoint one or more professionally qualified persons to oversee operations in their area. Those professionals could, for example, be Members of the British Society of Soil Scientists, or similar relevant professional institution. If there is pollution confirmed in Court, then that court has the option to restrict, penalise or withdraw the permitted supervision. That is, in effect, what Land Research set up with a percentage of the gate fee used to fund the supervision (Butterworth 2009a).

10.3 Results

10.3.1 Reverse Franchise Growth

Land Research developed a group of farmers by franchising them with a monitored package of organised discipline, marketing support and advice on compliance with regulations—and then gave them control over the franchise, i.e., they owned it. The individual farmers lost nothing in terms of sovereignty over their own land, but as a group, they could credibly bid for larger contracts than they could as individuals. The growth was significant as shown in the following Table 10.1.

1. The variation in monetary values per tonne reflected different wastes available at different times.
2. 1998 was restricted by an outbreak of BSE in the UK.

Table 10.1 Table of growth of field trails

Year ending March	Group turnover Net of VAT in UK £'000	Estimated tonnes total
1994	6.5	370
1995	12	700
1996	26	1300
1997	70.4	3500
1998	38	1950[a]
1999	94.6	4500
2000	14.4	1000
2001	28.5	3000
2002	68.6	30,000
2003	161	85,000
2004	167	80,000
2005	208	100,000
2006	288	131,000
2007	342	160,000
2008	500	250,000

[a]*Notes on table*

3. The group continued to expand after 2008, and most farms in the group were still operating in 2022.
4. Not only was the group growth significant, but its safety record in the face of significant over-regulation, was blemishless. Further, the farms involved verbally (with some evidence) recoded the following changes in cropping and soil structure. So also, were the following factors in Sects. 3.2–3.6.

10.3.2 Reducing Pollution of Soils and Groundwater

Over the period of the above table, the organisation's supervisors, plus the significant efforts of the statutory regulator, found no evidence of pollution.

10.3.3 Reducing Cultivation Energy

As compost applications increased year on year and soil organic matters rose, so did cultivation energy fall, especially on heavy clay soils. Some farms reported a reduction in excess of 50%. In particular, the use of power-driven cultivation equipment such as power harrows became discontinued. This reduction improved timelines and was one of the causes of more consistent and higher crop yields.

It generally took 4–7 years, depending on soil type, initial organic matter level, and compost quality and application rates, for the purchase and application of mineral fertilisers to fall to zero.

10.3.4 *Increase in Soil Biological Activity and Biodiversity up the Food Chain*

Farmers generally reported increase in wildlife and biodiversity. In some cases, observations included reference to increases in field mice and red kites.

10.3.5 *Reducing Irrigation Need*

Figure 10.7 shows the basic construction of a sub-surface reservoir of organic Carbon. Compost of any quantity can be put into the soil reservoir, mixed with enough of the local soil to create enough soil strength to support the field traffic (machinery) used in that situation. The proportion will vary according to the local soil and the constituents in the compost, particularly particle size, but as a guide, a 50/50 mix.

10.3.6 *Increasing Crop Yields*

All farms reported that, as soil organic matters increased, so did crop yield increased and become more consistent.

10.4 Reclaiming Desert

It is widely accepted that sewage, subject to tests for heavy metal levels, can and does make a very useful material to assist in reclaiming desert and arid soils (Labrecque 1995; Butterworth 1997; Müller Da Silva 2011). Further, in temperate climates, we can take construction waste "fines" (the sandy materials passing through a screen) with the right chemical composition, add a suitable compost (again with the right chemical composition), and produce a potentially productive soil. Suppose we apply this thought to reclaiming erosion-depleted upland soils or to making the deserts of the arid areas of the world productive, as many used to be? As an example, Malta is a relatively small island with a maximum distance coast to coast of only 7 times its airport runway length. Malta has around 400,000 total resident population and around 2 million tourists a year. Despite this population pressure, over 70% of the

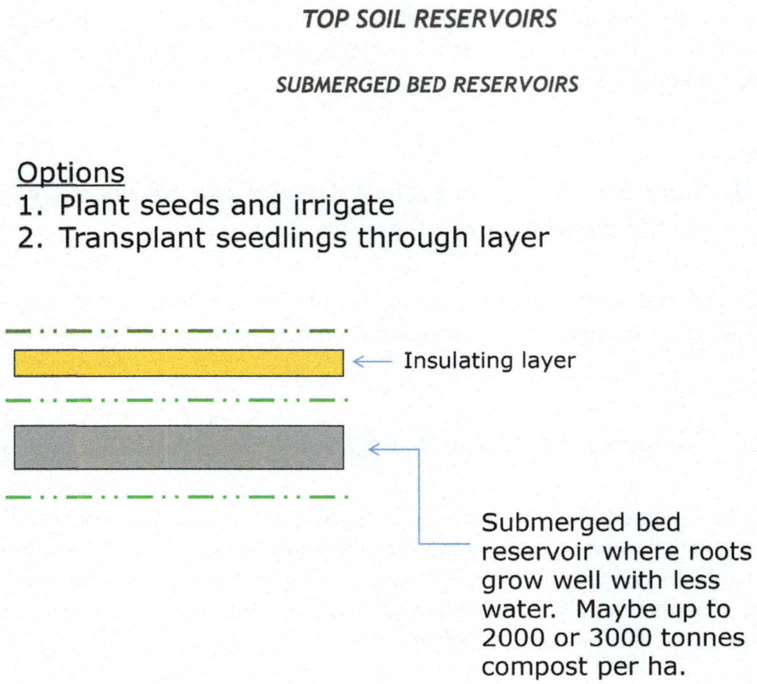

Fig. 10.7 Reservoirs at the crop roots

nation's waste output is construction waste. The island was almost certainly covered in much more vegetation historically, and yet much of the surface of the island, that which is not covered in concrete, is eroded back to the bare rock or, at best, very thin soils except in a limited amount of agricultural arable soils (which are still quite shallow and mostly poor in organic matter). That bare rock could be recovered back to vegetation and trees by using crushed and screened (or just screened) construction waste plus composted organic wastes and biosolids (the large particles from screening compost could be used for wind-protection bunds). The same approach could be used in the UK to reclaim uplands denuded of soil by over-grazing in the past; similarly, in most of the countries of the world.

A simplistic view of composting MSW with minimal separation and separating the product into reasonably high and possibly quite low-quality outputs can be achieved with relatively simple mechanical screens. These may be used, for example, at one extreme for food production and, as aerobic landfill, tree planting for windbreaks at the other. The arithmetic is straight forward. In approximate terms for a semi-arid area where there is some rainfall:

- 500 tonnes of "waste" will make around 350 tonnes of compost.
- 350 tonnes of compost per ha will hold around 2000 tonnes of water per ha (equivalent to 200 mm of rain/irrigation).

Incidentally, if biodiesel is needed, then the following example in the UK is worthy of note:

- 1 ha of crop will probably make 1 tonne of biodiesel and 200 L of bioglycerol.

The problem with water in hot situations is that it evaporates easily and often with no benefit to crop growth. The worst evaporation loss comes from using sprinkler irrigation which maximises evaporation before the water hits the ground, let alone be used by the crop. While it may be important to remember that this way of application may have effects on leaf temperatures and surface layer temperatures, it is still this maximising of evaporation loss which prompts an alternative thinking. If it were possible to make fewer applications, maybe only one, of enough water to grow the crop and hold it near the crop roots, then that evaporation could be dramatically reduced (Fig. 10.8). The concept of Top Soil Reservoirs, TSRs (as in Fig. 10.7) is (Butterworth 1997; Marshall 2019) to accept that water loss by translocation is necessary to crop growth but to limit the loss of water that has not been through the plant (Butterworth 1997, 1998a, b; Marshall 2019).

This concept indicates that the very top layer of soil is not much use in desert soils in the sense that it will often be too high a temperature for root and biological activity, lose water quickly and have its organic matter rapidly oxidised. However, it does have a real value as an insulating layer. So, in any particular situation, there is a task which is to design to plan for a stable top layer and design and a crop establishment programme which will get crop roots down into the reservoir area. In practice, what comes first is the design of the top soil reservoir (TSR) zone itself, i.e., the zone in which productive root growth will occur. Figure 10.9 shows the basic

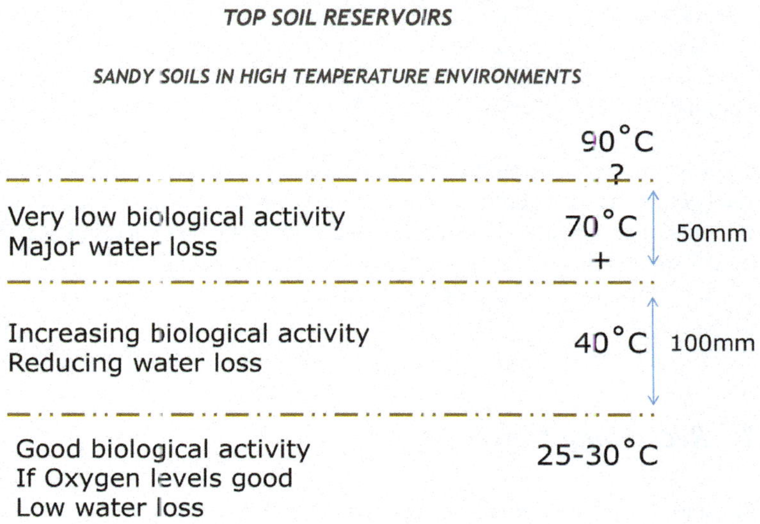

Fig. 10.8 Possible and likely soil temperatures in sandy soils in a hot climate

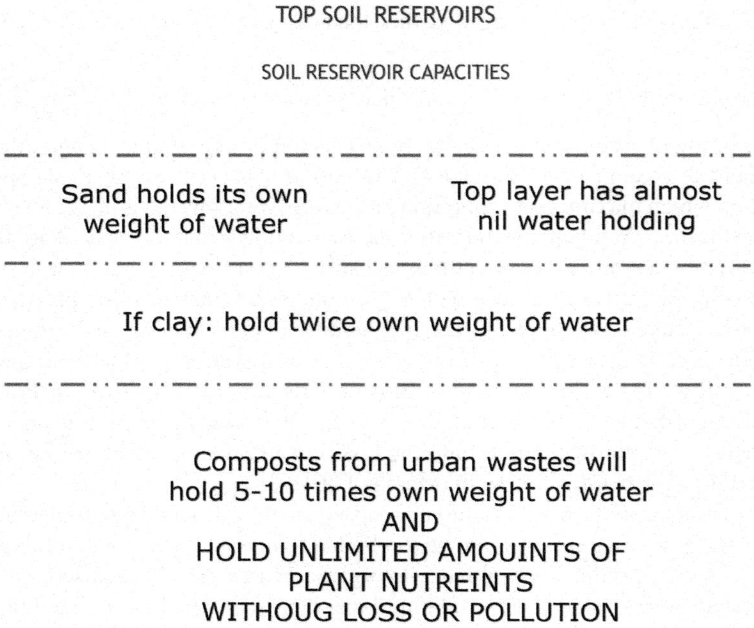

Fig. 10.9 Water holding capacities

thinking on water holding capacities. However, this leaves the question of how much organic matter to add per ha. Sand is useful in all soil structures to allow movement of air and water, some clay will help colloidal capacities, and therefore, the mixture of the two plus some silt (half way between the two in particle size), i.e., "loam" being advantageous. What is available locally will vary. Section 4.3 looks at the depth and construction of the TSR—the reservoir itself where the productive root growth will occur.

Many crop roots will go down to 2 m, and the design may not only allow for that but deliberately construct that depth. That is certainly expensive and would demand very large amounts of compost. Generally, the mix of the zone will vary, but as a starting point for the sake of this discussion, say 50% local material and 50% compost. That will vary in practise depending on the calculation of what the crop needs to grow to harvest.

10.4.1 Reclaiming Desert

Reducing and reversing desertification may not be as difficult as it might seem. Not, that is, if urban waste is available. Compost made from urban wastes will hold 5–10 times its own weight of water. Dressings of 2500–5000 tonnes per hectare, therefore, may hold enough rainfall or irrigation, if needed to grow a crop. Oxidation of the organic matter will vary enormously, up to 100%, but it can be limited to less than 10% per annum by cropping, cultivation techniques and waste/compost technology. To reclaim hot desert, possibly double the inputs of wastes. In deciding application rates, there needs to be recognition that there is no humus to start off with and the higher temperatures will increase the oxidation rate of the organic matter. This rate will depend on ambient soil temperatures and on cultivation techniques, and these, in turn, will affect crop species choice. Again, it is Fig. 10.7 which links soil organic Carbon and water-retaining capacity, and Fig. 10.3 shows an option for low grade compost or compost contaminated with trash such as plastic bottles. This can be used as hedgerow boundaries or forestry plantings. In hot climates, this can be used for planting energy crops such as Jatropha, as mentioned earlier.

10.5 Halting Global Warming

10.5.1 The Soil as a Processor, Bio-active Carbon

What the above figures and text indicate is that when we want to feed a crop (to make it grow), if we are in an organic-based system, then we feed the soil, and it is the soil fungi which feed the crop. As the fungi (mycorrhizae) feed off organic material of Carbon-based molecules, then having enough of the right sort of organic matter is logical. As this text has already discussed, adding dead plant and animal matter manures, or any other organic material—and many inorganic materials too—to soils, feeds the soil microorganisms which, when they die, form this complex mixture of hydrocarbons, carbohydrates and proteins which form the black, tarry material we call humus. That humus, in turn, feeds the mycorrhizae which feed the crop. This biologically active soil, then, is a very capable, bio-active global processing factory. We can use it to store Carbon.

Most organic Carbon will have come, at some stage, from green-leaved plants which will have taken Carbon dioxide out of the atmosphere in order to grow. Therefore, building a reservoir of organic Carbon will remove Carbon dioxide from the atmosphere and improve soil fertility and enable more food production per ha. However, as this text shows, soil organic matter will oxidise away, especially if near the surface and mechanically cultivated. "Near the surface" will vary according to soil type and conditions. For example, with a dry sand rapid oxidation will occur probably down to 100 or 150 mm (4–6 in). In a clay, this will be reduced to may be 50 mm. Therefore, logically, the best thing to do with compost is to plough it in, preferably below 150 mm. How much? There is no limit. There is no risk to groundwater from Nitrogen leaching if 10,000 tonnes per ha, or more, were ploughed in.

Putting it simply, any plant or animal tissue, dead or alive, contains the sort of organic matter that the soil microorganisms can turn into the black, tarry material which soil scientist call "humus" and is the bio-active Carbon sink (BACS). Why the label of "bio-active"? Because of the above Figs. 10.4, 10.5, and 10.6. Organic matter in the soil is the basis of soil fungi activity; the fungi (or mycorrhizae) feed on the organic matter, in turn feed the plant, crop, or trees they are symbiotic with, which in turn takes Carbon dioxide out of the atmosphere and drives the cycle round the Closed Loop (Butterworth 2022).

10.6 Further Development

10.6.1 Laboratory Study of Recycling a Microplastic

10.6.1.1 The Trial Material

The material in this microplastic trial originated from the manufacture of white electrical components such as light switches, plugs, and sockets. In its discarded form, i.e., recognisable switches and sockets, if such were buried in bio-active soils, they would still be unaltered in decades, maybe hundreds of years' time. However, such components can be ground up and are used as blast media (grinding dusts) and low-aggression polishers. The original raw material is described as an "amino moulding powder" and its production involves: Amino Moulding Powder is manufactured using approximately: Formaldehyde (29%), Urea (40%), Paper Pulp (30%), Pigments, and stabilisers (1%).

The composite material contains 22–24% Nitrogen nutrient of potential value as a Nitrogen fertiliser. After grinding, the smallest granules so produced can be used, for example, in the polishing of helicopter blades at servicing. The screened size is 180 microns down to zero. This is a very low diameter (i.e., it has a very large surface to weight ratio) material. Initial pot trials indicate that it can potentially be used in the Horticultural or Agricultural industry for its Nitrogen content in addition to the benefits gained by the organic Carbon (from the paper pulp) improving soil structure (Butterworth 2009c, 2014).

Initial conclusions from this small-scale trial show that when a waste can be ground fine enough so that the surface-to-volume ratio is high enough, and there are some nutrients including Nitrogen, then this may allow the soil mycorrhizae to digest the material. This trial certainly showed that there may be many materials for which it would not normally be financially viable to grind to a micro-granule size, but may be so ground for other commercial reasons, and then such materials could be used as organic fertilisers if the nutrients are present.

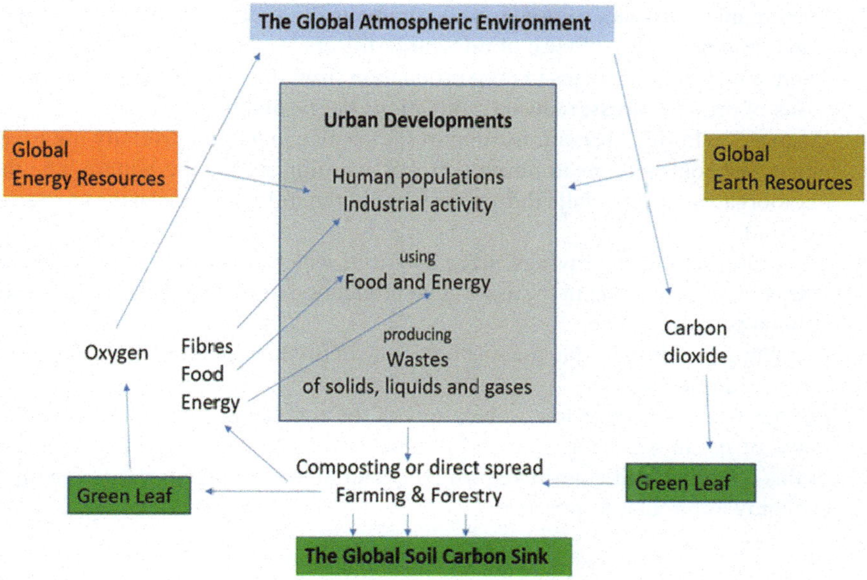

Fig. 10.10 Flow chart—relationship of urban development to the wider environment

10.7 Discussion

There are many references in the scientific literature to global warming and of possible contributions to stabilising climate change. Much of this is summarised in the on-going series of IPCC reports. This research and development work involved the safe recycling in the region of over 5 million tonnes of municipal and urban wastes, both solids and liquids, to farmland. The organisation of the farm operation was based on the science of composting and the closed conduit provided by the soil mycorrhizae. Many of the observations were made by farmers, not given to scientific recording in detail, in a complex situation. Nevertheless, the high volume of supervised fieldwork does confirm the technology and a number of conclusions, including an indication of the scale of the contribution that can be made to reducing the pollution and effects on global warming by recycling urban wastes to proximity land. A summary flow chart is shown in Fig. 10.10.

10.8 Conclusions

1. In terms of financial and environmental costs, in at least some cases, may be most, it is better to use urban wastes as fertilisers for farm and forestry than to use energy to produce mineral fertilisers.

2. The organic fertilisers so produced do dramatically reduce pollution of groundwater compared with the use of mineral fertilisers, especially Nitrogen.
3. There are clear advantages in crop production from the use of these composts in terms of drought stress, reduced cultivation energy, and better yields.
4. Crop and wildlife observations support the science in that biodiversity improvements and contributions to limiting global warming start at the bottom, i.e., the microorganisms, especially the mycorrhizae of the soil at the base of the terrestrial pyramid.
5. There are significant savings in energy use, especially in the production of Nitrogen fertilisers and in the associated production of Carbon dioxide and other pollutants.
6. Regulation needs to enable the safe recycling of urban wastes on proximity farm, forestry, and wild land.
7. Further work is needed to co-operate with the regulators to monitor as wide a range of urban wastes as safe and productive as possible to enable expansion of this approach and contribute positively and quickly to making the municipal economy as circular as possible.

References

Albrecht WA (1919–1970) The Albrecht papers, Acres USA: Charles Walter Books

Allen MF (1996) (1955) The ecology of arbuscular mycorrhizas: a look back into the 20th century and a peep into the 21st. Mycol Res 100:769–782

Berg B, McClaugherty C (2008) Plant litter: decomposition, humus formation, carbon sequestration. Springer, Berlin. https://doi.org/10.1007/978-3-540-74923-3

Bhadouria R, Upadhyay S, Tripathi S, Singh P (2022) Urban ecology and global climate change. Wiley and Sons, UK

Butterworth B (1986) The straw manual. Spon, London

Butterworth B (1997) Sewage solution for low-water farming. Far Eastern Agriculture, London, September/October

Butterworth B (1998a) Reservoirs at the crop roots. The Vegetable Farmer, London, pp 31–32

Butterworth B (1998b) Waste in the next millennium. Am Soc Agric Biol Eng 1998:11–12

Butterworth B (1999) A top idea that holds water. Water and Effluent Treatment News, 6 October, Big Issue, pp 4–5

Butterworth B (2000) Managing the soil rumen and ion exchange. Arable Farming, Ipswich, p 30

Butterworth B (2002) Nitrate nonsense. Landwards, Institution of Agricultural Engineers, Early Summer

Butterworth B (2008) Clamping down on compost. Am Soc Agric Biol Eng 2008:13–14

Butterworth B (2009a) Reversing global warming for profit. MX Publishing, London

Butterworth B (2009b) How to make on-farm composting work. MX Publishing, London

Butterworth B (2009c) The technology of waste, biofuels and global warming in viable closed loop, sustainable operations. Energies 2:1-x. https://doi.org/10.3390/en20x000x. Accessed via http://www.mdpi.com/1996-1073/2/4/1192

Butterworth B (2012) AD and TAD in focus, Organics Recycling: Summer 2012.

Butterworth B (2014) Socially integrated, eco-mimic reversal of global warming. J Technol Innov Renew Energy 58–62. 3 March 2014. Accessed via https://doi.org/10.6000/1929-6002.2014.03.02.3

Butterworth B (2015) Busting the myths. CIWM J 28–30
Butterworth B (2021) Surviving climate change. Amazon 2022. https://www.amazon.co.uk/Survival-Sustainable-Energy-Wastes-Shale-ebook/dp/B01H63EQX0
Butterworth B (2022) Survival—sustainable energy, wastes, shale gas, and the land. Amazon, 2016 and second edition 2022 in digital at https://www.amazon.co.uk/Survival-Sustainable-Energy-Wastes-Shale-ebook/dp/B01H63EQX0
Collins J (2001) Good to great. Random House Business Books, London
DEFRA (2021) UK statistics on waste. 15 July 2021
EPA, USA (1998) An analysis of composting as an environmental remediation technology. EPA530-R-98-008
Gellings CW, Parmenter EC (2004) Energy efficiency in fertiliser production and use, efficient use and conservation of energy. Encyclopaedia of life support systems. UNESCO
Hoonveg V, Bhada-Tata P (2012) What a waste—a global review of solid waste management. World Bank, No. 15. https://siteresources.worldbank.org/INTURBANDEVELOPMENT/Resources/336387-1334852610766/What_a_Waste2012_Final.pdf
IPCC, UN (2017) IPCC sixth assessment report, climate change: mitigation of climate change
IPCC, UN (2022) Climate change 2022: mitigation of climate change. Working Group 111, IPCC
Kinsey N (1999) Hands on agronomy. Acres USA
Klein A (2018) More than 60 prescription drugs are getting into river foodchains. New Scientist, London. 6 Oct 2018. https://www.newscientist.com/article/2184420-more-than-60-prescription-drugs-are-getting-into-river-foodchains/
Mishra P, Singh N, Shrama C, Pathak AK (2020) Landfill emissions and their impact on the environment. Int J Eng Res Technol 9(8):617–622. https://doi.org/10.17577/IJERTV9IS080187
Labrecque M et al (1995) Effect of wastewater sludge on growth and heavy metal bioaccumulation of two Salix species. Plant Soil 171(2):303–316
Marshall M (2019) Finding the real El Dorado. New Scientist, pp 26–29. 19 Jan 2019
Müller Da Silva PH (2011) Applying sewage sludge to Eucalyptus grandis plantations: effects on biomass production and nutrient cycling through litterfall. Appl Environ Soil Sci 2011:710614. https://core.ac.uk/reader/46675449
Ntostoglou E, Khatiwada D, Martin V (2021) The potential contribution of decentralized anaerobic digestion towards urban biowaste recovery systems: a scoping review. Sustainability 13:13435. https://doi.org/10.3390/su132313435
Richmond E et al (2018) A diverse suite of pharmaceuticals contaminates stream and riparian food webs. Nat Commun 9:4491. https://www.nature.com/articles/s41467-018-06822-ww
Singh UK, Kumar M, Chauhan R, Jha PK, Ramanathan AL, Subramanian V (2008) Assessment of the impact of landfill on groundwater quality: A case study of the Pirana site in western India. Environ Monit Assess 141:309–321. https://doi.org/10.1007/s10661-007-9897-6
USEPA (2018) An analysis of composting and an environmental remediation technology. 2, EPA530-R-98-008, April 1998
World Bank (2019) World's population will continue to grow and will reach nearly 10 billion by 2050. Data Blog, 8 July 2019
World Water Quality Alliance (2021) World water quality assessment: first global display of a water quality baseline. A consortium effort by the world water quality alliance—towards a full global assessment. Information document annex for display at the 5th session of the United Nations Environment Assembly, Nairobi 2021

Chapter 11
Transitioning Urban Agriculture to a Circular Metabolism at a Neighbourhood Level

Sharmila Jagadisan and Joy Sen

Abstract We are living in an era of human dominance which has caused a pervasive influence on the planets' ecosystem and climate. The consequences of unbalanced urbanization have made cities disproportionately susceptible to a range of environmental hazards such as soil degradation, loss of bio-diversity and pollution which directly affected fertile soils and natural areas that lose their functionality. These crises not only affect the quality of life and health of our community but also play vital access with respect to the affordability of water, food and energy. The context of climate change amplifies the multiple socio-economic urban challenges which are prompting the re-emergence of an urban metabolism that encapsulates holistic and circular imperatives to achieve sustainability fragrances. Over the last few years, there has been an increasing interest in cities balanced urban development which is inextricably linked to the liveability of urban areas along with food, water and energy security. Urban metabolism has now symbolized the big visionary ideas and fostered new ways of thinking about how cities can be made sustainable as they are purely heterotrophic in nature which makes them dependent on energy and materials imported from the supporting system. The contribution of urban agriculture is to create more resilient and sustainable cities as they have the capacity to integrate streams such as water, waste and energy and reduce their negative ecological footprint to zero. This chapter is an attempt to answer the question of how urban cultivation can improve metabolic processes and recuperate urban sustainability through planning and architectural interventions. How few buildings integrated agriculture concepts like edible landscapes, school gardens, community gardens, indoor agriculture and rooftop production by building integrated interventions at a neighbourhood level can push forward the development of material and resource cycling in cities.

Keywords Circular economy (CE) · Climate change · Neighbourhood · Recycling · Sustainability · Urban agriculture

---

S. Jagadisan (✉)
School of Architecture, VIT (Vellore Institute of Technology), Vellore, Tamil Nadu 632014, India
e-mail: sharmila.jagadisan@vit.ac.in; sjag005@gmail.com

J. Sen
Department of Architecture and Regional Planning, Indian Institute of Technology Kharagpur, Kharagpur, West Bengal 721302, India

R. Bhadouria et al. (eds.), *Urban Metabolism and Climate Change*,
https://doi.org/10.1007/978-3-031-29422-8_11

11.1 Introduction

The concept of urban metabolism is the buzz word in the recent years in planning and other disciplines as cities have become the driving force of human health, well-being and economic development. As conceived by Wolman (1965) and Odum (1996), urban metabolism model helps to describe and analyse the materials and energy flow meandering within cities through the metaphor of living organisms as they maintain a continuous exchange of matter and energy within their environment to enable operation, growth, and reproduction (Wolman 1965). The emerging grand challenge of continued urbanization accompanied by rapid population growth, climate change and degradation of ecosystem services has led to profound impacts on the epoch of the Anthropocene with unknown dimensions. The World Economic Forum 2018 report shows that resource extraction has increased multi-fold between 1900 and 2015. Global extraction of materials has almost tripled in the past 40 years from 26.7 billion tonnes in 1970 to 84.4 billion tonnes in 2015. This projection is predicted to be doubled between 170 and 184 billion tonnes by 2050 (World Economic Forum 2018). As a consequence, it has raised concerns on how to manage ecological resources more sustainably. Growing awareness that the linear model which has the traditional pattern of *produce-use-dispose* has accelerated the hunger for greater resources made our planet unsustainable. There were two main approaches adopted and identified as part of study of urban metabolism. The first is based on the work of Odum that deals with the concept of urban metabolism in terms of energy equivalents and mainly solar energy equivalent or energy (Odum 1996) and the other approach/method follows a broader approach expressing the different resources flows through the city: water, materials and nutrients, in terms of mass fluxes.

In scientific terms, one should say that cities are complex with multifunctional systems consisting of several overlapping components (people, infrastructure, services, etc.) interacting with each other unfortunately in a silo or in a fragmented way (Paola 2021). If we believe that the future of humanity is irrevocably linked to the city, then we have to address the totality of human needs by adopting a holistic approach of integrating energy, material and waste flows (Bhadouria et al. 2022). So, there is an urgent need to break this compartmentalization and develop a regenerative paradigm at different scales (macro, meso and micro) in order to establish an environmentally healthy, restorative relationship between cities and the ecosystem services. According to André (2019), our World's economy is only 9% circular today. This means out of 92.9 billion of all metals, minerals and fossil fuels that enter the economy, only 9% are recycled every year leaving a massive "Circularity Gap" that needs to be addressed (André 2019). These alarming statistics exposes how very few resources cycled back into our economy and uncover how much material goes wasted beyond recovery. This reflects how deeply this linear system (refer Fig. 11.1) is still ingrained in our every day's life. Our linear model is "no longer fit for purpose", and it is a threat to both people and the planet (Novotny, 2012). Transitioning to a circular economy paradigm is therefore a means to an end that provides an actionable way to mitigate the associated climate impacts. The report says that out of total

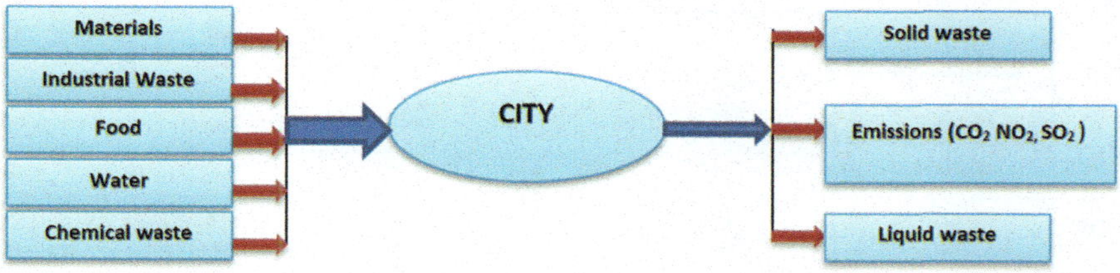

Fig. 11.1 Linear model of urban metabolism

anthropogenic biggest contributor (67%) of global GHG (greenhouse gas) emissions is associated with material management.

11.2 Urban Metabolism and Circular Economy (CE)

Urban metabolism is a multi-disciplinary and integrated platform which allows us to identify and understand material and energy flow in cities, and at the same time emphasizes the linear dimension of urban economies. Urban metabolism can be defined as "*the sum total of the technical and socio-economic processes that occur in cities, resulting in growth, production of energy and elimination of waste*" (Kennedy et al. 2008). The importance of having clarity is of utmost importance to understand the concept of urban metabolism. Many cities continue to be dependent on their surrounding hinterlands for their survival and function. The current pattern of resource extraction and expansion of cities has altered the pattern of livelihood of rural settlements, loss of species and forest diversity, greater loss of agricultural land and increase in carbon footprints and pollution (Bansal et al. 2022). Unregulated solid waste management and disposal practices pose a huge threat to public health and the environment. There is an urgent need to embrace a more viable method so that global value chains achieve improved resource efficiency in terms of utilization and waste emissions (OECD 2021). The Sustainable Development Goals (2030) popularly referred to as Global goals encourage nations to have a paradigm shift towards a cyclical and restorative model which focuses on higher-value material loops and waste prevention for social, economic and environmental benefit.

The concept of Circular Economy (CE) gained visibility since the dawn of industrialization where it could be applied to modern economic systems and industrial processes (Lucertini and Musco 2020). A circular economy necessitates substantial transformation where there are no wastes associated with CE and there are only secondary raw materials ready for a new life process. To achieve sustainable outcomes, both the concepts of urban metabolism and circular economy have to be adopted simultaneously to promote an economic system that is regenerative by design (refer Fig. 11.2). In a circular economy framework, the flow of biological materials can be reintegrated into the biosphere, whereas technical material flows

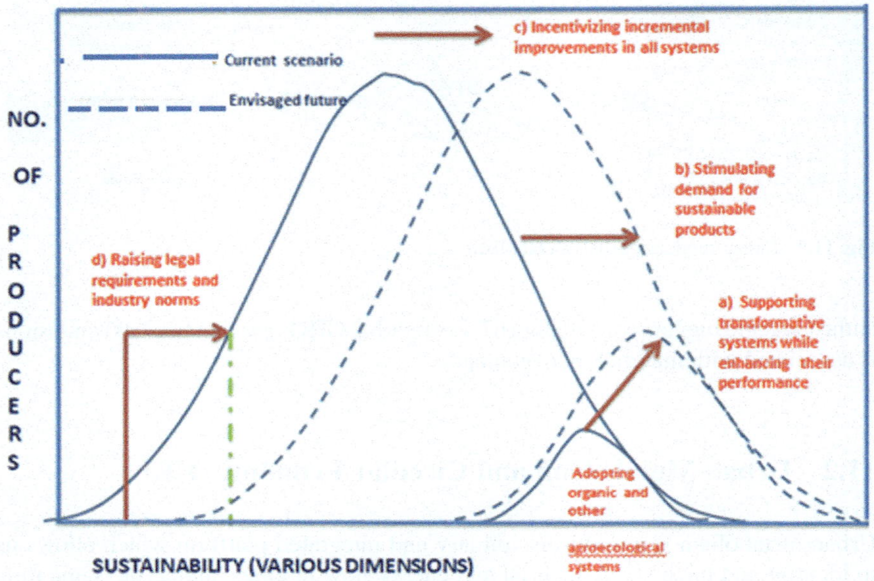

Fig. 11.2 Comparative analysis of traditional and futuristic sustainable practices

can be enhanced, reused and recycled without harming the environment. Therefore, the circular economy model aims at promoting the reuse and recycling of materials in downstream cycles, minimizing wastage and increasing resource efficiency. As Giulia and Francesco stated that "*Within the vast urban metabolism and circular economy frameworks, a better understanding of interactions, trade-offs and cause-effect relationships between different resource flows can also lead to more flexible and resilient territories and contribute to building up new synergies between cities and rural hinterlands*" (Lucertini and Musco 2020). Most importantly, the objective of shifting towards a circular economy is to slow depletion of scarce natural resources, reduce the impact of environmental damage from extraction and to decrease pollution from the processing, use and end-of-life of virgin materials.

11.3 Urban Agriculture and Its Role in Urban Metabolism

Due to intense urbanization and extension of undefined city's boundaries, humankind faces a plethora of complex environmental challenges such as diminishing agriculture land and green spaces, loss of soil fertility, social security problems, vulnerability in global food supply chains which results in economic disparity and inequality among urban inhabitants which further results in illness and stress syndromes within our community. According to Wirsenius (2003), the food and agriculture system is among the largest anthropogenic activities in terms of appropriation of land and biological primary production, as well as alteration of the grand biogeochemical cycles of carbon, water and nitrogen which has led to ecological vulnerabilities (Wirsenius 2003). In order to have long-term food security of cities and to improve the liveability of cities, we need to facilitate the importance of urban agriculture with the context of nature-based solutions (NBS) in circular cities of the future (D'Ostuni 2021). We need to concentrate more on global sustainability strategies which integrates innovative solution and focus on smart development. Adapting to the idea of circular city where relocalization of the food system, reuse of natural and organic disposals and a narrowing of the cities' food sheds is considered to be systematic perspective to solve existing challenges.

Circular urban agriculture aims to make maximum use of resources through recycling and reuse (Ekins 2019). Developing local food economy, high-tech indoor agriculture and local food hubs will be the strategies to solve the biggest challenges facing the world of feeding growing urban populations. It is estimated that vertical farming is expected to reach a value of $7.3 billion globally by 2025 (Pierce 2021). In order to mitigate severe food crisis in future, we are in need of taking necessary pre-step for building adaptive capacities where we focus more on the recovery of agricultural knowledge which will play a crucial role for people to build psychological and social resilience in period of crisis. Integrating urban agriculture within the context of food systems approach to support urban economic and ecological system is very critical at this stage.

In the context of urban agriculture, the concept of circularity focuses on agricultural production and engagement using minimal amounts of external inputs, closing nutrient loops, regenerating the soil mitigating the negative discharges to the environment (de Boer 2018). In recent years, population grows exponentially and food supply to ever-growing urban areas follows the trend of linearity as the resources per person tend to be very low in urban areas while the generation of potentially reusable wastes exerts on the natural environment by human species tends to be larger (Kapoor et al. 2020). There is an urgent need to develop the idea of circularity converge in urban centres in order to balance the demand for circularity and the supply of inputs (D'Ostuni and Zaffi 2021).

From a global perspective, urban agriculture appears to be progressive as they are capable to deliver a suite of social, economic and environmental benefits (Delind 2011). For example, urban agriculture has emerging concept to connect people to their food production and increase agricultural literacy by promoting healthy food diets. Michelle Obama's kitchen garden has been at the forefront of a healthy eating initiative where it has developed as a vibrant mechanism in many schools that have recently promoting food gardens for pleasure, community building or subsistence (Gibbs et al. 2013). The rural fringe of Chinese cities has long history of farming and urban agriculture has a strong tradition as a strategy for their survival, and you can still find people growing their own particularly horticultural production (Tan et al. 2005).

11.4 Circular Economy in Indian Agriculture: Rethinking Growth and Prosperity

As per the World Bank report for the first time in history half of the global population (55%), approximately 4.2 billion inhabitants live in urban areas. This trend is expected to double its current size by 68%, which means nearly 7 of 10 people in the world will live in cities (World Bank 2021). The National Commission on Population (NCP) in India predicts that in the next 15 years (i.e. by 2036), about 38.6% of Indians (approximately 600 million) will live in urban areas. The United Nations, too, highlights that India's urban population size will nearly double between 2018 and 2050, from 461 to 877 million (Rumi 2021). The perceived impact of accelerated urbanization in the rural hinterland is leading to reduced farmlands, accompanied by a shift in land use and often involving fertile soils and agricultural lands being repurposed for commercial and residential uses. As per the UN Economic and Social Affairs, "*Food-related CO_2 emissions could double by 2050 without changes to current unsustainable food systems and consumption patterns*" (UN 2021). From the city branding perspective, cities are competing with each other in order to attract affluent users, new residents, investors, tourists, etc. To accommodate core services and amenities for these new inhabitants, the sizeable number of people in the neighbourhood is being displaced, and places of consumption have been built to serve the new user's needs. The effect of this type of policy creates spatial inequality where city centres have become thriving places for affluence and marginalized communities near the peripheries. Overall, our current economic growth model seems to be unsustainable, unstable and unfair.

By embarking on a circular economy transformation, India can directly move towards an effective system of positive, regenerating and value-creating development and can create direct economic benefits in different sectors like construction and cities, food and agriculture and vehicle manufacturing (Foundation et al. 2016). The circular economy is aimed at preserving and enhancing natural capital by minimizing the usage of natural resources; keeping the products and materials in use at a minimum, and eliminating the risks associated with wastes and negative externalities such as water, soil and air pollution (Foundation et al. 2016). By 2050, food demand is likely to increase between 59 and 98% which will be an undeniably major challenge for agricultural markets. In the light of ever-growing demand, food production could be increased by agricultural intensification by adopting new methods like precision farming, etc (Pearson et al. 2010). The growth of agriculture has increased the chemical burden of natural ecosystems through more intensive use of chemical fertilizers and synthetic pesticides which over time has destroyed the soil health and degraded the quality of land for cultivation (MIT 2021). The above challenge provides an opportunity for the development of a circular economy (CE) using innovative technologies to redesign the existing production systems in agriculture with respect to waste management strategies such as recovery and reuse of waste products. Therefore, while increasing agricultural production, circular economy requires the adoption of a closed loop system where it maintains the balance between production and the restoration of nature which work towards the long-term economic and environmental sustainability.

Current research indicates that agriculture and the agri-food industry are considered to be the second-largest source of material footprint accounting for 20.1 billion tonnes, and a carbon footprint of 6.5 billion tonnes of carbon dioxide (CO_2) equivalent or the fourth-largest behind other industries like housing, consumables, etc. (Bauer et al. 2016). The circular economy represents a strategic sustainable approach for lower resource demand by minimizing agricultural waste and reducing the negative environmental impacts while improving economic performance (Kuisma and Kahiluoto 2017; Stegmanna et al. 2020).

Agriculture can be defined as "*the art and science of cultivating the soil, growing crops and raising livestock. It includes the preparation of plant and animal products for people to use and their distribution to markets*" (National Geographic 2007). Crop production comprises the following activities:

(i) processes,
(ii) soil nutrient stock and
(iii) nutrient supply associated with the production of arable crops, including produces (fruits and vegetables) horticulture and grasslands.

As Brunner and Rechberger states that crop production relies on the most intensive farming as they consume more natural resources and has the highest aggregate level of water and energy consumption. Globally, agriculture consumes more than 70% of fresh water. For example, approximately 1000 L of water are required to produce 1 kg of cereal and 43,000 L to produce 1 kg of beef (Pimente et al. 2004). These types of activities account for more than 90% of water-related issues such as water stress and loss of bio-diversity. So, in future more research efforts have to be initiated to develop a comprehensive and coherent strategy which is needed to improve the resource efficiency and sustainability of crop production. Understanding the importance of a circular economy is the need of an hour.

11.5 Benefits of Circular Economy

A circular economy alters cities existing systems and helps them to emerge as circular cities where they adopt regenerative and adaptive urban ecosystems to increase urban resource security for future generations (Williams 2021). The major goals underpinning this process are to close the resource loops and improve resource productivity and make waste obsolete in order to diminish the ecological footprint. Circular cities offer spaces to transform and allow cities to adapt to a broad range of shocks and stresses in the wider landscape and enable transformation in the local system (such as local food banks, recycling wastes and reusing infrastructural systems (e.g. grey water recycling) and encourage urban inhabitants to build stronger social capital (Madanipour 2018).

11.6 Why Circular Economy Is the Need of an Hour?

According to the United Nations' Committee on World Food Security, food security is defined as the means that all people, at all times, have physical, social and economic access to sufficient, safe and nutritious food that meets their food preferences and dietary needs for an active and healthy life (IFPRI 2021). As per the standards of FAO, 1800 kilocalories are the threshold value for a healthy life. If an adult continues to eat less for a very long duration, then it is considered to be starvation (Dodgson 2017). Food prices at the international level reached unprecedented levels and the UN warns certain countries to find means to insulate themselves from potential food shortages and food price spikes. In many regions, the supply of food demand is far below the demand quotient. The reasons for this are very diverse, for instance, natural conditions such as climate or the nature of the soil. Basically, food production needs resources that are not available in surplus quantities in the current situation. Increasing inequality in the supply of food accompanied by the unavailability of land for food production has an ecological dimension (UN 2015). Agricultural production is highly dependent on land, water and energy. Rising demand for cultivation has created incentives where forests have been razed to make spaces and it is estimated about 27% of the total loss between the years 2001 and 2015 is due to intensive large-scale farming and ranching (FAO and UNEP 2020). One approach to reducing the impact of environmental problems is to orient ourselves towards the principles of sustainable agriculture that uses soil preserving methods and acts without synthetic chemical agents. Figure 11.2 shows the comparison between traditional and futuristic sustainable practices.

A global environmentally sustainable and resource-efficient agriculture requires a shift to fully organic agriculture within a properly designed food system (Eyhorn et al. 2019). Does organic agriculture alone sustain the food demand of a growing population? The review paper done by Jonathan Wachter which was published in Nature Plants demonstrates that organic farming systems produce yields that average 10–20% less than conventional agriculture they are more profitable and environmentally friendly and more nutritious as they contain less or no pesticides and provide greater social benefits than their conventional counterparts (Reganold and Wachter 2016). In addition, organic waste which comprises a variety of waste such as dried plant materials, animal manure can be converted into fertilizers which are rich in nitrogen, potassium and other nutrients long been used as a source of fertilizer for agriculture which minimizes the cost and resource demand external inputs of synthetic fertilizers (AgriPortal 2016). Wastewater from irrigation and storm water run-off, and animal production can be reused for pastures and plant production if treated properly. Solid biomass and waste from plant and animal waste are considered as major sources of energy and can be used directly or modified to produce biofuels (used as transportation fuel and electricity). Biomass is a versatile renewable energy source that is abundant and can reduce reliance on fossil-based fuels and waste in the landfill.

Though the concept of the circular economy remains in its infancy, it is becoming increasingly important as it offers an avenue to sustainable growth if we adopt an even more comprehensive integration of circular economic practices in agriculture. One example of a successful implementation of a circular economy is a farm in Uganda which has been developed as small-scale mixed farming (where a range of plants and livestock' are grown on the farm) where waste has been repurposed and reused in the following ways:

(a) It is observed that livestock waste is fed to maggots as a direct feed which can be used as feed for fish and other animals, with the insects' waste and nutrient-rich aquaculture wastewater has been used to fertilize and irrigate crops.
(b) Biogas for cooking has been generated using an anaerobic digester which processes the livestock manure which is again a source of renewable energy.

We are all aware that one approach to tackling food security is to adopt small-scale food production. It is considered as an indigenous, traditional system which could play a substantial role in providing the base for food availability as it has a tendency to be in harmony with natural closed cycles. One of the recent experiments is completed by one of our VIT Vaial faculty where they attempted to cultivate mushroom by converting agricultural residues into protein biomass through solid state fermentation.

11.7 Case Study: Conversion of Agricultural Residues into Protein Biomass by Milky Mushroom Fungus *Calocybe indica* Var. Apk2 Through Solid State Fermentation by Priyadharshini Bhupathi and Krishnamoorthy Akkana Subbaiah (Bhupathi and Subbaiah 2020)

Most of the agricultural residues generated as waste have a rich content of ligno-cellulosic compounds whose handling and disposal are poorly managed and often cause significant environmental problems due to their chemical structure and decomposition properties. It is estimated that 1.3 billion tonnes/year have been generated globally. The demand for agro-residues such as rice straw, wheat straw, maize and sorghum stalks, rice husk, coir pith, leaves and bark are produced in large quantities and left unattended in the fields after harvests or have been used as fodder or material landfill or burnt. They have been used as an alternative for woody cellulosic fibres due to their low cost compared to plant fibres.

> "*Calocybe indica*" P&C is native to India and was first reported by Purkayastha and Chandra in 1974 (Purkayastha 1974.). The technology for commercial cultivation and the variety APK2 has been introduced first from Tamil Nadu Agricultural University, Coimbatore, India (Krishnamoorthy 1998). Mushroom culture offers an excellent means for the recycling of agro-wastes presently available in the country (Sohi 1988). Alam et al. have used 30% maize powder to supplement paddy straw substrate in order to increase mushroom yields. More promisingly, supplements like soybean and cotton seed cake gave the highest absolute mushroom yields (64.8% and 59.2% increased biological efficiency over control). Converting

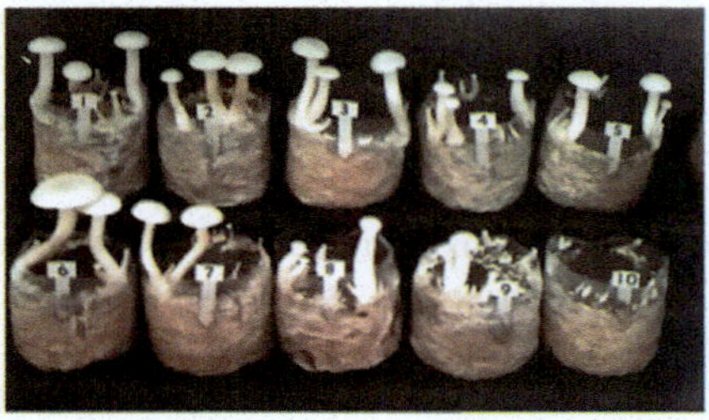

Fig. 11.3 Illustrating the growth of mushroom on various substrates (Bhupathi and Subbaiah 2020)

lignocellulosic agricultural and forest residues into protein-rich mushrooms is one of the most economically viable and sustainable biotechnology processes to address world food demand, especially proteins (Bhupathi and Subbaiah 2020).

The present study proved that the paddy straw and maize stalks as substrates gave significantly higher yields in mushroom cultivation (356.5 and 354.3 g per bed, respectively) followed by sorghum stalks and vetiver grass (Fig. 11.3).

The utilization of agro-wastes helps in reducing the wastes, converting them into mushroom protein and vitamins. Thus, the use of waste can provide more food, more jobs, better family income and improved living standard, curb global warming and clear up the crop residues on road sides and forest margins. This demonstrates how circular economy is an alternative to the traditional linear economy where it eliminates the whole concept of waste and promotes sustained economic growth.

11.8 Conclusions

The practice of utilizing organic wastes for urban agriculture plays a significant role in society. It provides benefits and creates a positive role to society in terms of health economic and ecological externalities (Olivier and Heinecken 2017). It provides additional benefit by demonstrating how food is produced and making people aware of the advantages of utilizing/recycling organic wastes to maintain sustainable urban livelihood. Utilization of organic wastes through urban agriculture can empower urban communities to not only secure their own sustainable food source, but also create a more uplifting environment and helps to establish self-reliant backyards gardens in planting their own crops. In future plans, the municipalities should take initiative to integrate the organic waste management in the food strategy for effective waste management.

11.9 Recommendations

We cannot achieve a healthy planet with healthy community without a fundamental transformation of moving towards a circular economy where it is emphasis on converting organic waste into compost, fertilizer or bioenergy. It makes the cities to have paradigm shift from "end consumer" to part of the solution. This chapter provides us an insightful look how the concrete choice of adopting for more regenerative food systems will be a critical component in nurturing our agrobiodiversity that contribute to the economic development. There are three ambitions of achieving a circular economy with respect to agriculture. Firstly, designing and marketing food products that appeal to people using more locally available and seasonal ingredients would increase cities' connection with local farmers and could help spark the transition to regenerative practices (Foundation 2019). Secondly, it improves our inventory management as it increases food traceability which has significant potential to protect consumers in terms of food recall and safety. Finally, the cities can be a hub of innovation by converting food by-products into compost and fertilizers which would definitely attract the peri-urban farmers which successfully linkup all the elements within food value within the local and regional food value chains.

References

AgriPortal T (2016) Organic farming. [Online]. Available at: https://agritech.tnau.ac.in/org_farm/orgfarm_manure.html

André G (2019) Youmatter [Online]. Available at: https://youmatter.world/en/is-circular-economy-the-key-to-fight-climate-change-a-circle-economy-report/

Bansal V, Jagadisan S, Sen J (2022) Water urbanism and multifunctional landscapes: case of Adyar River, Chennai, and Ganga River, Varanasi, India. In: Urban ecology and global climate change. John Wiley & Sons, pp 85–103

Bauer SE, Tsigaridis K, Miller R (2016) Significant atmospheric aerosol pollution caused by world food cultivation. Geophys Res Lett 43(10):5394–5400. https://doi.org/10.1002/2016GL068354

Bhadouria R, Upadhyay S, Tripathi S, Singh P (2022) Urban ecology and global climate change. Wiley-Blackwell

Bhupathi P, Subbaiah KA (2020) Conversion of agricultural residues into protein biomass by Milky Mushroom Fungus Calocybe Indica var. Apk2 through solid state fermentation. Madras Agric J 107:17–20. https://doi.org/10.29321/MAJ.S.000438

D'Ostuni M, Zaffi L (2021) Nurturing cities: pathways towards a circular urban. In: Conference on World Heritage and Design for Health, Napoli-Capri

de Boer IavIM (2018) Circularity in agricultural production. [Online]. Available at: https://www.wur.nl/nl/show/Circularity-in-agricultural-production.htm

Delind L (2011) Are local food and the local food movement taking us to where we want to go? Or are we hitching our wagons to the wrong stars? Agricultural human values, pp 273–283

Dodgson L (2017) New health advice says men and women should only eat 1800 calories a day—here's the idea behind it [Online]. Available at: https://www.businessinsider.in/science/health/new-health-advice-says-men-and-women-should-only-eat-1800-calories-a-day-heres-the-idea-behind-it/articleshow/62272656.cms. Accessed 22 March 2019

Ekins PDTDP (2019) He circular economy: what, why, how? OECD/EC Workshop, Paris

Eyhorn F, Muller A, Reganold JP, Frison E, Herren HR, Luttiholt L, Mueller A, Sanders J, Scialabba NEH, Seufert V, Smith P (2019) Sustainability in global agriculture driven by organic farming. Nature sustainability 2:253–255
FAO & UNEP (2020) UN environment programme [Online]. Available at: https://www.fao.org/3/ca8642en/ca8642en.pdf
Foundation EM (2019) Cities and circular economy for food. Ellen MacArthur Foundation. https://ellenmacarthurfoundation.org/cities-and-circular-economy-for-food
Foundation EM, ClimateWorks & UNCTAD (2016) Circular economy in India: rethinking growth for long-term prosperity [Online]. Available at: https://ellenmacarthurfoundation.org/circular-economy-in-india
Gibbs L et al (2013) Expanding children's food experiences: the impact of a school-based kitchen garden program. J Nutr Educ Behav 45:137–146
IFPRI (2021) Food security [Online]. Available at: https://www.ifpri.org/topic/food-security
Kapoor V et al (2020) Ecological economics of an urban settlement: an overview. In: Verma P, Singh P, Singh R, Raghubanshi AS (eds) Urban ecology: Emerging patterns and socio-ecological systems. Elsevier, pp 91–110
Kennedy C, Cuddihy J, Engel-Yan J (2008) The changing metabolism of cities. J Ind Ecol 11:43-59. https://doi.org/10.1162/jie.2007.1107
Krishnamoorthy ASMMMTNVA (1998) APK2 milky mushroom-Extn. Bulletin, TNAU, India
Kuisma M, Kahiluoto H (2017) Biotic resource loss beyond food waste: agriculture leaks worst. Resour Conserv Recycl 124:129–140
Lucertini G, Musco F (2020) Circular urban metabolism framework. One Earth 2:138-142
Madanipour A (2018) Temporary use of space: urban processes between flexibility, opportunity and precarity. Urban Stud 55:1093–1110
MIT (2021) Mission 2014: feeding the world [Online]. Available at: http://12.000.scripts.mit.edu/mission2014/background
National Geographic (2007) The art and science of agriculture [Online]. Available at: https://education.nationalgeographic.org/resource/agriculture
Novotny V (2012) Water and energy link in the cities of the future—achieving net zero carbon and pollution emissions footprint. In: Water/energy interactions of water reuse. IWA Publishing, London
Odum H (1996) Environmental accounting: emergy and environmental decision making. Wiley-inderscience, New York
OECD (2021) Towards a more resource-efficient and circular economy. OECD, Italy
Olivier DW, Heinecken L (2017) The personal and social benefits of urban agriculture experienced by cultivators on the Cape Flats. Development Southern Africa, pp 168–181
Paola P (2021) Urban metabolism: the city as a living organism is circular by nature [Online]. Available at: https://www.renewablematter.eu/articles/article/urban-metabolism-the-city-as-a-living-organism-is-circular-by-nature
Pearson C, Pilgrim S, Pretty J (2010) Urban agriculture: diverse activities and benefits for city society. Routledge
Pierce M (2021) NASA research launches a new generation of indoor farming [Online]. Available at: https://www.nasa.gov/directorates/spacetech/spinoff/NASA_Research_Launches_a_New_Generation_of_Indoor_Farming
Pimentel D, Berger B, Filiberto D, Newton M, Wolfe B, Karabinakis E, et al. (2004) Water resources: agricultural and environmental issues. BoiScience 54:909–918
Purkayastha R, Chandra A (1974) New species of edible mushroom from India. Trans Br Mycol Soc 62:415–418
Reganold JP, Wachter JM (2016) Organic agriculture in the twenty-first century. Nat Plants 2:15221
Rumi A (2021) Urban futures: managing India's urban transition in 2021 [Online]. Available at: https://www.orfonline.org/expert-speak/managing-india-urban-transition-2021/
Sohi H (1988) Mushrooms cultivation in India—recent research findings. Indian Phytopath 313–326

Stegmanna P, Londo M, Junginger M (2020) The circular bioeconomy: its elements and role in European bioeconomy clusters. Resour Conserv Recycl X 6:100029. https://doi.org/10.1016/j.rcrx.2019.100029
Tan MX, Li HX, Lu C (2005) Urban land expansion and arable land loss in China: a case study of Beijing–Tianjin–Hebei region. Land Use Pol 22:187–196
UN (2015) Climate change and food security: risks and responses. Food and Agriculture Organization of the United Nations
UN (2021) UN economic and social affairs [Online]. Available at: https://www.un.org/development/desa/dpad/publication/un-desa-policy-brief-105-circular-agriculture-for-sustainable-rural-development/
Williams J (2021) Circular cities: what are the benefits of circular development? Sustainability 13(10):5725
Wirsenius (2003) Efficiencies and biomass appropriation of food commodities on global and regional levels. Agric Syst 77:219–255
Wolman A (1965) The metabolism of cities. Sci Am 213:179–190
World Bank (2021) Urban development [Online]. Available at: https://www.worldbank.org/en/topic/urbandevelopment/overview#1
World Economic Forum (2018) Circular economy in cities evolving the model for a sustainable urban future. World Economic Forum, Switzerland

Chapter 12
Eight years to Go, to Meet the SDG Targets: Waste Management as Enabler and Enabled

G. Venkatesh

Abstract Waste management has evolved from the earlier five-step hierarchy to include many more R's—reclaim, repurpose, remediate, renovate, replenish, revere nature, being a few of them. It can play a key role in the alleviation of, and the simultaneous adaptation to the repercussions of climate change. Waste valorisation, which is gradually entrenching itself, in both principle and practice, can go a long way in directly and indirectly enabling humankind to get closer to several sustainable development goals (SDGs) targets and perhaps overachieve in some respects. Value creation by adopting the R's wherever, however, whenever and by whosoever possible, is a *sine qua non* for achieving the SDGs by year-2030 and continuing in the same vein thereafter, when the world will have to grapple more perceptibly with the repercussions of climate change. It is clear that we cannot avert climate change now. We can, at best, alleviate the intensity of its repercussions, though unfortunately not uniformly all over the world. This chapter posits waste management (urban and otherwise) in the scheme of things related to the sustainable development goals (SDGs), as both enablers and enabled. The exposition introduces readers to the multi-dimensionality of sustainable development, and thereby efficient, value-generating waste management in a circular economy/bio-economy.

Keywords Circular bio-economy · Climate change · Recover · Recycle · Reduce · Remediate · Reuse · Sustainable development goals (SDGs) · Urban metabolism · Valorisation · Wastes · Waste management

12.1 Introduction and Motivation

In a take-make-use-dispose economy, one either dispose of waste irresponsibly to the environmental media or resorts to the so-called 'end-of-pipe' treatment. Awareness of resource scarcity (abiotic depletion) and the adverse impacts of anthropogenic

G. Venkatesh (✉)
Department of Engineering and Chemical Sciences, Karlstad University, Universitetsgatan 2, 65188 Karlstad, Sweden
e-mail: Venkatesh.govindarajan@kau.se

R. Bhadouria et al. (eds.), *Urban Metabolism and Climate Change*,
https://doi.org/10.1007/978-3-031-29422-8_12

activities on the environment that supports them, gave birth to the popular waste management hierarchy—reduce, reuse, recycle and recover (energy). Landfilling was considered to be the last resort. The sought-after gradual transition to a circular bio-economy, a paradigm for the future, according to Venkatesh (2021), has added more R's to the afore-named set of waste management approaches (and ideals)—reclaim, restore, remediate, rainwater-harvest, repair, renovate, refuse, replenish, replace and revere nature (Prasad 2016). The author would further like to add 'repurpose', 'reconsider' and 'restrict' to the mix.

This chapter takes the form of a general discussion, positing waste management in the scheme of things related to the sustainable development goals (SDGs) (United Nations-Department of Economic and Social Affairs 2015). Needless to add, the references to published literature are by no means all-encompassing. While readers can be motivated by this discussion to model waste management strategies and policies, linked to the SDGs (and specific targets), no metrics/indicators or decision-making approaches have been recommended by the author. This exposition introduces readers to the multi-dimensionality of sustainable development, and thereby of efficient waste management as an enabler of the same.

12.2 Discussion

Needless to state, there are complementarities and conflicts among the 17 SDGs, and the targets are defined under each one of them (Venkatesh 2022a, b). Trade-offs are indispensable to ensure that one does not upset the applecart, by missing the wood for the trees. In other words, one can, at best, focus on optimising improvement across the swathe of targets. Further, different towns, cities, regions and countries will start from different initial states, on a continuous journey towards moving targets they set for themselves. Sustainable waste management approaches, while contributing to (enabling) some SDGs, are enabled by the progress made towards the attainment of others. The discussion is structured in order of the relevant SDGs. The official definitions of the SDGs (United Nations—Department of Economic and Social Affairs 2015) have been truncated in length, for the sake of brevity of the sub-subheadings.

12.2.1 SDG 1—End to Poverty in All Its Forms Everywhere

The seven targets defined under SDG 1 (United Nations—Department of Economic and Social Affairs 2015) can be combined thus—*create sound policies, mobilise resources, build resilience and implement social protection systems to eradicate extreme poverty, provide access to the poor and the vulnerable to basic services, microfinance,* etc. Waste management, in a way, has been sustaining the livelihoods of the poor in the developing world, enabling many to stay afloat above the extreme poverty line—currently defined as USD 1.25 per day. The reference here is to the informal recycling sector which governments in developing countries may be averse to dismantling completely. Umair et al. (2015) posit the e-waste recycling sector in Pakistan is such an enabler. However, Ojha et al. (2014), in a case study of the Mirzapur district of the state of Uttar Pradesh in north India, reported that the average daily income for rag-pickers in the district (people who make a living by collecting recyclables from dumping grounds or open landfills and selling the same to middlemen who sell them forward to recycling units) was about INR 60. While this was the reported average, there were some who earned more than USD 1.25 per day (which in 2014 was equivalent to INR 88). Going by the mean value, however (a metric which is commonly used to describe the situation, for decision-makers), the average rag-picker took home approximately 85 US cents per day. Prihandoko et al. (2021), in their case study of the Piyungan landfill in Yogyakarta (Java, Indonesia), have observed that the minimum daily earning of rag-pickers is USD 1.68. They further state that 80% of these rag-pickers are keen on adding to their incomes by supporting the implementation of any composting and incineration technologies in the future, to decrease the loading on the landfill and enable a more-complete value-recovery (as fertilisers and energy, in this case) from the wastes. While these two are instances from two densely populated Asian countries, these rag-pickers are representative of the millions in developing world countries in South America, Africa and Asia, who make a living in this fashion. The 'repair-and-reuse culture' which prevails in developing countries plays a key role in extending the lifetimes of products (decreasing waste generation in the process) and providing employment to several poor people who are able to imbibe certain basic repairing skills (Govind and Mahongnao 2021).

A question to be answered here is if modernising the informal waste recycling sector is sustainable. Is not there a risk that many people who are barely managing to stay above the extreme poverty line will sink into the abyss of penury again? However, progressive thinking would entail socially-inclusive policy-making to create new jobs in the circular bio-economies of the future and help pull more and more individuals (and their families) out of extreme poverty. Social entrepreneurship, however, supported by top-down policy-making, can show the way.

12.2.2 SDG 2—Food Security, Improved Nutrition, Sustainable Agriculture, End to Hunger

The very first priority in the waste management hierarchy—reduce—is evoked in an Irish case study (Oldfield et al. 2016), in which they differentiate between avoidable food waste and unavoidable food residues along the food chain. Food waste reduction in households (as reported in Wikström et al. (2016), in Europe, one-third of food waste along the value chain, happens in households), can be motivated by generating awareness and bringing about a change in consumer behaviour which is largely influenced by the prevailing social norms (Di Sorrentino et al. 2016). The question that is likely to be posed here is—how can a decrease in food waste in European households solve the problem of hunger in the world? As most downstream food waste happens in the richer, developed countries of the world which also incidentally rely a lot on imported foodstuffs from the developing world nations, a substantial reduction in food waste downstream, are likely to free up more food for domestic consumption in the exporting countries. Additionally, the western world countries may also think of exporting any surplus of their locally-cultivated fruits and vegetables for instance (a surplus that may come about when food wastage in households is considerably decreased), at reduced prices to countries striving to combat hunger and feed their populations.

With time, the stress on arable land to continue feeding humankind (directly and indirectly by feeding livestock) will be exacerbated. It can be alleviated by reducing food losses all along the value chain. There is another vital non-substitutable resource—phosphorus—that may not be available as easily, affordably and abundantly in the future, as it has been heretofore. While minimisation of phosphorus losses (waste reduction in other words) along the entire chain from phosphate mining to food/feed is recommended by Verstraete et al. (2016), it has been noted by Venkatesh (2018) that phosphorus recovery from wastewater will become run-of-the-mill in the future and may be mandated legally by policy-makers. Mihelcic et al. (2011) are of the view that in 2050, 1.68 million tonnes of phosphorus (tantamount to 20% of the global phosphorus demand) can potentially be recovered from brown wastewater (yellow (urine) + black (faeces)), globally. There are numerous publications (Antonini et al. 2011; Talboys et al. 2016; Tian et al. 2016; McConville et al. 2017; Simha et al. 2017; Simha and Ganeshpillai 2017; Panagiotou et al. 2018; Taddeo et al. 2018) focussing on phosphorus recovery from wastewater upstream (by separating urine at-source) and in wastewater treatment plants (WWTPs).

SDG 2 also refers to 'sustainable agriculture'. Phosphorus apart, nitrogen can be extracted as urea from source-separated yellow wastewater (putting coconut wastes to good use as adsorbents in Simha et al. (2017) and alkaline dehydration in Simha et al. (2020). Valorisation of bio-wastes (from households, forests, fields, fisheries and industries) in bio-refineries (ubiquitous entities in circular bio-economies of the future—Venkatesh (2021); Kumar et al. (2020); Kopsahelis and Kachrimanidou (2019); Maina et al. (2017); Vea et al. (2017); Mengal et al. (2018)) to yield resource inputs for the agricultural sector, can entrench a steady stream of biochar as soil amendment (Dahal et al. 2018; Yang et al. 2020). Vermicomposting (Sharma and Garg 2019) can be done in households with the intention of circularising their food waste as bio-fertilisers to farmers. In regions experiencing water scarcity, the reuse of wastewater (which is adequately-treated) for irrigation purposes will have to entrench itself firmly, as advised in case studies focussed on Italy, Spain, Greece, Tunisia, the Gaza Strip and Egypt (Shomar et al. 2010; Agrafioti and Diamadopoulos 2012; Ben Brahim-Neji et al. 2014; Cirelli et al. 2012; Abdel-Shafy et al. 2017; Lavrnič et al. 2017; Libutti et al. 2018; Stathatou et al. 2015).

12.2.3 SDG 3—Healthy Lives and Well-Being for All at All Ages

While e-waste recycling enables the Pakistani workers to earn between USD 2–3 per day and that indeed may augur well for SDG 1, Umair et al. (2015) express concern about the health and safety aspects, which go abegging. They have pointed out that the workers do not use any protective gear, suffer from breathing problems and skin disease and endure improper ventilation in the recycling shops. Child labour is common too, and the blood lead (Pb) levels of children are dangerously high, owing to the leakage of Pb from the e-waste into soil and water. Rag-pickers in the developing world are also exposed to a plethora of health and safety hazards when they foray for recyclables in garbage dumps. A recent publication that focusses on rag-pickers in the city of Lucknow in the State of Uttar Pradesh in India (Ojha et al. 2014) had chosen another city in the same state for their case study, lists respiratory, eye and skin problems, fever, cough and cold, dysentery and diarrhoea, road accidents, animal bites and frost bite (Santoshi and Kiran 2022).

Citing from Venkatesh (2019a), *'…at the ship-breaking yards in Alang in the western-Indian State of Gujarat, a yard-owner brushed aside a question from this author concerning deaths of workers who were not provided gloves, helmets or masks while working at heights above the ground and blow-torching the steel bodies of ships, some of which had been delivered to the yard with inflammable oils inside. He said that if one worker dies, his jobless brother from a faraway eastern state comes over to replace him, and thereby the yard does not fall short of labour'*. Recycling is indeed a good thing, be that e-waste imported into the country (Umair et al. 2015), recyclables jettisoned in garbage dumps (Ojha et al. 2014), or steel from ships that have reached the end of their lifetimes and have to be scrapped (Venkatesh 2019a) if it contributes to SDG 1. However, the virtue of waste management should not be overshadowed/offset by the adverse impacts to the health and well-being of the actual value-creators—the rag-pickers and workers, in this case—it tends to bring in its wake.

While recycling treated wastewater for irrigation is good for SDG 2, one must be wary of the presence of pharmaceuticals, alkyl phenols, polar pesticides, illicit drugs, oestrogens, pathogens and toxic heavy metals in it (Carr et al. 2011; Köck-Schulmeyer et al. 2011). These are often not completely removed during the treatment processes and must not be recirculated back into the agricultural soils and thereby the food chain (Therefore, the descriptor 'adequately-treated' was used in sub-Sect. 2.2).

12.2.4 SDG 4—Inclusive, Equitable Quality Education and Lifelong Learning Opportunities

People will not know about the R's of waste management unless they are enlightened about the looming challenges facing humankind—resource scarcity, environmental degradation, climate change and the like and motivated to contribute to surmounting them (Govind and Mahongnao 2021). Reverence for nature is easily imbibed by school-going children. While it is necessary to ensure that all children the world over are able to go to school, it is also important to teach them about the R's (referred to Hartley et al. 2015).

Academic researchers may be tasked with the responsibility of educating the unaware and sceptical sections of the population about the long-term benefits of a circular bio-economy, in which what were hitherto wastes are rightly looked upon as valuable (valorizable) resources (Viaggi 2015; Bikse et al. 2019). Universities around the world must and will start offering new post-graduate courses in the bio-based circular economy to jumpstart this paradigm shift, toeing China's line (Nibbi et al. 2019; Luttenberger 2020). Education—pre-primary, primary, technical, vocational and tertiary—thus will continue to play a vital role in positing waste management as a value-creator and an enabler of sustainable development.

12.2.5 SDG 5—Gender Equality and Women Empowerment

In the developing world, resource management and conservation (and thereby waste handling) in households are usually undertaken by women (Bureau of International Information Programmes—United States Department of State 2012). This responsibility often extends to the community level, and in rural areas, encompasses agriculture, forestry and animal husbandry. Their experiences, therefore, are vital to policy-making for sustainable development.

A study conducted in New Zealand (Desrochers et al. 2019) which may very well apply to other parts of the world (both developed and developing) concluded that girls and women tend to be environment-friendlier than men and have a lower greenhouse gas (GHG) footprint vis-à-vis the males of the species. Thus, the researchers attributed to traits of kindness, conscientiousness, honesty and emotionality. It follows that if women are empowered in society, industry and government, as leaders, experts, educators and innovators, the R's of waste management in a circular (bio-) economy (Prasad 2016) will be applied to a greater extent. In the western world though, women are already empowered, thanks to higher levels of education and gender parity, and play a crucial role as sustainable consumers (Bureau of International Information Programmes—United States Department of State 2012), making most of the 'green' decisions for the households, optimising resource use, and minimising waste generation.

12.2.6 *SDG 6—Availability and Sustainable Management of Water and Sanitation for All*

Figure 12.1, in which it has been assumed that the source of raw water and the sink for treated wastewater are the same (which is often the case in many parts of the world) is at once illustrative of the role of a well-developed water-and-sanitation utility (SDG 6) servicing consumers who are also environmentally-aware and cooperative, can play in uncovering sustainable and greener energy alternatives (SDG 7, refer sub-Sect. 2.7), contributing to the zero-hunger goal (SDG 2) assisting in combating climate change (SDG 13), aiding sustainable consumption (SDG 12) and maintaining the carrying capacities of the water bodies in question (SDG 14).

Envisaging the downstream of the water-and-sanitation system (Fig. 12.1) as a 'multi-output refinery' recovering value (bio-energy, bio-fertilisers and metals, non-metals and chemicals when industrial wastewater streams are subjected to treatment at-source) from wastewater can even recast the water utilities as for-profit enterprises. This incentivises investments made in developing the downstream (wastewater transport) of the urban water cycle (Johansson and Venkatesh 2017). The profits can be channelled back to progressive sustainable development (supporting research projects in academic circles; SDG 9) and/or lead to making the water and wastewater services increasingly more affordable for the consumers (applicable in the developing world countries) (Kapoor et al. 2020a, b).

The words 'reduce', 'reuse in a cascade' and 'rainwater-harvest' appear below the 'consumer' step in Fig. 12.1. Rainwater which is not collected (harvested) can be looked upon as 'wasted water' which wends its way through stormwater pipelines to the WWTPs. Likewise, 'reduce leakage' appears under the 'water distribution' step. All these are waste minimisation approaches—not just of water, but also of the resources consumed to provide these services. It is consumer demand that drives the utility in a way, and every effort made by the consumers to optimise their water usage (and thereby wastewater discharges) will go a long way to improve sustainability (SDG 12).

12.2.7 *SDG 7—Access to Affordable, Reliable, Sustainable and Modern Energy for All*

Waste valorisation can contribute to SDG 7, by uncovering sustainable alternatives to fossil fuels. Biodiesel can be produced from, *inter alia*, activated sludge (Severo et al. 2019), waste cooking oil (Loizides et al. 2019), spent coffee grounds (Atabani and Al-Rubaye 2020) and palm oil mill wastes (Quayson et al. 2020). Hydrogen is heralded as fuel for the future, and researchers (Sharma and Li 2010; Teng et al. 2010; Ren et al. 2015; Baeza et al. 2017) have touted wastewater as a valorizable starting material for the production of bio-hydrogen. Production of biogas/bio-methane (ultra-pure biogas in other words) by anaerobically digesting activated wastewater sludge, food waste

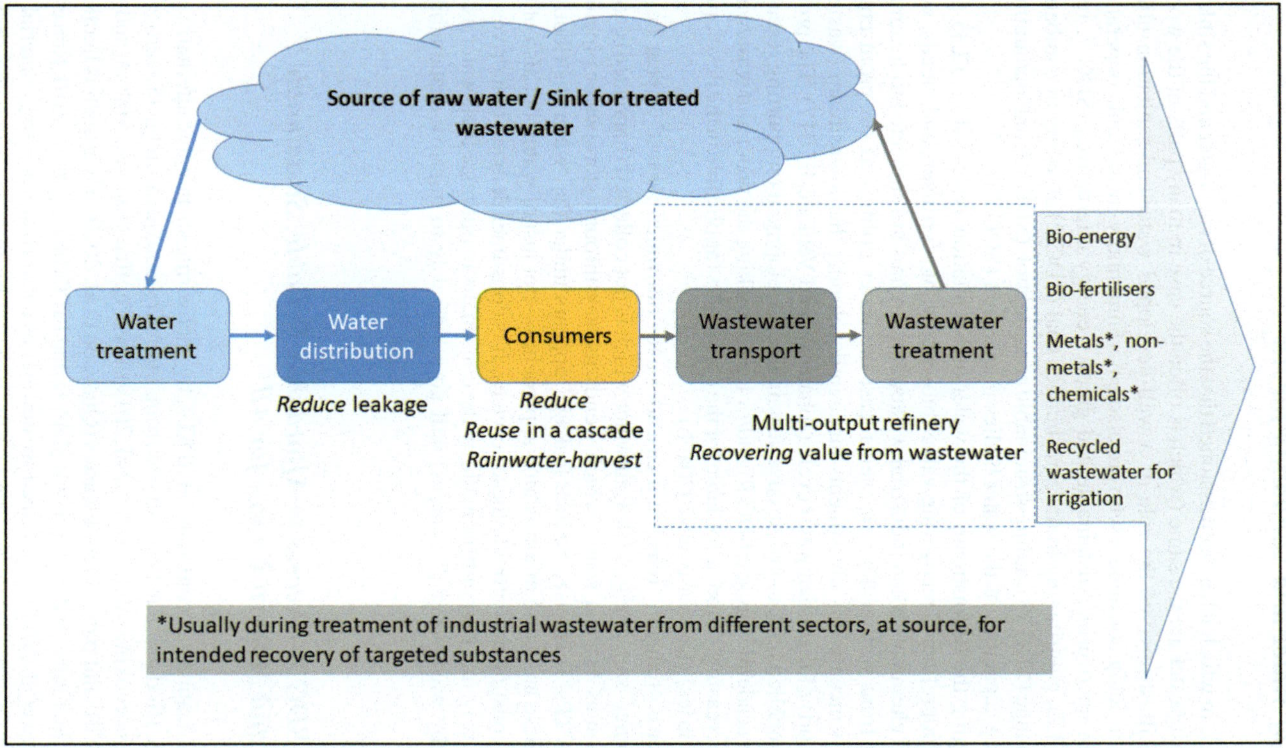

Fig. 12.1 Achieving the sanitation targets in SDG 6 helps SDGs 2, 7, 12, 13 and 14 automatically

and other easily digestible biomass is well entrenched in several parts of the world. However, the unharnessed potential is huge, at the time of writing. For instance, if all the bio-solids in South African WWTPs could be valorised to biogas, bio-hydrogen and bioethanol, about 7% of the country's electricity needs could be satisfied (Stafford et al. 2013). In New Zealand, reportedly, energy recovery from wastewater (biogas being one of the recovered energy carriers) can easily be sextupled, according to Heubeck et al. (2011), while Duan et al. (2020) are of the view that maximising the bio-methane output by anaerobically digesting all the human wastes in China, and using it to substitute coal for electricity generation, can potentially reduce Chinese GHG-emissions by 142 kt daily.

Pellets produced from biomass (wastes, wherever possible to be availed of), serve the dual purpose of ensuring energy security (both in developed and developing countries) and environmental sustainability. Crude palm oil effluent is one such valorizable waste, according to Sadhukhan et al. (2020). Wastes like plastics and paper are combusted in combined-heat-and-power (CHP) plants in many countries of the world (Karlsson et al. 2018). While being affordable and reliable for sure, combusting plastics for energy generation is far from 'clean'. As noted by the authors in Karlsson et al. (2018), CHPs in Sweden for instance are keen on decreasing the quanta of waste plastics in their respective input fuel-mixes and divert more of them thereby for recycling.

As several researchers testify, the potential exists, and countries of the world must 'get down to brass tacks', to convert the reported estimates of bio-waste-derived energy generation to reality. While bio-wastes (materials) are valorizable in bio-refineries and WWTPs to solid, liquid and gaseous carriers of energy, (waste) heat can be recovered from the wastewater flowing to the WWTPs, using water-source-heat-pumps (Gu and Wang 2013; Sitzenfrei et al. 2017; Spriet and Hendrick 2017), and utilised for purposes like district heating. In thermal power plants, cooling the spent steam in the condenser enables the power plant to maintain its net electricity output (some of the generations is for captive consumption by the pump that pressurises the condensed steam onward to the boiler). It is also a way of recovering useful heat energy which may be put to good use—for district heating (Xu et al. 2018) and for greenhouses in the horticultural sector (Yu and Nam 2016), obviating the need for fossil fuels for these purposes and bringing about a net reduction in the GHG-emissions (refer Fig. 12.2). Waste heat recovery can also entail upgrading some of the high-temperature heat to high-exergy electricity (Blaise and Feidt 2019; Tang et al. 2020).

12.2.8 SDG 8—Sustained, Inclusive, Economic Growth, Decent Work, Productive Employment

While informal recycling in cities of the developing world does lift some families out of extreme poverty (sub-Sect. 2.1), and in a small but significant way also contributes

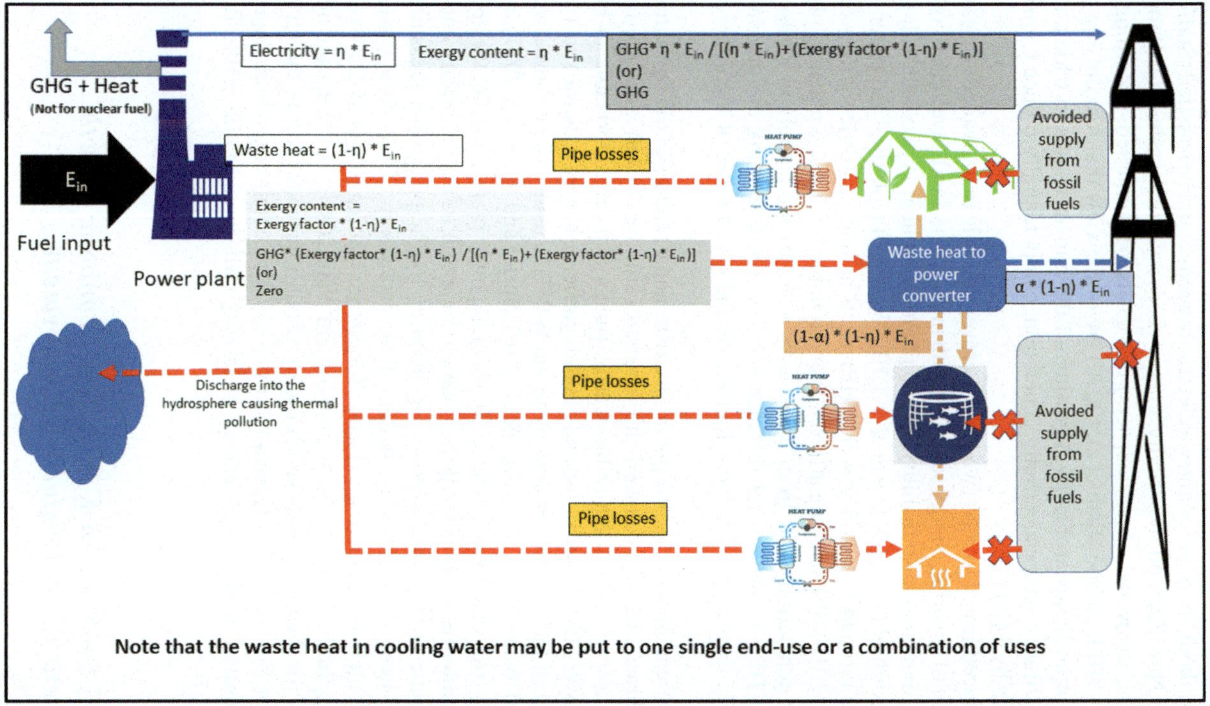

Fig. 12.2 Utilisation of the waste heat in power plant cooling water streams for electricity generation, greenhouses, fish farms and district heating (top-to-bottom on the right-hand side of the Figure). (α and η are conversion efficiencies, and the specific GWP-100 factors have been indicated for exergy-based allocation and 100%-allocation to the electricity output)

to economic growth by recycling value (abiotic materials like metals and plastics usually) back into the anthroposphere, one perhaps may not be able to label it as 'decent work' or 'productive employment' per se. This is because the actual 'value-creators' in this case do not earn in proportion to the value they create, and the health risks they are exposed to. This is a hotspot in the value chains of several products, with operating nodes in the developing world (Venkatesh 2019a, b), which needs to be highlighted and redressed by local governments. However, waste valorisation, in general, must and will entrench itself in the circular economies of the future. As observed by Mengal et al. (2018), the bio-based Industries Joint Undertaking in the EU is working towards facilitating private investment in new bio-refineries. These, while valorising bio-wastes and competing in the process, with non-bio-based products, will create several 'decent' and productive employment opportunities.

12.2.9 SDG 9—Resilient Infrastructure, Sustainable Industrialisation and Innovation

Waste collection, handling and valorisation ought to go hand in hand with industrialisation, if the latter needs to be labelled as sustainable. Here, the reference is both to wastes generated during industrial processes, at points-of-sale of products supplied by the industries and post-consumption. Change is the only constant, and innovation is key. This may be about doing different things (technological breakthroughs, product- and process-related 'disruptive' innovations) or doing things differently ('creative destruction' of the conventional approaches).

Innovations targeted at recovering resources from waste take time to entrench themselves in economies and deserve top-down support. In sub-Sect. 2.6, the downstream of the water utility has been labelled as a 'multi-output refinery' (refer Fig. 12.1). While nutrients and bio-energy are derived from the bio-based wastes in the wastewater, researchers have proven that at-source treatment of industrial wastewaters can be done to recover different metals, non-metals and chemical compounds—silver, palladium, gallic acid, sodium, potassium and magnesium salts, copper, chromium, chlorine, zinc sulphide, gold, indium, caustic soda, aluminium, calcium fluoride, lead, uranium, cadmium and polyhydroxyalkanoates (bio-plastics in other words) (Venkatesh 2018).

Necessity is the mother of invention. Technological breakthroughs which enable resource-recovery and happen in R & D labs in industries, research institutes and universities, often serendipitously. Knowledge about them must be disseminated globally (of course without compromising the rights of the innovators to hold patents and avail of the advantages thereof), and once the desiderata of SDG 17 (sub-Sect. 2.16) are in place, the rewards of this knowledge-sharing can be reaped on a larger scale (Venkatesh 2021).

12.2.10 SDG 10—Reduction of Inequality Within and Among Countries

Waste management (with a focus on valorisation) will aid in the reduction of inequality within and among countries, indirectly by contributing to the other SDGs:

(a) Sustaining the space for informal recyclers (SDG 1, sub-Sect. 2.1) to function (with some attention given to health and well-being issues (refer to sub-Sect. 2.3), stay above the extreme poverty line and rise up gradually in the process. Creating more employment opportunities in a green economy focussed on the collection, handling and valorisation of wastes (sub-Sect. 2.8).
(b) Aiding electrification of far-flung areas in the developing world, by availing of waste-derived energy options (bio-waste as far as possible) and bringing electricity to the lives of people who have not experienced its benefits thus far (sub-Sect. 2.7).
(c) Spurring investments in sanitation in the developing world, motivated by the potential of WWTPs to be multi-output refineries (sub-Sect. 2.6).
(d) Eliminating hunger (sub-Sect. 2.2) by way of reducing food waste, recovering nutrients from wastewater and producing bio-fertilisers from bio-wastes in bio-refineries and bridging the gap between the vast swathes of the human population who struggle to get three square meals a day, and the relatively smaller numbers who waste the surplus that they are able to afford.

12.2.11 SDG 11—Inclusive, Safe, Resilient and Sustainable Cities and Human Settlements

The adjectives 'resilient' and 'sustainable' are relevant here from a waste management perspective. The targets include the provision of basic services to all inhabitants, and access to sanitation facilities can be mentioned here. As mentioned in sub-Sect. 2.6, when the downstream of the water utility is envisaged as a 'multi-output refinery' generating a lot of value from wastewater (also refer Fig. 12.1), city administrations are incentivised to expand sanitation services to the entire population in phases—'sustainably' and 'inclusively'. Another target is sustainable public transportation, and as discussed in sub-Sect. 2.7, bio-methane (or biogas), biodiesel and bio-hydrogen in the future—all derived from bio-wastes in bio-refineries—can become the fuels of choice, replacing slowly but surely, and eventually completely, fossil-gasoline and fossil-diesel (exemplified by Sweden, as discussed in Venkatesh (2022a, b)). In this regard, Kapoor et al. (2020a, b) would like to point to the open burning of agro-wastes in India, which accounts for a sizable fraction of the country's greenhouse gas (GHG) emissions. The tremendous potential they hold for conversion to biogas that can be supplied as a sustainable transport fuel in the towns and cities of the country must not be wasted.

Resilience entails being able to uncover a host of alternatives to fall back on, to supply the essential needs of citizens, in the event of disruptions. Waste valorisation is a promising option that must be availed of, to adapt to the fallouts of climate change. Also targeted is a reduction in the per-capita environmental impact, and all the R's of waste management working in tandem towards sustainable consumption (refer the next sub-section) will speed up the process towards it.

12.2.12 SDG 12—Sustainable Consumption and Production Patterns

Sustainable production entails in the context of waste management:

(a) Improving the efficiency of usage of resources (and thereby reducing waste generation)
(b) Recycling the in-process wastes that are invariably generated
(c) Increasing the recycled content in the products supplied
(d) Designing products for recyclability, durability and energy efficiency in the use phase (to decrease energy losses—also a form of waste to be avoided)
(e) Setting in place incentivised take-back schemes to efficiently recycle materials back into the anthroposphere
(f) Utilising waste-derived energy (preferably bio-waste-derived)—electricity, heat and solid/liquid/gaseous fuels—in production processes wherever possible.
(g) Reusing wastewater in a cascade within the plants, to the greatest extent possible; and rainwater harvesting if possible.

Sustainable consumption, in the context of waste management entails, *inter alia:*

(a) Reducing/optimising one's material footprint and minimising waste of all kinds (Oldfield et al. 2016; Wikström et al. 2016)
(b) Prioritising bio-based alternatives
(c) Repurposing creatively and extending the usable/functional lifetime of products
(d) Cooperating with waste-collection-and-valorization initiatives undertaken by entrepreneurs (Loizides et al. 2019)
(e) Supporting reuse by donating to second-hand shops
(f) Availing of the repair-and-reuse option to extend the lifetimes of products (Govind and Mahongnao 2021)
(g) Handling the end-of-life of products responsibly
(h) Reusing water in a cascade at home and rainwater harvesting if possible (sub-Sect. 2.6).

It follows that the entrenchment of the R's of waste management in production upstream and consumption downstream will contribute significantly to sustainable development (SDG 12 in this case).

12.2.13 SDG 13—Combating Climate Change and Its Repercussions

Climate change is attributed directly to the emission of greenhouse gases (GHGs) from anthropospheric activities. While circularising bio-wastes—from households, industries, points-of-sale, WWTPs (sub-Sect. 2.6), fields/farms and forests—and valorising them into forms of energy to supplant fossil fuels (sub-Sect. 2.7) are a sure-fire strategy to have GHG-emissions on a leash, bio-materials and bio-fertilisers produced in bio-refineries will contribute by avoiding the production of conventional GHG-emitting alternatives. In all these cases, however, an environmental life-cycle assessment needs to be carried out to determine if net GHG reductions are achieved by means of such replacements. Having more alternatives (courtesy the circularity paradigm and the realisation that 'waste' is a misnomer, the hidden value of which needs to be harnessed) to take recourse to, will impart resilience to societies and economies, as they try to adapt to any inevitable repercussions of climate change.

As more and more changes are incorporated in the journey towards SDG 12 (12 of several strategies outlined in sub-Sect. 2.12), the targets of SDG 13 (and those of SDGs 14 and 15) will progressively seem nearer at hand.

12.2.14 SDG 14—Conservation of Oceans, Seas and Marine Resources

In sub-Sect. 2.7, the author referred to waste heat recovery from the cooling water discharged from thermal power plants. In the absence of any heat recovery, the heat content of the cooling water is transferred to the hydrosphere—into marine or freshwater ecosystems. This leads to thermal pollution, and habitat destruction, as the dissolved oxygen content in the water bodies, decreases with rising water temperature and this tends to affect the aquatic biota (Rosen et al. 2015). Thus, waste heat recovery, while aiding the journey towards SDG 7 and SDG 13, also contributes to SDG 14.

While thermal pollution of water bodies (both freshwater and marine) is to be avoided, the issue which has been dominating headlines in the recent past and also currently is marine pollution by waste plastics. A four-fold approach is called for, to tackle this problem.

(a) Refusing and reducing the use of plastics and being willing to pay more for bio-alternatives (La Fuente et al. 2022) (SDG 12, and a part of sustainable consumption) and assuming the responsibility to deposit waste plastics in designated bins/containers
(b) Replacing plastics with bio-alternatives (SDG 12, sustainable production; SDG 9, industrial innovation)

(c) Restricting the use of plastics by bans and introducing fees (Senturk and Dumludag 2021) (top-down policy decisions, which may even be based on international agreements, encompassed by SDG 17)
(d) Remediating the oceans by recovering the biodiversity-destroying plastic wastes from the hydrosphere (a necessary part of SDG 14—Caruso 2015; Oluniyi Solomon and Palanisami 2016; Auta et al. 2017; Cho 2009) and subsequently harnessing energy from them, as recycling them will be quite challenging (SDG 7, though this energy cannot be deemed to be 'clean', as also mentioned earlier in sub-Sect. 2.7)

Remediating the oceans of the world necessitates global partnership (SDG 17; sub-Sect. 2.16) and innovative responsible initiatives undertaken by governments of countries that have coastlines and economies bolstered by proximity to the oceans/seas (South Korea for instance as gathered from Cho (2009)) and by non-governmental organisations (Plastics News 2021).

12.2.15 SDG 15—Conservation of Terrestrial Ecosystems, Forests and Biodiversity

Just as the atmosphere can be 'remediated' by capturing and utilising/sequestering carbon dioxide (Peplow 2022) (as one of the strategies for SDG 13); and the oceans can be remediated by recovering waste plastics therefrom (SDG 14; sub-Sect. 2.14), soils—be they on arable land, in forests or on land in close proximity to industrial complexes—and the groundwater beneath them can be remediated by removing toxic (and other undesirable, though non-toxic) wastes therefrom. Soil and groundwater remediation is an advanced field of practical research, which will serve the purpose, when it comes to SDG 15 (Grifoni et al. 2022). Remediation technologies can restore soils and groundwater partially to their pre-contaminated levels, restoring a good deal of their carrying capacities. Soil amendment and fertilisation—in reclaimed forests and agricultural land—can be sustainably done by utilising bio-fertilisers and biochar from bio-refineries, and sludge from WWTPs (SDG 2, sub-Sect. 2.2; SDG 6, sub-Sect. 2.6).

It may read like a cliché, but the more consumers try to minimise their material footprints by reducing, reusing, recycling, repurposing and refusing (SDG 12; sub-Sect. 2.12), the degree of circularity in an economy can be considerably enhanced. While a developing bio-economy may be characterised by a slightly greater dependence on arable land and forests, attempts made to circularise it will strike a good balance between economic development and conservation of terrestrial ecosystems (and thereby the biodiversity which they sustain).

12.2.16 SDG 17—Revitalised Global Partnership for Sustainable Development

For waste valorisation to entrench itself in a significant way, and become pervasive, as observed by Venkatesh (2021) that cites from a host of publications from literature (not referred to here), the following are indispensable:

(a) Stronger private–public partnerships
(b) Collaborations among 'conventional and non-conventional entities' in the economy
(c) Robust institutional structures at local and regional levels
(d) Strong governance to ensure transparency, decent work and productive employment (SDG 8; refer sub-Sect. 2.8).
(e) Effective policy-mix of 'carrots and sticks' (penalties/fees and incentives/subsidies)
(f) Revamp of international quality standards to accommodate products valorised from wastes and encourage and facilitate global trading of the same.
(g) Systemic-level changes through the joint global efforts of researchers, technology centres, industries, the primary sector, new entrepreneurs, consumers, civil society and governments and relentless innovation (SDG 9; sub-Sect. 2.9) are indispensable.

12.3 Summary and Conclusions

This commentary has taken recourse to published literature, but as mentioned in Sect. 12.1, they account for an extremely small sub-set of extant waste management literature.

As the degree of urbanisation keeps increasing, cities and towns evolve into ever-larger centres of consumption of a plethora of resources, and consequently, as generators of huge quantities of wastes [refer to the other chapter (Chap. 9) by this author, in this collection, related to urban metabolism). While waste generation is inevitable, the management of the same is an art, science and business at the same time, involving a variety of stakeholders—producers and manufacturers, consumers, governments, media and academic researchers. Each one of us generates wastes and thereby can try to incorporate the R's of waste management consciously and responsibly into our lives. It can play a key role in the alleviation of, and the simultaneous adaptation to the repercussions of climate change. Waste valorisation, which is gradually entrenching itself, in both principle and practice, can go a long way in directly and indirectly enabling humankind to get closer to several SDG targets, and perhaps overachieve in some respects. As shown in Fig. 12.3, waste management envisaged as value creation will enable SDGs 1, 2, 6, 7, 8, 10–15, while it will be enabled by SDGs 3, 4, 9, and 17, and must be handled with care when it comes to SDG 2. SDG 8 can be considered to enable and be enabled, at the same time—mutually reinforcing in other words.

The author has excluded SDG 16 from the mix, considering it to be not so relevant in a waste management context.

The take-away messages from this chapter can be listed hereunder:

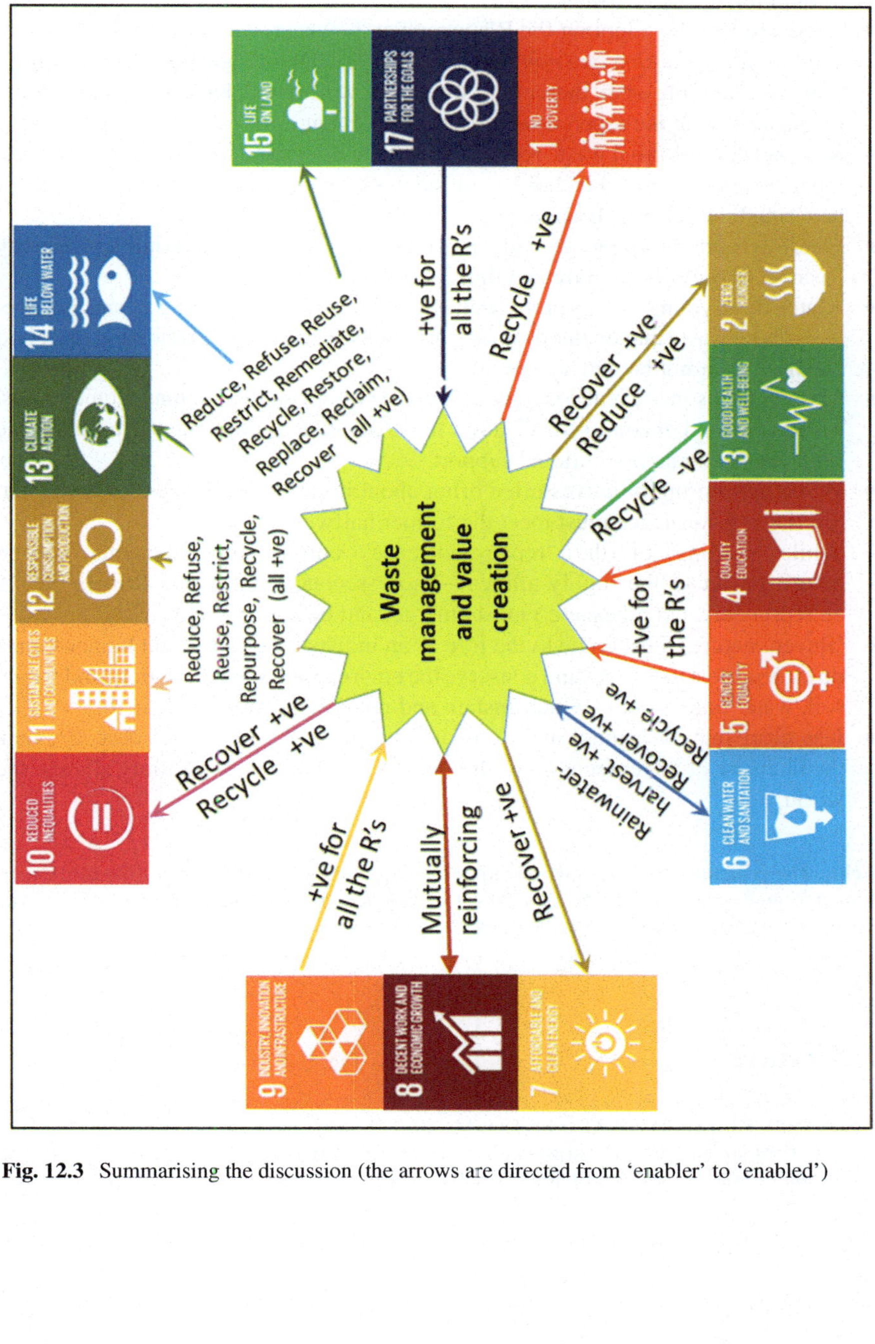

Fig. 12.3 Summarising the discussion (the arrows are directed from 'enabler' to 'enabled')

- '*Waste not, want not*' is a popular slogan all readers may have heard or read before and perhaps just not given much thought to.
- However, what is inevitably wasted must hereafter, be looked upon as a resource with 'hidden value' that has heretofore not been appreciated. A steady overhaul in thinking is indispensable.
- Value creation by adopting the R's wherever, however, whenever and by whosoever possible, is a *sine qua non* for achieving the SDGs by year-2030, and continuing in the same vein thereafter, when the world will have to grapple more perceptibly with the repercussions of climate change.
- Rag-pickers (informal waste recyclers) must be integrated into the waste management system, as vital stakeholders of the waste management system. They could be termed the 'Green Brigade'.
- Landfills and dumping grounds must be renamed as 'Material and Energy Recovery Facilities' and treated thus.
- 'Catch them young' works pretty well when it comes to developing robust perspectives about waste management among urbanites. The power of education and good teachers must not be underestimated.
- Creative ideas in the realm of resource conservation and waste management must be boosted and encouraged. Ideation must advance to implementation, with the right financial and institutional support.
- While bottom-up initiatives often bring about impressive changes for the better, top-down governance must meet the former halfway.
- While availing of the 'repair-and-reuse' culture must become second nature—even to the highly-affluent—more second-hand shops (run by social entrepreneurs, churches, etc.) must emerge, and be patronised.
- 'Revere nature' must come to the fore as an indispensable 'R', and humans must do their utmost to respect and conserve the environmental media—flora and fauna, soil, water bodies etc.—which sustain and nourish the anthroposphere.
- It is clear that we cannot avert climate change now. We can, at best, alleviate the intensity of its repercussions, though unfortunately not uniformly all over the world.

Acknowledgements I dedicate this paper to the memory of my late wife, Varshita. Thanks to Linnea Casemyr, graduate student at the Department of Engineering and Chemical Sciences, Karlstad University, whose coursework from 2021 was useful for sub-Sect. 2.14.

References

Abdel-Shafy HI, El-Khateeb MA, Shehata M (2017) Blackwater treatment via combination of sedimentation tank and hybrid wetlands for unrestricted reuse in Egypt. Desalin Water Treat 71:145–151. https://doi.org/10.5004/dwt.2017.20538

Agrafioti E, Diamadopoulos E (2012) A strategic plan for reuse of treated municipal wastewater for crop irrigation on the Island of Crete. Agric Water Manag 105:57–64. https://doi.org/10.1016/j.agwat.2012.01.002

Antonini S, Paris S, Eichert T, Clemens J (2011) Nitrogen and phosphorus recovery from human urine by struvite precipitation and air stripping in Vietnam. Clean Soil Air, Water 39(12):1099–1104. https://doi.org/10.1002/clen.201100036

Atabani AE, Al-Rubaye OK (2020) Valorization of spent coffee grounds for biodiesel production: blending with higher alcohols, FT-IR, TGA, DSC, and NMR characterizations. Biomass Convers Biorefinery 12:577–596. https://doi.org/10.1007/s13399-020-00866-z

Auta HS, Emenike CU, Fauzia SH (2017) Distribution and importance of microplastics in the marine environment: a review of the source, fate, effects, and potential solutions. Environ Int 102:165–176

Baeza JA, Martinez-Miro A, Guerrero J, Ruiz Y, Guisasola A (2017) Bio-electrochemical hydrogen production from urban wastewater on a pilot scale. J Power Sources 356:500–509. https://doi.org/10.1016/j.jpowsour.2017.02.087

Ben Brahim-Neji H, Ruiz-Villaverde A, Gonzalez- Gomez F (2014) Decision aid supports for evaluating agricultural water reuse practices in Tunisia: the Cebala perimeter. Agric Water Manag 143:113–121 https://doi.org/10.1016/j.agwat.2014.07.002

Bikse V, Lusena-Ezera I, Volkova T, Rivza B (2019) European bioeconomy policy and new opportunities for bio-based business development. Int Multi Sci GeoConference Surveying Geol Mining Ecol Manag, SGEM 19:317–325. https://doi.org/10.5593/sgem2019/5.3/S21.040

Blaise M, Feidt M (2019) Waste heat recovery and conversion into electricity: current solutions and assessment. Int J Thermodyn 22(1):1–7. https://doi.org/10.5541/ijot.487951

Bureau of International Information Programs, United States Department of State (2012) Women and the Environment. In: Global women's issues: women in the world today, extended version. https://opentextbc.ca/womenintheworld/. Accessed 31 Aug 2022

Carr G, Potter RB, Nortcliff S (2011) Water reuse for irrigation in Jordan: perception of water quality among farmers. Agric Water Manag 98(5):847–854. https://doi.org/10.1016/j.agwat.2010.12.011

Caruso G (2015) Plastic degrading microorganisms as a tool for bioremediation of plastic contamination in aquatic environments. J Pollut Eff Cont 3(03):3. https://doi.org/10.4172/2375-4397.1000e112

Cho D-O (2009) The incentive program for fishermen to collect marine debris in Korea. Mar Pollut Bull 58:415–417. https://doi.org/10.1016/j.marpolbul.2008.10.004

Cirelli GL, Consoli S, Licciardello F, Aiello R, Giuffrida F, Leonardi C (2012) Treated municipal wastewater reuse in vegetable production. Agric Water Manag 104:163–170. https://doi.org/10.1016/j.agwat.2011.12.011

Dahal RK, Acharya B, Farooque A (2018) Biochar: a sustainable solution for solid waste management in agro-processing industries. Biofuels 12(2):237–245. https://doi.org/10.1080/17597269.2018.1468978

Desrochers JE, Albert G, Milfont T, Kelly B, Arnocky S (2019) Does personality mediate the relationship between sex and environmentalism? Personality Individ Differ 147:204–213. https://doi.org/10.1016/j.paid.2019.04.026

Di Sorrentino P, Woelbert E, Serenella ES (2016) Consumers and their behaviour: state of the art in behavioural science supporting use phase modelling in LCA and eco-design. Int J Life Cycle Assess 21(2):237–251. https://doi.org/10.1007/s11367-015-1016-2

Duan N, Zhang D, Khoshnevisan B, Kougias PG, Treu L, Liu Z, Lin C, Liu H, Zhang Y, Angelidaki I (2020) Human waste anaerobic digestion as a promising low-carbon strategy: operating performance, microbial dynamics and environmental footprint. J Cleaner Prod 256:120414. https://doi.org/10.1016/j.jclepro.2020.120414

Govind M, Mahongnao M (2021) Recycling of municipal solid waste in India: empirical findings from some select colonies in Delhi. Afr J Sci Technol Innov Dev 13(7):899–912. https://doi.org/10.1080/20421338.2021.1979169

Grifoni M, Franchi E, Fusini D, Vocciante M, Barbafieri M, Pedron F, Rosellini I, Petruzzelli G (2022) Soil remediation: towards a resilient and adaptive approach to deal with the ever-changing environmental challenges. Environments 9(2):18. https://doi.org/10.3390/environments9020018

Gu Y, Wang S (2013) Research on the energy-saving and economic analysis of wastewater source heat pump. Energy Educ Sci Technol Part a: Energy Sci Res 31(2):1167–1170

Hartley BL, Thompson RC, Pahl S (2015) Marine litter education boosts children's understanding and self-reported actions. Mar Pollut Bull 90:209–217. https://doi.org/10.1016/j.marpolbul.2014.10.049

Heubeck S, De Vos RM, Craggs R (2011) Potential contribution of the wastewater sector to energy supply. Water Sci Technol 63(8):1765–1771. https://doi.org/10.2166/wst.2011.115

Johansson E, Venkatesh G (2017) Wastewater treatment plants as energy producers: comparison of status quo in Sweden and India. Vatten 73(4):151–160

Kapoor R, Ghosh P, Kumar M, Sengupta S, Gupta A, Kumar SS, Vijay V, Kumar V, Kumar Vijay V, Pant D (2020a) Valorization of agricultural waste for biogas based circular economy in India: a research outlook. Bioresour Technol 304:123036. https://doi.org/10.1016/j.biortech.2020.123036

Kapoor V, Tripathi S, Devi RS, Srivastava P, Bhadouria R (2020b) Ecological economics of an urban settlement: an overview. In: Verma P, Singh P, Singh R, Raghubanshi AS (eds) Urban Ecology: Emerging patterns and socio-ecological systems. Elsevier, pp 91–110. https://doi.org/10.1016/B978-0-12-820730-7.00006-9

Karlsson J, Brunzell L, Venkatesh G (2018) Material-flow analysis, energy analysis and partial environmental-LCA of a district-heating combined heat and power plant in Sweden. Energy 144:31–40. https://doi.org/10.1016/j.energy.2017.11.159

Köck-Schulmeyer M, Ginebreda A, Postigo C, Lopez-Serna R, Perez S, Brix R, Llorca M, Alda MLD, Petrović M, Munne A, Tirapu L, Barcelo D (2011) Wastewater reuse in Mediterranean semi-arid areas: the impact of discharges of tertiary treated sewage on the load of polar micro pollutants in the Llobregat river (NE Spain). Chemosphere 82(5):670–678. https://doi.org/10.1016/j.chemosphere.2010.11.005

Kopsahelis N, Kachrimanidou V (2019) Advances in food and by-products processing towards a sustainable bioeconomy. Foods 8(9):425. https://doi.org/10.3390/foods8090425

Kumar PS, Saravanan A, Jayasree R, Jeevanantham S (2020) Food industry waste bio-refineries. In: Refining biomass residues for sustainable energy and bioproducts: technology, advances, life cycle assessment, and economics, pp 407–426. https://doi.org/10.1016/B978-0-12-818996-2.00018-1

La Fuente CIA, Tribst AIL, Augusto PED (2022) Knowledge and perception of different plastic bags and packages: a case study in Brazil. J Environ Manage 301:113881. https://doi.org/10.1016/j.jenvman.2021.113881

Lavrnič S, Zapater-Pereyra M, Mancini ML (2017) Water scarcity and wastewater reuse standards in Southern Europe: focus on agriculture. Water Air Soil Pollut 228(7):251. https://doi.org/10.1007/s11270-017-3425-2

Libutti A, Gatta G, Gagliardi A, Vergine P, Pollice A, Beneduce L, Disciglio G, Tarantino E (2018) Agro-industrial wastewater reuse for irrigation of a vegetable crop succession under Mediterranean conditions. Agric Water Manag 196:1–14. https://doi.org/10.1016/j.agwat.2017.10.015

Loizides MI, Loizidou XI, Orthodoxou DL, Petsa D (2019) Circular bioeconomy in action: collection and recycling of domestic used cooking oil through a social, reverse logistics system. Recycling 4(2):16. https://doi.org/10.3390/recycling4020016

Luttenberger LR (2020) Waste management challenges in transition to circular econom—case of Croatia. J Clean Prod 256:120495. https://doi.org/10.1016/j.jclepro.2020.120495

Maina S, Kachrimanidou V, Koutinas A (2017) A roadmap towards a circular and sustainable bioeconomy through waste valorization. Curr Opin Green Sustain Chem 8:18–23. https://doi.org/10.1016/j.cogsc.2017.07.007

McConville JR, Kvarnström E, Jonsson H, Karrman E, Johansson M (2017) Is the Swedish wastewater sector ready for a transition to source separation? Desalin Water Treat 91:320–328
Mengal P, Wubbolts M, Zika E, Ruiz A, Brigitta D, Pieniadz A, Black S (2018) Bio-based industries joint undertaking: the catalyst for sustainable bio-based economic growth in Europe. New Biotechnol 40:31–39. https://doi.org/10.1016/j.nbt.2017.06.002
Mihelcic JR, Fry LM, Shaw R (2011) Global potential of phosphorus recovery from human urine and faeces. Chemosphere 84(6):832–839. https://doi.org/10.1016/j.biortech.2016.04.094
Nibbi L, Chiaramonti D, Palchetti E (2019) Project BBChina: a new master program in three Chinese universities on bio-based circular economy; from fields to bioenergy, biofuel and bio-products. Energy Procedia 158:1261–1266. https://doi.org/10.1016/j.egypro.2019.01.416
Ojha VK, Tripathi V, Krishna V (2014) Role of rag-pickers in solid waste management: a case study of Mirzapur, U.P. Int J Environ Technol Manag 17(2–4):134–142. https://doi.org/10.1504/IJETM.2014.061803
Oldfield TL, White E, Holden NM (2016) An environmental analysis of options for utilising wasted food and food residue. J Environ Manage 183:826–835. https://doi.org/10.1016/j.jenvman.2016.09.035
Oluniyi Solomon O, Palanisami T (2016) Microplastics in the marine environment: current status, assessment methodologies, impacts and solutions. J Pollut Eff Cont 4(2):1000161. https://doi.org/10.4172/2375-4397.1000161
Panagiotou E, Kafa N, Koutsokeras L, Kouis P, Nikolaou P, Constantinides G, Vyrides I (2018) Turning calcined waste egg-shells and wastewater to Brushite: phosphorus adsorption from aqua media and anaerobic sludge leach water. J Clean Prod 178:419–428. https://doi.org/10.1016/j.jclepro.2018.01.014
Peplow M (2022) The race to recycle carbon dioxide. Nature 603:780–783
Plastics News (2021) Ocean CleanUp says redesigned collector can remove plastics from the Pacific. Plastics News 32(32):0004. Crain Communications, Inc.
Prasad MNV (2016) Recovery of resources from bio-waste for pollution prevention. In: Environmental materials and waste: resource recovery and pollution prevention, pp 1–9. https://doi.org/10.1016/B978-0-012-803837-6.00001-9
Prihandoko D, Budiman A, Setyono P, Fandeli C, Budiastuti MTS (2021) Appropriate technology for municipal solid waste management based on wastepreneurship implementation. ASEAN J Chem Eng 21(1):113–123
Quayson E, Amoah J, Rachmadona N, Morita K, Darkwah L, Hama S, Yoshida A, Kondo A, Ogino C (2020) Valorization of palm biomass waste into carbon matrices for the immobilization of recombinant Fusarium heterosporum lipase towards palm biodiesel synthesis. Biomass Bioenerg 142:105768. https://doi.org/10.1016/j.biombioe.2020.105768
Ren H-Y, Liu B-F, Kong F, Zhao L, Ren N (2015) Hydrogen and lipid production from starch wastewater by co-culture of anaerobic sludge and oleaginous microalgae with simultaneous COD, nitrogen and phosphorus removal. Water Res 85:404–412. https://doi.org/10.1016/j.watres.2015.08.057
Rosen MA, Buluca CA, Mastorakis NE, Brindusa CC, Bulucea CA, Jeles AC (2015) Evaluating the thermal pollution caused by wastewaters discharged from a chain of coal-fired power plants along a river. Sustainability 7(5):5920–5943. https://doi.org/10.3390/su7055920
Sadhukhan J, Dugmore TIJ, Matharu A, Martinez-Hernandez E, Aburto J, Rahman PKSM, Lynch J (2020) Perspectives on "game changer" global challenges for sustainable 21st century: plant-based diet, unavoidable food waste biorefining, and circular economy. Sustainability 12(5):1976. https://doi.org/10.3390/su12051976
Santoshi KM, Kiran UV (2022) Prevalence of health problems of rag-pickers due to various hazards at Lucknow city. Hum Factors Healthc 2:100023. https://doi.org/10.1016/j.hfh.2022.100023
Senturk G, Dumludag D (2021) An evaluation of the effect of plastic bag fee on consumer behaviour-Case of Turkey. Waste Manag 120:748–754. https://doi.org/10.1016/j.wasman.2020.10.042

Severo IA, Siqueira SF, Deprá MC, Maroneze MM, Zepka LQ, Jacob-Lopes E (2019) Biodiesel facilities: what can we address to make biorefineries commercially competitive? Renew Sust Energ Rev 112:686–705. https://doi.org/10.1016/j.rser.2019.06.020
Sharma K, Garg VK (2019) Vermicomposting of waste: a zero-waste approach for waste management. In: Sustainable resource recovery and zero waste approaches, pp 133–164. https://doi.org/10.1016/B978-0-444-64200-4.00010-4
Sharma Y, Li B (2010) Optimizing energy harvest in wastewater treatment by combining anaerobic hydrogen producing biofermentor (HPB) and microbial fuel cell (MFC). Int J Hydrogen Energy 35(8):3789–3797. https://doi.org/10.1016/j.ijhydene.2010.01.042
Shomar B, El-Madhoun F, Yahya A (2010) Wastewater reuse for alfalfa production in the Gaza Strip. Water, Air and Soil Pollution 213(1–4):105–119. https://doi.org/10.1007/s11270-010-0371-7
Simha P, Ganesapillai M (2017) Ecological sanitation and nutrient recovery from human urine: How far have we come? A review. Sustain Environ Res 27(3):107–116
Simha P, Zabaniotou A, Ganesapillai M (2017) Continuous urea–nitrogen recycling from human urine: a step towards creating a human-excreta-based bio–economy. J Clean Prod 172:4152–4161. https://doi.org/10.1016/j.jclepro.2017.01.062
Simha P, Karlsson C, Viskari E-L, Malila R, Vinnerås B (2020) Field testing a pilot-scale system for alkaline dehydration of source-separated human urine: a case study in Finland. Front Environ Sci 8:570637. https://doi.org/10.3389/fenvs.2020.570637
Sitzenfrei R, Hillebrand S, Rauch W (2017) Investigating the interactions of decentralized and centralized wastewater heat recovery systems. Water Sci Technol 75(5):1243–1250
Spriet J, Hendrick P (2017) Wastewater as a heat source for individual residence heating: a techno-economic feasibility study in the Brussels Capital Region. J Sustain Develop Energy, Water Environ Syst 5(3):289–308
Stafford W, Cohen B, Pather-Elias S, von Blottnitz H, van Hille R, Harrison STL, Burton SG (2013) Technologies for recovery of energy from wastewaters : applicability and potential in South Africa. J Energy South Afr 24(1):15–26
Stathatou P-M, Gad F-K, Kampragou E, Grigoropoulou H, Assimacopoulos D (2015) Treated wastewater reuse potential: mitigating water scarcity problems in the Aegean Islands. Desalin Water Treat 53(12):3272–3282. https://doi.org/10.1080/19443994.2014.934108
Taddeo R, Honkanen M, Kolppo K, Lepisto R (2018) Nutrient management via struvite precipitation and recovery from various agro-industrial wastewaters: process feasibility and struvite quality. J Environ Manage 212:433–439. https://doi.org/10.1016/j.jenvman.2018.02.027
Talboys PJ, Heppell J, Roose T, Healey JR, Jones DL, Withers PJA (2016) Struvite: a slow-release fertiliser for sustainable phosphorus management? Plant Soil 401:109–123. https://doi.org/10.1007/s11104-015-2747-3
Tang S, Wang C, Liu X, Su G, Tian W, Qiu S, Zhang Q, Ruiheng L, Bai S (2020) Experimental investigation of a novel heat pipe thermoelectric generator for waste heat recovery and electricity generation. Int J Energy Res 44:7450–7463. https://doi.org/10.1002/er.5465
Teng S-X, Tong Z-H, Li W-W, Wang S-G, Sheng G-P, Shi X-Y, Liu X-W, Yu H-Q (2010) Electricity generation from mixed volatile fatty acids using microbial fuel cells. Appl Microbiol Biotechnol 87(6):2365–2372. https://doi.org/10.1007/s00253-010-2746-5
Tian X, Wang G, Guan D, Li J, Wang A, Li J, Yu Z, Chen Y, Zhang Z (2016) Reverse osmosis brine for phosphorus recovery from source separated urine. Chemosphere 165:202–210
Umair S, Björklund A, Pettersen EE (2015) Social impact assessment of informal recycling of electronic ICT waste in Pakistan. Resour Conserv Recycl 95:46–57. https://doi.org/10.1016/j.resconrec.2014.11.008
United Nations—Department of Economic and Social Affairs: Sustainable Development (2015) The 17 goals. https://sdgs.un.org/goals. Accessed 30 Aug 2022
Vea EB, Romeo D, Thomsen M (2017) Biowaste valorisation in a future circular bioeconomy. Procedia CIRP 69:591–596. https://doi.org/10.1016/j.procir.2017.11.062
Venkatesh G (2018) Recovery of different types of resources from wastewater—a structured review. Vatten 74(1–2):65–82

Venkatesh G (2019a) Social LCA—an introduction. The What, the Why and the How. Bookboon, Copenhagen, Denmark. ISBN: 978-87-403-2841-7

Venkatesh G (2019b) Critique of selected peer-reviewed publications on applied social life cycle assessment: Focus on cases from developing countries. Clean Technol Environ Policy 21(2):413–430. https://doi.org/10.1007/s10098-018-1644-x

Venkatesh G (2021) Circular bio-economy—paradigm for the future: systematic review of scientific journal publications from 2015 to 2021. Circ Econ Sustain 2:231–279. https://doi.org/10.1007/s43615-021-00084-3

Venkatesh G (2022) Not merely public transport but green too—from footprint to toe-print. Sci Rep 59(1):48–49. http://nopr.niscair.res.in/handle/123456789/58785

Venkatesh G (2022a) A brief analysis of SDG 3—Good health and well-being—and its synergies and trade-offs with the other sustainable development goals. Problemy Ekorozowju 17(2):133–140

Verstraete W, Clauwaert P, Vlaeminck SE (2016) Used water and nutrients: recovery perspective in a '*panta rhei*' context. Bioresour Technol 215:199–208. https://doi.org/10.1016/j.biortech.2016.04.094

Viaggi D (2015) Research and innovation in agriculture: beyond productivity? Bio-based and Appl Econ 4(3):279–300. https://doi.org/10.13128/BAE-17555

Wikstrom F, Williams H, Venkatesh G (2016) The influence of packaging attributes on recycling and food waste behaviour—an environmental comparison of two packaging alternatives. J Clean Prod 137:895–902. https://doi.org/10.1016/j.jclepro.2016.07.097

Xu ZY, Mao HC, Liu DS, Wang RZ (2018) Waste heat recovery of power plant with large-scale serial absorption heat pumps. Energy 165(B):1097–1105. https://doi.org/10.1016/j.energy.2018.10.052

Yang Z, Liu Y, Zhang J, Mao K, Kurbonova M, Liu G, Zhang R, Wang W (2020) Improvement of biofuel recovery from food waste by integration of anaerobic digestion, digestate pyrolysis and syngas bio-methanation under mesophilic and thermophilic conditions. J Cleaner Product 256:120594. https://doi.org/10.1016/j.jclepro.2020.120594

Yu MG, Nam Y (2016) Feasibility assessment of using power plant waste heat in large scale horticulture facility energy supply systems. Energies 9:112. https://doi.org/10.3390/en9020112

Chapter 13
Emerging Approaches for Sustainable Urban Metabolism

G. Gupta, R. Shrivastava, J. Khan, and N. K. Singh

Abstract Extensive urbanization with rapidly increasing human population leads to changes in urban environment specially land use pattern, energy budget, over-exploitation and depletion of natural resources, degradation of water, air and soil quality, and waste generation including wastewater and greenhouse gases. Further, increasing human activities and extensive development leads to more encroachment on land and water resources which cause more imbalances in urban environment. Therefore, more sustainable approaches need to be adapted in changing scenario for urban planning specially in developing green cities with sustainable resource utilization and better waste management practices. Advancement in technologies for waste generation reduction, waste treatment, recycling of waste, and energy conservation approaches could be helpful in upkeeping the better urban environment and to ensure sustainable development. Urban resources management is currently being included into a method for urban spatial planning. The urban harvest approach (UHA), which is founded on the idea of urban metabolism, seeks to improve resource management by utilizing cutting-edge technologies, ending urban cycles, and harvesting urban resources. Economic expansion, the rise in living standards, and the general development of civilization all have an impact on the environment, and it is now clear that this impact is distorting the natural equilibrium. Novel technologies must be critically reviewed to meet the complex needs necessary for a balanced and sustainable future.

Keywords Climate change · Resources · Sewage · Urban environment · Urbanization · Waste disposal

G. Gupta · R. Shrivastava · J. Khan · N. K. Singh (✉)
Environmental Science Discipline, Department of Chemistry, Manipal University Jaipur, Dehmi Kalan, Sanganer, Jaipur, Rajasthan 303007, India
e-mail: naveenenviro04@gmail.com

R. Shrivastava
Poornima College of Engineering, Jaipur, Rajasthan 302022, India

R. Bhadouria et al. (eds.), *Urban Metabolism and Climate Change*,
https://doi.org/10.1007/978-3-031-29422-8_13

13.1 Introduction

Human activities related to urbanization have a crucial impact on the environment. Half of the global human population lives in cities, and it is predicted to increase up to 66% by 2050 (United Nations 2014). Extensive urbanization has resulted in unsustainable development activities around the world, and cities are intensifying with ever-increasing population and increasing demands for water, food, shelter and proper sanitation (Edenhofer et al. 2014; Shao et al. 2020). The depletion of the resources that were nearest and easiest to get in the past have been inhibited by exponential urbanization (Tainter 2000). Technological and infrastructure advancements, on the other hand, have fueled increases in the resource use and waste generation in urbanized areas (Kennedy et al. 2007; Monstadt 2009). More burden on the capitals land, infrastructure, housing stock and resources is being brought by unprecedented urban growth and demands (Clark et al. 2019; Ford et al. 2019). For example, one of the most vulnerable urban areas to climate change is London (Ford et al. 2019), because of the freshwater supply issues, flooding, heat waves, and poor ambient air quality. Especially over the past two centuries, the world's population, industrialism, urbanization, and technical advancement have all led to increased resource demands on a global scale. Emerging urbanization project comprises designing the built-in gray infrastructure as well as the surrounding green and blue space (Erell et al. 2015; Wu et al. 2019). The carrying capacity of the global ecosystem has received decreasing amounts of attention since the industrial revolution. Ponting (2007) described the ecosystems and societies can collapse as a result of irreparable environmental damage. It has been widely understood over the past decade that present utilization rate is higher than the rate of sustainable utilization due to overexploitation (Arrow et al. 2004).

Sustainable urban development is a significant challenge in a world of cities and may be the most important present and future environmental issue (Girardet 2003). It is essential to know about urban metabolic systems to understand its functions and dynamics (Decker et al. 2000; Girardet 2003). In this chapter, we discussed about urban planning sustainability framework (UPSUF) with emphasis on the environmental issues of extensive urbanization. Assessment of urban evaluation sustainability combines benefits from natural spaces inside the bounds of cities and serves as the foundation for the sustainability evaluation (Tan et al. 2020). Definition of a waste is not well understood, even though it is commonly accepted that waste management services are necessary services that must be offered in every society. Understanding waste is an individualized idea since one thing considers to be waste might be a resource for another. Therefore, it is crucial to establish a clear definition of what constitutes waste. Solid waste generation is a routine consequences of human existence and development while improving human life in urbanization process and needs safe disposal. Initially, the main intention of solid waste management (SWM) strategies was to remove waste from places that were close to residential areas to protect public health. Municipal areas have a broad range of solid waste sources specially generated from domestic, commercial and industrial activities (Saseanu et al. 2019;

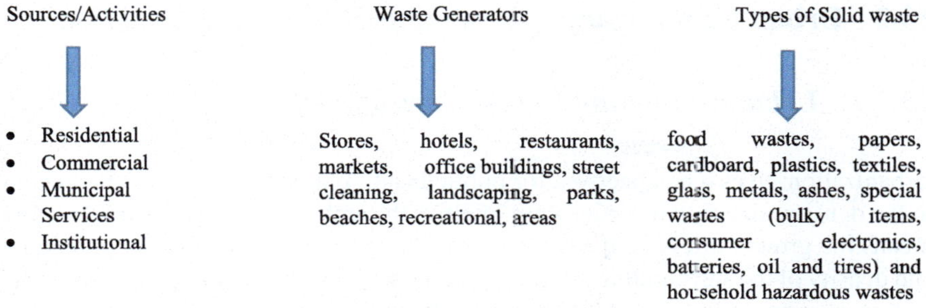

Fig. 13.1 Types of municipal solid waste (MSW) and their sources

Bhadouria et al. 2022). Solid waste disposal is regulated by various provisions and laws in several Asian countries and is classified into different categories (Fig. 13.1).

Urban metabolism (UM) and circular economy (CE) have both recently attracted a lot of attention worldwide. The amount of waste generation and consumption of natural resources are, in fact, critical problems which needs appropriate solutions with extensive urbanization (Longato et al. 2019). Urban metabolism is being investigated as a viable tool for assessing and improving resource utilization and waste disposal practices in a sustainable manner at local and regional level (Ulgiati and Zucaro 2019). Recent years have seen the development of numerous material and energy recovery methods, which are now included into contemporary systems (Jouhara et al. 2018). Water resources have been depleted in many areas of the nation because of excessive domestic, agricultural and industrial use of the nation's water supply, which has not only caused a decline in water quality. The country's water crisis is not due to one single cause; rather, several factors, including population growth, unaccounted water, and dwindling groundwater supplies due to farmers' excessive groundwater extraction, inadequate pricing, and insufficient investment in treatment facilities at the federal, state, and local levels have all contributed to the decline in water quality (Sharma et al. 2014). The impending water crisis and degrading water quality are now apparent in several parts of the country. We need to act decisively to secure our water assets and use them sensibly. To refocus solid waste management systems toward sustainability, there are now active global activities. Asian nations are heavily involved in this transformation. Nevertheless, the level of concern about sustainability varies from country to country and is related to economic condition. It is vital to assess the situation and create a strategy that might aid Asian nations in developing solid waste management practices.

13.2 Urban Planning and Sustainable Resources Utilization

13.2.1 Urbanization and Urban Environment

Urbanization gives a perspectives for public health planning and relates to changes in the density, size, and heterogeneity of cities through time. All in all, urbanization means the growth of cities. It mainly depends upon the base size of the population and density of the surrounding areas which directly depends on the "pull" and "push" factors of urban migration (Godfrey and Julien 2005). Where push factors include migration of population due to any natural disaster, civil disturbances, etc., and pull factors include the upward mobility of family for the modern lifestyle. Cities especially urban areas are growing rapidly. In this manner, urbanization of a particular area mainly depends on geographical changes or land use patterns (Dewan et al. 2009). United Nations Environment Program (UNEP) reported that the composition of natural resources by the urban population will be increased about 125% by 2050 from 40 billion tons in 2010 (Zucaro et al. 2022). Although urban area covers only 0.4 to 0.9% of the total global land area but is responsible for more than three-quarters of the global final energy and high rate of CO_2 emissions like living organisms as well as the city's population requires a large amount of food, energy, etc., to live (Guan et al. 2018). Some factors are responsible for rapid urban growth; these are macroeconomic factors (globalization, integration, etc.) and microeconomic factors (increased living standards). Rapid urbanization in several nations is closely correlated with the displacement of communities, which imposed large-scale expropriation and transfer of land (Huang et al. 2018; Lu et al. 2018). The urban environment can be understood by three different concepts: the physical environment, urban resource infrastructure, and the social environment. The value and interactions that have been shared by the makers of social groups come under the social environment (Morello et al. 2000; Ji et al. 2020).

13.2.2 Changes in Land Use Pattern

There are several factors that contribute to rapid urbanization. Densification or urbanization is mostly caused by naturally occurring population expansion and population migration from rural to urban areas (Weber and Puissant 2003; Chakraborty et al. 2015). This growing population is now becoming a burden on the natural resources, so there is an urgent need to balance the requirements of land for future needs. To maintain the environmental qualities in the fringe area of expanding cities, it is important to preserve and prevent agricultural land for future purposes (Farroq and Ahmad 2008). Using the land for urbanization and related activities other than farming is part of the erosion of natural ecosystems by human civilization since ancient times. Visible consequences of land use change after urbanization are the spatial expansion of the built-up area (which means land cover characteristics), along with changes in urban land use patterns (Hassan 2017; Mansour et al. 2020). Due to the overpopulation in urban areas, most of the agricultural lands have now been used for industrialization. To define the specific pattern of urban growth, the term "sprawl" was given earlier. Since then, this term urban sprawl is used to define the drawbacks and impacts of urban land use change in the developing world. Urban land use change is often labeled "land depletion" (Frenkel 2004; Shetty and Reid 2014), which is a somewhat imprecise concept because the land (quantity) will not be reduced by changing its use. Concepts of city, sprawl, land depletion, and land occupation demonstrate the linkages of urban land use changes that have negative effects. It is estimated that the average annual growth rate of built-up areas in the developing countries is about 3.6%, while the average in the industrialized countries is only 2.9% (Haase et al. 2014). Poor land management leads to inadequate services, infrastructure, and corresponding transport facilities in urban areas, and the cost of addressing these problems in future can be very high (Moore et al. 2003). Monitoring urban development for sustainable urban development is required to prevent urban sprawl and the subsequent unsustainable development of any city in future (Newman et al. 2022). Where urban planning for effective use of resources and allocation of infrastructure initiatives is a key issue that contributes to the field of effective land use planning in cities (Button 2002; Liu et al. 2014).

Land use and land cover change (LUCC) program aimed to inform diverse initiatives to forecast upcoming environmental changes on a range of scales, from local to worldwide (Turner et al. 2021). Remote sensing techniques have shown their importance in mapping urban land use/land cover, urban growth trends, and monitoring land use/land cover changes (Pathan et al. 1993). To assess urban development trends, monitoring land use change is essential for planning studies at the local and regional levels. These studies should be based on accurate and up-to-date planning of land use information. Therefore, urban planners need a mechanism to effectively detect, monitor, and analyze changes in urban land use patterns. Change detection is the process where a different state of geographic features is identified by observing them at different times (Singh 1989). With the advancement of digital processing, analysis, and availability of multispectral imagery in digital form, remote sensing has emerged as a new tool for land use change detection. Numerous automatic change detection techniques have been developed, i.e., image rationing (Howarth and Boasson 1983), including image differencing (Quarmby and Cushnie 1989), post-classification comparison (Howarth and Wickware 1981), principal components analysis (Byrne et al. 1980), change vector analysis (Almutairi and Warner 2010), GIS-assisted change detection (Turker and Derenyi 2000; Turker and Asik 2002), and direct multi-date classification (Douml 2011). Only insufficient insight into the dynamics of urban land use changes (even about population growth) because there are different types of urban land use change with fairly diverse impacts (Chen et al. 2001). Hence, the only quantity of the land converted into urban uses directly affects the land use and land cover features. Knowing these aspects is not only a comprehensive understanding of land use change dynamics and their knock-on effects on environmental quality, but it can also be used as a basis for urban planning and management. Digital land use and land cover data with significantly better resolution and quality for more advanced monitoring land use, development, and siting patterns, as well as the dynamics of land use change with variances in urban design, etc. (Schneider and Woodcock 2008).

13.2.3 Resource Availability and Utilization

Generally, urban areas consume between 70 and 75% of the world's natural resources (Li et al. 2021) and expected population growth, and subsequent resource utilization within the urban system is difficult to meet the sustainability approaches. Of course, sustainable developmental goals cannot be to maintain already deteriorated urban areas, however, it could be implemented for urban to rural settlements with appropriate integration of sustainable natural resources utilization (Simon et al. 2012). Demand for goods and services is increasing in metro regions due to population growth and explosion. Ignore the diversity that needs so much attention, i.e., attractive urban systems, and detrimental environmental changes related to its continued growth with limited resources of earth planets can cause intractable problems like overuse of fossil fuels, mining effects, urban sprawl, overexploitation of resources, and generation of solid waste in near future (van Leeuwen et al. 2012). Availability of resources is limited by its limited usability, environmental, and social concerns due to their exploration, processing, storage, and transportation. Urbanized population may subtract more resources for well-being in developing cities (Lederer et al. 2014; Fletcher et al. 2021). For sustainable resource utilization, we need to develop more adequate strategies to get more goods and services with minimum resource use. It implies development of sustainable cities with conserving resources in a strategic management to satisfy the needs of present as well as for future generation. Cities develop like self-organizing organisms and strive to improve their efficiency by attracting resources for more functions. Lotka's maximum energy principle states that, regardless of efficiency, when resources are abundant, maximum energy is attained through competition utilizing the resources, however, in resource scarce condition, efficiency increases through collaboration with different components of the urban environment (Sirmon and Hitt 2003).

13.2.4 Encroachment on Land and Overexploitation of Water Resources

Urbanization is the process where due to the activity of human beings, green space is converted into black space. In India due to the uncontrolled overpopulation in the urban areas, natural resources are continuously exploited inadvertently (Kalnay and Cai 2003). Urban encroachment which is a result of urban sprawl is responsible for some uncontrolled activity such as changes of land use patterns, loss of open space and encroachment on surface water resources, and disturbance of ecologically sensitive habitats. Irresponsible encroachment of land mainly reflects the unplanned human activity which destroys green space and leads to environmental pollution. The process of conversion of largely agricultural land into an urban area due to urbanization or rapid urban sprawling is known as urban encroachment (Morello et al. 2000). Land degradation can be defined as where soil loses its functions, which poses a threat to the terrestrial ecosystem as well as to agricultural production. This includes the changes in the topography of soil structure, physical changes in soil structure, and chemical changes like acidification, deposition of heavy metals, and salinization, which directly affect the fertility rate of soil and production of land (Asif 2014). Using poor quality water for irrigation purposes leads to main problems like soil salinization and acidification due to the presence of metals and other contaminants. In addition to this due to the use of fertilizers especially nitrogen and phosphorus fertilizers, manures of livestock not only degraded the quality of soil but also pollute the surface water by leaching excess chemicals like nitrogen as nitrate (Balkhair and Ashraf 2016). Other chemicals like pesticides and herbicides also cause the use of heavy metals deposition on land that affects the soil quality but also affects the water bodies like rivers, groundwater, etc.

Industrial waste caused due to industrialization also contributes to the chemical degradation of valuable land resources. Dumping of urban wastes and coal ash (obtained from thermal power plants) into the agricultural lands digging landfills for construction purposes has detrimental effects to the environment and the agricultural as well as residential areas. The quality and quantity of water resources are continuously affected by anthropogenic factors including industrialization and urbanization process which also includes farming through unscientific disposal of solid and liquid wastes. A huge amount of organic and synthetic effluents is leached into the underground water. Due to the urban sprawling, natural water resources are continuously contaminated (Balkhair and Ashraf 2016). The problem was significantly worsened by the overuse of ground water caused by population growth. Surface runoff following each rainfall event is affected by anthropogenic activities and, on the other hand, has an impact on groundwater recharge.

Wetlands which are known as the "kidney of the earth" are the ecosystem that acts as an intermediate zone between the land and vegetation characteristics (Brinson and Malvárez 2002). Wetlands are highly productive areas with more variety of flora and fauna; hence, they are very essential components of the ecosystem needed for conserving the biodiversity. Wetlands are adjoining areas of terrestrial and aquatic ecosystems with water throughout the year; however, land degradation and poor waste management practices directly reduce the number of wetlands due to their encroachment. These wetlands commonly occur near the low-lying areas that receive freshwater near the edges of ponds, lakes, streams, etc. (Cao et al. 2021). Due to this urban encroachment, ecological functions are continuously deteriorating the wetlands, for example, frequent flood disasters and loss of soil and water equality (Jones 2002; Herath 2004). Apart from this, mineral exploration and using raw materials for construction purposes and industrialization are the common activities which is also responsible for severe losses of natural forest areas. The heavy pressure of this urbanization on natural biodiversity is a common cause of changes in land use patterns which is a great concern to save the ecosystem.

13.2.5 Impact on Water and Ambient Air Quality

The unexpected growth of population degrades the quality of water and causes an increase in nutrients and microbial loads (Maillard and Santos 2008; Ghosh et al. 2014). Therefore, these urban areas adversely affect the environment (Grimm et al. 2008; Hanjra and Qureshi 2010; Breuste and Quereshi 2011). Land use change due to urbanization has changed the flow patterns of wastewater which directly affects the aquatic life as it affects the river water quality (Astaraie-Imani et al. 2012). Urbanization reflects demographical changes, resources overuse, and waste emission and disposal. Cities are the main and common root cause of the different environmental problems that are related to air and water quality. Today, the main problem of this urban sprawl is an inadequate supply of safe drinking water which is being contaminated by disposal of improper discharge of sewage and industrial effluents into water resources. In recent years, the disposal of wastewater into water bodies including rivers, pond, and lake has been increased (Grimm et al. 2008).

Due to overcrowding in many parts of India, drinking water supply is now reached in scarce condition. It is predicted that drinking water will be scarce resource due to more urbanization and change in the climatic condition over the next century (Fraga et al. 2012). Household effluents and industries continuously deteriorating the quality of water due to organic and inorganic nutrients inputs which are continuously increasing in the water resources. The presence of these nutrients in water also increases the growth rate of bacteria in river water. Due to the overexploiting activities of humans, a large area of a water body is continuously disturbing that also affects the aquatic biodiversity (Zhang et al. 2007). Human activities come under the eutrophication process or encroachment on a water body. At present many water bodies are getting encroached by humans through urbanization and solid waste disposal practices leading to water crisis, monsoon flooding, waterlogging, squeezing and pollution of rivers, lake and water borne diseases (Davis and Glick 1978). Over the last few years, due to urban sprawling, air pollution has also become a serious problem that has been continuously faced by most of the countries across the globe because of poor maintenance of vehicles, and improper installations of industries near residential areas, most developing countries are suffering from serious outdoor air pollution (Ravindra et al. 2003).

Air pollution has increased globally because of the industrial revolution and rising vehicles demand with more modernization. These activities cannot be regulated because they are directly related to the overall development of human beings and improving life standard (Nandan et al. 2017). The common air pollutants which come directly from industrialization are mainly sulfur dioxide (SO_2), nitrogen dioxide (NO_2), and suspended particulate matter (Reddy et al. 2005). Particulate matter with a size less than 10 μm is known as a respirable suspended particulate matter which penetrates the respiratory system. Respirable suspended particulate matter (RSPM) can be further divided into ultra-thin particles with less than 2.5 μm (Fenger 1999; Goswami and Baruah 2008). RSPM particles are found in the air in the form of dust, smoke, and liquid droplets and exists in the air for a long time in suspended conditions (Goswami and Baruah 2008). Most of the areas of Northern India face serious health risks due to the presence of particulate matter (10 μm) in the air.

Most common anthropogenic sources of air pollution are domestic fuel burning, industrial emissions, transportation activities, and thermal power plants. In India, especially Delhi, about 67% of total air pollution is caused due to vehicle activity which is 300 metric tons per day. Almost more than 50% of the cities have moderate to poor air quality, increased amount of fog reduces the temperature range, especially during the winter months (Pasricha et al. 2003). Tall buildings use air conditioners (ACs) and hindered dispersal of this pollution into the air which also reduces the quality of air (Wang et al. 2018). Motor vehicles also contribute a major role to air pollution by emitting greenhouse gases. Almost 90% of the total air pollution emission emitted due to transportation. Diesel particulates also play an important role and are known to be carcinogenic. Petrol driven vehicles contribute more pollutant load as compared to diesel driven vehicles. Emission from the vehicles enters the air as air pollutant in the form of particles and gases. Common air pollutants including SO_2, NO_2, and suspended particulate matter are found in the range of over 600 mg/m^3 (Misra et al. 1993). In Delhi, amount of SOx is now decreasing, but in Mumbai, the amount of sulfur content in the air is continuously increasing. This is due to the excessive use of vehicles. In megacities, amount of GHG emissions is also increasing where common sources are transportation, burning of fossil fuels, burning of agriculture waste, etc. Greenhouse gas footprints of Delhi, Kolkata, Chennai, and Mumbai are 38,633.20 Gg, 14,812.10 Gg, 22,090.55 Gg, and 22,783.08 Gg, respectively. A high concentration of metals in the air also affects the vegetation through their deposition on surface of crop plants (Singh and Kumar 2006).

13.3 Waste Generation and Management

Waste can take on many different forms, and there are numerous ways to characterize waste. Among the features frequently used in the classification of waste are their physical properties, biodegradability, source of origin, and environmental impact (Dixon and Jones 2005; Demirbas 2011). Some examples of wastes generated in the urban systems are given below:

13.3.1 Wastewater

Wastewater volumes have been rising in many areas due to urbanization, increasing population growth, improving living standards, and economic growth. The quantity of wastewater is growing in proportion to the expanding urbanization and residential water supply. According to Central Public Health and Environmental Engineering Organisation (CPHEEO), Govt. of India estimates, about 70–80% of the entire amount of water supplied for domestic use, return in the environment as sewage discharge. Most of the wastewater comes from more populated states like Uttar Pradesh, Maharashtra, West Bengal, Delhi, and Gujarat (CPCB 2007). Sewage and industrial wastewater is characterized by their physico-chemical and microbiological properties (Ritter 2002; Kesari et al. 2021). Therefore, domestic and industrial sources of wastewater need safe disposal, and after proper treatment, however, most of the wastewater generated is being discharged with any treatment into the aquatic bodies which lead to water contamination (Jassby et al. 2018). Wastewater may contain suspended particles, dissolved organic waste, and other toxins. A soluble fraction of about 63% of phosphate has been reported in wastewater (Warwick et al. 2013). Wastewater treatment and its reuse are one of the most important requirements of the present situation, despite the rapidly depleting freshwater supply and rising freshwater demand. Around 70% of the freshwater, which is derived from rivers and groundwater sources, is used for agriculture (Kesari et al. 2021).

13.3.2 Emission of Air Pollutants and Greenhouse Gases

It is essential to have a complete understanding of sources and pattern of greenhouse gas (GHG) emission and air pollutants before putting into place the proper climate and air quality mitigation strategies. In recent years, more emissions inventories have been developed for large regions including India (Streets et al. 2003; Zhang et al. 2011). Emission inventories are being incorporated for air quality and climate model simulations to better understand global pattern of air pollution and climate change. The largest amount of greenhouse gases released by human activity specially by burning of fossil fuel. The percentages of CH_4, N_2O, and other trace gases in greenhouse gas emissions are roughly 10.1, 7.0, and 2.8%, respectively (Lyu et al. 2021).

13.3.3 Vehicular Emission

Air pollution is a serious environmental issue in urban areas, as a majority of population is exposed to poor air quality (Alam and Khan 2020). Transportation is a significant contributor to air pollution in many nations across the globe due to the numerous automobiles that are already on the road. In urban areas, particularly in large cities, vehicle-related air pollution has turned into a significant issue. Vehicle pollution is beginning to appear through symptoms including cough, headaches, nausea, eye irritation, and numerous bronchial and visibility issues (Manisalidis et al. 2020). Major concern is the detrimental effect of air pollution on the environment and public health, and as a result, regulations on vehicle emissions are getting stricter globally. By implementing these regulations, petrol and diesel engines should emit fewer hydrocarbons (HCs), CO, NO_2, or NOx, and particulate matter (PM) (Winkler et al. 2018). Pollution caused by motor vehicles can be reduced by utilizing cutting-edge techniques, alternative energies, and government policies. These technologies needed to be applied properly if they were to considerably improve the condition of the environment. The current analysis is in favor of the environmental sustainability initiatives made by the automotive industry.

13.4 Sustainable Approaches for Green Cities

Sustainable growth is very important for the growth of green towns which is directly related to eco city. A green society should be able to meet present needs without compromising the ability of future generations to meet their own needs (Sanchez Rodriguez et al. 2018). A stable society means stable and sustainable life to increase efficiency of energy use and mitigate the impacts of developmental activities. Without proper design of cities, another cost is applied to maintain the design of buildings such as energy inefficiencies, emissions, and construction waste (Barrico and Castro 2016). To reduce the energy consumption, some efficient solar techniques and innovative building architecture are used. Compatible consumption increases the use of daylight to decrease the electricity demand. In renewable electricity sources like solar power, photovoltaic cells can be used in low emission technologies like electric motors (Hossain et al. 2018). Green cities are those that work to minimize their negative effects on the environment by reducing waste, lowering emissions, improving recycling, increasing housing density while preserving open space, and supporting the growth of locally owned, sustainable companies (Ghorab et al. 2016).

13.4.1 Green Building and Cities

Green city planning approach provides an opportunity to sweep away the existing urban planning and development. Today to reduce the ecological and environmental impacts, it is challenging to create and enable urban planning strategies. The main principle of green city planning includes energy utilization, management of the natural water ecosystem, management of solid waste, sewage waste management, and rainwater harvesting (Franco et al. 2021). Sustainable urban planning has been proposed to using low carbon energy resources, regulations and policies, natural ecosystem, and develop the sustainable projects including low interest loan and tax rebates. An integrated plan is also proposed that has the potential to use affected land use areas with the coordination of other urban activities and by mitigating negative impacts (Gennadievna-Pahomova et al. 2017). Sustainability means "lasting" that means the use of natural resources in such a way that can be used in future without any environmental degradation, and to restore a harmonious balance between nature and humans (Choi 2010; Franco et al. 2021). Green city planning is mainly to create a balance between the natural and built environment to ensure the long-term survival of both. Natural ecosystem, newer technologies, plans, science, and engineering shall be used for lasting development needs for green cities. This principle of green city planning can be achieved by proper and systematic land use planning. The planning of green cities offers opportunities to bring sustainability between economic and social activities. Land use planning gives the best opportunity to combat climate change by giving zero-carbon development. For planning a green city, it is also required to understand natural elements like topography, microclimate and river catchments, etc. These green cities have improved the conditions of life of humans for better life-like community and sociability, health, education, park, and open spaces which are also taken care of through green building technologies.

13.4.2 Safe Water Supply and Sanitation

Availability of water resources is now dropped from past years, i.e., from 13,000 to 6000 cubic meters per year. Most of the areas even still do not get pure drinking water, and two billion people are still there who are lacking access to proper sanitation (Pokhrel and Viraraghavan 2004). From twentieth century, as the population has increased, the water thirst has also increased. At present, safe drinking water has become a top priority at all levels. The government is taking many initiatives like investments in pipelines and pumps but this is not good enough to fulfill the water supply. The problem of safe and clean water supply in developing countries is now a major issue. India has a very limited daily water supply in comparison with the industrialized world. In India, depending on the region, it ranges from 16 to 300 L per day (Bartram et al. 2014), whereas for developed countries, it is about 100–600 L per day. Population which is served water supply without any pipeline receiving a low amount of water. In many countries, the water stress is mainly due to the poor distribution efficiency and inequalities in service provision. In many cases, water loss occurs due to the inappropriate management of water. Supply and reduction in water losses can be done through proper planning over administrative and financial issues as well as a technical and operational issue. In most developing countries, water supply is intermittent in place of continuous supply. This intermittent supply leads to inequality in water supply and loss of water pressure (Mahgoub et al. 2010). This intermittent supply also leads to high level of water contamination. Sanitation is also worst as compared to water supply, as almost 2.6 billion people are living without proper sanitation. Due to urbanization, as the housing demands and infrastructure services increase, sanitation services also increase; therefore, it is a must to focus on some master plan.

13.4.3 Renewable Energy Use (E-Vehicle)

Renewable energy is a form of energy that is derived from natural resources and cannot be depleted over time. More use of renewable energy resources decreases the CO_2 emission in the environment and produce clean air quality for the population. Renewable energy produces around 18% of total global energy consumption (Demirbaş 2006) including solar, wind energy, biofuels, geothermal energy, etc. Renewable energy sources (RESs) use indigenous resources to provide energy sources with less amount of greenhouse gases and air pollutants (Singhal 2007). These renewable energy resources converted natural energy into useful energy. India's population of more than a 1028 million is growing continuously, and using fossil fuels as energy resources is becoming scary. Use of fossil fuels also causes different global environmental issues (Golait et al. 2009). As the demand increases, this is satisfied by the use of coal, oil, and petroleum as non-renewable energy resources, which is not a permanent solution and also gives many environmental issues; therefore for sustainable development all over the world, we must switch from non-renewable energy resources to renewable energy resources. For the above reasons, now, the use of RES and technologies is playing an important role in the sustainable development of India. For the economic development in every sector of the Indian economy, some new renewable energy resources had been found. The government of India has already made some plans and laws for the effective utilization of RES. In urban areas due to over urban sprawling, air pollution through vehicles is also contributing a major role to polluting the environment. By using electric vehicle (E-vehicle), these environmental issues can be controlled which plays a significant role in the reduction of CO_2 especially in the transport sector as it works on "zero emission". In E-vehicle, hydrogen is an important future energy source, which is used to convert renewable energy sources for transport sector (Emadi et al. 2008; Hawkins et al. 2012). In the electricity market, E-vehicle constitutes a new category of consumers as it directly plays an important role in CO_2 reduction.

13.4.4 Zonation (Residential, Industrial, and Eco-Sensitive Zone)

Zoning is a tool to regulate the sustainable environment and for land use planning. The main purpose of zoning is to regulate and control land by ensuring its complementary uses. It also can stimulate or slow down the development to conserve that particular area. Zoning is a series of laws and regulations which allows land use for

any use of a particular piece of property (Kuriakose and Kylasam 2016). By residential zoning specific blocks provided to the individuals with specifications of lot size and floor area. Zoning for the residential area is a framework to develop the residential area in such a manner that can be used in future with sustainability. In which specific development is allowed with some significant planning, especially in urban areas (Gupta 2006; Dai et al. 2012). This residential zoning toolkit includes rules and specific development with the proper design and quality of public spaces. The zoning baseline allows for site planning and development rights. Industrial zoning provides land for commercial and industrial activities separately in the urban areas which can be subdivided into light industrial zone and heavy industrial zone (Dai et al. 2012). Light industrial zoning means it includes the manufacturing process without using heavy machinery; however, heavy industrial zones include industries that use high intensity and energy loss, for example, plants for oil and natural gas. So basically, zoning provides the rules and regulations for particular industries for their development (Kuriakose and Kylasam 2016).

Differentiating eco-sensitive zones serves as a "shock absorber" for the protected areas around national parks and sanctuaries. It differentiate transitional space between the locations with more restrictions on human activities and those with less restrictions (Deb et al. 2014). The National Board for Wildlife has determined that, unless specially required differently, activities in eco-sensitive zones are compulsorily prohibited. A sensitive corridor, connectivity, and biologically significant patches that are necessary for landscape linkage should be included in eco-sensitive zones when they extend over 10 km width. The primary objective of establishing any eco-sensitive zone is to regulate certain activities in and around national parks and wildlife sanctuaries in order to reduce the negative impacts to conserve the protected areas with their biodiversity.

13.4.5 Recycling and Recovery of Resources

Due to urbanization and industrialization, now, waste management is a significant concern for all of us. Management of this waste also gives significant impacts on the environment based on the methods that are adopted to treat them like incineration, compositing, landfilling, etc. Improper waste management not only contaminates water and soil but also affects the whole ecosystem and gives many health risks (Banerjee et al. 2013). The other problem of this waste management is to search and identify the new landfill location and incinerate which also affects the local environment. So at present, a few major "green" waste management options have been identified that are environmentally friendly and energy savings:

Recycling—Due to urban sprawling as the population is growing in urban areas, the high amount of waste generation is also creating a problem for the environment. So, recycling is a process where waste from raw materials and destroyed items is reduced and converted into new products to prevent air and water pollution (Nakatani and Hirao 2011). Recycling is utilized to reduce the potential risks to human health and the environment from the potential depletion of natural resources. Pre-consumer materials where waste is generated during the manufacturing processes are again regenerated. Post-consumer materials are generated by households, commercial, and through industrialization (DeVoy et al. 2021). **Integrated Solid Waste Management (ISWM)**—ISWM system wherein developing countries have some efficient techniques and models is developed to recycle waste of urban environment. The cost of recycling depends on the cost of collection and sorting. It is also estimated that waste that is generated is about 51% and can be recycled by using a cost-efficient strategy. The ever-growing population in an urban area also increases the pressure on natural resources (either minerals or metals); therefore, the demand for using these resources is continuously increasing. And recycling is important for the sustainability of these natural resources. Preventive strategies and upgraded recycling systems give benefits over the exploitation of natural resources (Kellenberg 2012). By applying green policies, a large amount of waste can be again reused as secondary raw material. In some cases, the wastes can be again reused like coal ash from the cement industry can be reused again. Domestic wastes can be treated or recycled in three ways, i.e., by recycling, the exportation of these wastes to the other countries, and disposal by this domestic waste can be reused again as the domestic economy which can be again reprocessed and consumed.

Recovery—It can be defined as "*any product or part of components that are not waste can be used again for same purposes*". In this process, three steps are followed, i.e., checking, dealing, and repairing for recovery operation by which the components that have become waste can be reused again. Reuse is better than recycling where waste products can be again recovered without any change in the physical properties of the material. It is an energy effective process as compared to recycling (Yokoo 2010). The waste and resources action program has developed a specific methodology for quantifying the impacts of reusing products. Recovery of any waste material gives potential environmental and socioeconomic advantages. They reduce the amount of CO_2 emission, generate wastes, and reuse and recover electric items, and there is an increasing proportion of energy efficient appliances (Kopsidas and Giakoumatos 2021). Recovery and reuse of any product require proper checking and transport to the particular area. The reusable product must be properly and well designed. Thus, by recovery of any product, it is possible to produce a new product at a very low cost. Renovation of any building gives a high amount of waste, and this waste (mixture of concrete, bricks, excavated soil, etc.) can be again reused for new construction projects like highways (Yokoo 2010). Reusing graywater from the city can be reused for agricultural fields, thus reducing the overall demand for freshwater.

13.4.6 Restoration of the Ecosystem

Ecosystem restoration is processed to reverse the degradation of an ecosystem like rivers, lakes, and oceans. In other words, it is a process to improve the capacity and production rate of the ecosystem for the future needs of society. This can be done by regeneration of natural ecosystem that is overexploited or by plantation (De Groot et al. 2013). The main objective of ecosystem restoration is to conserve biodiversity for its sustainable use by contributing to the improvement of food and water security. An ecosystem is a socio-ecological area that is used for multiple purposes by which economic issues can be managed (Strassburg et al. 2020). Ecosystem restoration is a compliment to conservation activities and provides many benefits with its sustainable use. It should be well planned at a broader scale and implemented by using the best available knowledge. In this context, there should be effective participation of local and indigenous people and stakeholders at all stages of the process. Public awareness, communication, and education are also important for complete ecological restoration. For ecological restoration, an action plan is made with four groups of activities, i.e., assessment of opportunities for ecosystem restoration. In this process, this is done to ensure the assessment of the proper area where ecosystem restoration is needed by implementing mapping, broad-scale ecosystem assessment, this can be done with various considerations (Aronson et al. 2020). These actions are assess the location, type, extent, and degree of the degraded ecosystem at regional, national, and local levels so that restoration initiatives can be taken. To identify the priority geographical areas, restoration strategies would be applied. Involve the local population, indigenous people, and stakeholders with full participation in the identification of areas where restoration is needed. Assess the multiple benefits of ecosystem restoration, by restoring the ecosystem like food and water security, health benefits, least health risks, etc., by identifying these benefits can be maximized by giving opportunities. Identifying options and reasons which is continuously deteriorating the ecosystem can be done with the help of local and indigenous people (De Groot et al. 2013). By improving the ecological environment, by providing social-economic incentives and appropriate mechanisms, ecosystem degradation can be managed.

13.5 Advances in Technologies for Urban Management

13.5.1 Sources Control Technology for Air Pollutant Management

13.5.1.1 Removing Pollutants from the Air

Recently, China has built an air filtration system that uses GHGs to pump air through a 100-m-tall chimney-shaped tower in Xi'an city. Separate studies have been done on chemicals that react with NO_2 to absorb it from the air in the Japan, England,

and Netherlands among other countries (Quarmby et al. 2019). Water cannons had recently been tried in Delhi, but they had little discernible impact on the air quality (Rizwan et al. 2013). More evidence is there that planting vegetation to urban surfaces, such as roofs and walls, can have a limited and localized improvement in quality of air (Currie and Bass 2008; Jayasooriya et al. 2017).

13.5.1.2 Switching to Less Polluting Cars

There are numerous technologies available to reduce emissions from conventional vehicles. Catalytic convertors devices reduce the danger of tailpipe emissions in the UK and the most of other nations across the world. The oil corporation cell has created a synthetic "drop in" replacement for diesel that does not require any engine modifications. This may have a good impact on particulate matter and NOx emissions, although it has yet to be rigorously assessed (Melony and Palma 2020). Instead, San Paulo has emphasized "flex" vehicles, which can run on a variety of fuels, usually a mixture of ethanol and petrol (Slovic and Ribiero 2018). Flex vehicles are usually combine with electric and internal combustion engines. Electric cars are the most prominent cleaner cars, and their share of the global vehicle fleet is increasing (Senecal and Felix 2019). The adoption of electric vehicles crosses the boundaries between strategies and technologies because the public's changing purchasing habits are the only way this new technology can improve air quality.

13.5.1.3 Roadside Barriers

Similar to congestion-charging zones, barriers at the side of the road that reduce noise also have an impact on the air quality. Artificial barriers are thought to be a potential air quality measure since they often lead in less pollution behind the barrier (Brantley et al. 2014; Tong et al. 2016). The most effective air quality barriers are those that include vegetation and man-made materials, while barriers built entirely of vegetation with dense foliage are also seen to hold promise (Abhijith et al. 2017).

13.5.1.4 Trees and Vegetation

In urban environments, trees and vegetation can lead to reduction in air quality when placed beside main road, tree canopies serve as a buffer zone and protect pollutants from their dissipation. The most effective tree species for enhancing air quality are conifers, which have been found to absorb more pollutants than broad-leaved trees (Yang et al. 2015). Several cities throughout the world are exploring with urban greening, including Singapore, Sydney, and Mexico City. Although additional study is needed to properly understand the effects of urban greening (Zupancic et al. 2016), data shows that urban design is important to their effectiveness. For instance, "urban canyons" are streets with tall buildings on either side. In these areas, green walls might be most beneficial.

13.5.2 Sewage Treatment Plants (STPs) and Effluent Treatment Plants (ETPs) for Wastewater Treatment

One of the biggest challenges to the sustainability of human civilization is the unregulated discharge of municipal and industrial wastes into the environment, which contaminates the life supporting system of the environment (Jhansi and Misra 2013). To ensure health and environmental safety, lower intolerable levels of public toward the environmental changes in the face of growing populations and economies, wastewater treatment, and safe disposal are crucial. About 76% of wastewater production is dumped directly into waterways without being treated. Even a sizable portion of the industrial sector lacks sufficient wastewater treatment facilities (ETPs). The high capital expenses associated with ETPs are the main cause of their dearth. Additionally, small-scale enterprises cannot afford to build their own recycling facilities (Yadav et al. 2019). Industrial units are choosing to use common effluent treatment plants due to the significance of wastewater recycling; consequently, all large-scale industries must be required to establish ETPs (Shah and Ruparelia 2022). The Central Pollution Control Board (CPCB) is in favor of using decentralized sewage treatment facilities due to high cost of creating a sewage network. For instance, all new housing estates in Chennai are required to establish their own STPs. Freshwater needs to be given at a reduced cost and should cost more than sewage-treated water under appropriate tariff structures. To encourage the use of recycled water, all these restrictions must be backed with education campaigns (Chopra and Das 2019). In a recent study, 67% of survey participants in Bengaluru city said they would be happy to purchase recycled water (Ravishankar et al. 2018). Wastewater treatment has a lot of promise, and customers and governments may be able to co-sponsor the construction of wastewater treatment plants (WWTP). In fact, because they internalize the unfavorable environmental externalities of urbanization, WWTPs are a pre-requisite to the general development of urban metropolises.

13.5.3 Rainwater Harvesting

Water harvesting has a strong tradition in India (Krishan 2011). Since ancient times, our ancestors have collected rainwater in a variety of structures to fulfill their basic water needs. These structures have continued to be lifelines for the local population till this day. However, these traditional techniques must be integrated with cutting-edge technology and scientific equipment. The Indian government should support the development of infrastructure for rainwater collection by offering subsidies. If rainwater collection is used in domesticated regions, there are major benefits. Collected and stored rainwater can be utilized for irrigation as well as for household activities, including washing, heating, and cooling. Central Ground Water Authority (CGWA) mandates that all residential groups, including schools, institutions, hotels, and companies install roof-top rainwater-harvesting systems on their land. Rainwater collection satisfies the need for non-potable purposes like bathing, flushing, cleaning, and watering gardens.

13.5.4 Solid Waste Management

Solid waste management (SWM) is becoming more important due to a variety of factors, including the concentration of people in urban areas, governmental initiatives, the development of newer technologies, and rising public awareness of the need of sanitation and hygiene (Kumar and Agrawal 2020). The municipal area has a wide range of sources for solid waste, including governmental and private institutions, households, and businesses. The SWM system manages also septic tank sludge, incinerator ashes, and sewage treatment plant sludge in some countries (Magutu and Onsongo 2011). The composition of solid waste is also impacted by the economy. In industrialized economies, there is a high proportion of recyclable materials (plastic and paper), whereas in low-GDP countries, there is a significant amount of organic material that degrades quickly. In urban areas, solid waste is collected at permanent stations or through house-to-house collection techniques. When using a fixed station system, which is common in developed countries, residents are required to dispose of their waste at the locations authorized by the local governments by a certain hour on a specific day of the week. Residents of the area are expected to dispose of their waste, when necessary, in the community bin that has been created or designated as a fixed station in developing countries (UNEP 2002). Solid waste is produced for recycling to reduce the volume and potential for pollution at landfill areas.

The cost-effectiveness of a recycling industry is a key factor in determining its sustainability, which is mostly influenced by a society's economic status. There is a systematic procedure in place in developed economies for the collection and processing of different recyclables, including papers, glassware, and metals, with both public and private participation. The thermal processing technique of incinerating the combustible portions of solid waste has long been established. In addition to recovering the heat produced during combustion, the aim is to decrease the amount of waste that needs to be dumped in a final place. The complicated technology needed for extensive burning and air pollution control drives up the cost of the process (Kokalj and Samec 2013). South Korea, Japan, Singapore, and Taiwan have all been constantly enhancing their SWM systems with the main objective of eliminating landfills from their systems (Shekdar 2009). The SWM systems have stabilized in these countries, which had rapid economic growth in the closing decades of the twentieth century, thanks to a number of legal measures and national subsidies. Reliable information is regularly obtained and used for system operations and SWM planning. For technical training, there are well-established facilities. Similar to this, there is a significant literature on many aspects of SWM. Due to the increasing expenses in the service industries, a higher emphasis is placed on the effective deployment of capital equipment as a labor replacement. Technologies are well established for tasks including collection, transportation, processing, and disposal, and a stable financing source is available to run the system.

13.6 Conclusion and Recommendations

The quick pace of population growth, urbanization, and industries have considerably lead to the degradation of water quality due to disruption of the natural hydrological cycle. The quality of water has been impacted due to inadequate management and rising worldwide demand, resulting in serious environmental impacts. Increased population and the creation of new communities are two factors that contribute to increased waste generation. The key components of an effective waste management system are social acceptance, cost-effectiveness, and environmental friendliness. Reducing gaseous emissions, solid leftovers, and pollution is the goal of the "zero-waste production" and "waste prevention" movements, which helps to safeguard the climate and environment. Waste management is a first step toward finding ecologically responsible solutions by using green technology. Recycling and composting are simple methods that are effective in reducing the amount of garbage produced and creating useful goods with multiple uses. The concept of trash valorization is appealing and growing in acceptance as waste residue generation rises quickly. The management of the environment in this area is just as crucial as other public services or infrastructure, without which modern man's existence would be very challenging. This is due to research demonstrating a direct connection between diseases including lung cancer, heart disease, cholera, and hepatitis and air, water, and land pollution. In addition, air and water pollution directly contributes to climate change and eutrophication. Smart technology can play a significant part in solving today's major problems and preparing the ground for a sustainable future. A clever strategy offers the chance to integrate knowledge, which is essential to addressing the major issues facing modern civilizations.

Acknowledgements The authors are thankful to Manipal University Jaipur, India for providing facilities and continuous encouragements.

References

Abhijith K, Kumar P, Gallagher J, McNabola A, Baldauf R, Pilla F, Broderick B, Di Sabatino S, Pulvirenti B (2017) Air pollution abatement performances of green infrastructure in open road and built-up street canyon environments—a review. Atmos Environ 162:71–86

Alam MS, Khan A (2020) The impact study of vehicular pollution on environment. IJSART 6:30–37

Almutairi A, Warner TA (2010) Change detection accuracy and image properties: a study using simulated data. Rem Sens 2:1508–1529

Aronson J, Goodwin N, Orlando L, Eisenberg C, Cross AT (2020) A world of possibilities: six restoration strategies to support the United Nation's Decade on Ecosystem Restoration. Restor Ecol 28(4):730–736. https://doi.org/10.1111/rec.v28.410.1111/rec.13170

Arrow K, Dasgupta P, Goulder L, Daily G, Ehrlich P, Heal G, Levin S, Mäler KG, Schneider S, Starrett D, Walker B (2004) Are we consuming too much? J Econ Perspect 18:147–172

Asif K (2014) Encroachment of agricultural land in urban fringe areas of Aligarh city, India–process and parameters. Asian Geogr 31:129–148

Astaraie-Imani M, Kapelan Z, Fu G, Butler D (2012) Assessing the combined effects of urbanisation and climate change on the river water quality in an integrated urban wastewater system in the UK. J Environ Manage 112:1–9

Balkhair KS, Ashraf MA (2016) Field accumulation risks of heavy metals in soil and vegetable crop irrigated with sewage water in western region of Saudi Arabia. Saudi J Biol Sci 23:S32–S44

Banerjee S, Aditya G, Saha GK (2013) Household disposables as breeding habitats of dengue vectors: linking wastes and public health. Waste Manage 33(1):233–239

Barrico L, Castro P (2016) Urban biodiversity and cities' sustainable development. In: Biodiversity and education for sustainable development, pp 29–42

Bartram J, Brocklehurst C, Fisher MB, Luyendijk R, Hossain R, Wardlaw T, Gordon B (2014) Global monitoring of water supply and sanitation: history, methods and future challenges. Int J Environ Res Public Health 11:8137–8165

Bhadouria R, Upadhyay S, Tripathi S, Singh P (2022) Urban ecology and global climate change. Wiley-Blackwell

Brantley H, Hagler G, Deshmukh P, Baldauf R (2014) Field assessment of the effects of roadside vegetation on near-road black carbon and particulate matter. Sci Total Environ 468–469:120–129

Breuste J, Qureshi S (2011) Urban sustainability, urban ecology and the Society for Urban Ecology (SURE). Urban Ecosyst 14:313–317

Brinson MM, Malvárez AI (2002) Temperate freshwater wetlands: types, status, and threats. Environ Conserv 29:115–133

Button K (2002) City management and urban environmental indicators. Ecol Econ 40:217–233

Byrne GF, Crapper PF, Mayo KK (1980) Monitoring land-cover change by principal component analysis of multitemporal Landsat data. Remote Sens Environ 10:175–184

Cao Y, Kong L, Zhang L, Ouyang Z (2021) The balance between economic development and ecosystem service value in the process of land urbanization: a case study of China's land urbanization from 2000 to 2015. Land Use Policy 108:105536

Chakraborty S, Das K, Roy U (2015) Concentrated or dispersed urbanization: a critical analysis of newly emerged census towns of West Bengal India in 2011. J Geogr Reg Plan 8:218–227

Chen D, Stow D, Daeschner S, Tucker L (2001) Detecting and enumerating new building structures utilizing very-high resolution imaged data and image processing. Geocarto Int 16:71–84

Choi E (2010) Green on buildings: the effects of municipal policy on green building designations in America's central cities. J Sustain Real Estate 2:1–21

Chopra V, Das S (2019) Estimating willingness to pay for wastewater treatment in New Delhi: contingent valuation approach. Ecol Econ Soc INSEE J 2:75–108

Clark J, Pratt N, Muirhead O (2019) How can sustainable drainage (SuDS) improve our environment? Westminster sustainable business forum (policy connect), discussion summary

CPCB (2007) Evaluation of operation and maintenance of sewage treatment plants in India 2007, control of urban pollution series: CUPS/68/2007. Central Pollution Control Board, India

Currie B, Bass B (2008) Estimates of air pollution mitigation with green plants and green roofs using the UFORE model. Urban Ecosyst 11:409–422

Dai X, Li Z, Lin S, Xu W (2012) Assessment and zoning of eco-environmental sensitivity for a typical developing province in China. Stoch Env Res Risk Assess 26:1095–1107

Davis AM, Glick TF (1978) Urban ecosystems and island biogeography. Environ Conserv 5:299–304

De Groot RS, Blignaut J, Van Der Ploeg S, Aronson J, Elmqvist T, Farley J (2013) Benefits of investing in ecosystem restoration. Conserv Biol 27:1286–1293

Deb S, Ahmed A, Datta D (2014) An alternative approach for delineating eco-sensitive zones around a wildlife sanctuary applying geospatial techniques. Environ Monit Assess 186:2641–2651

Decker EH, Elliott S, Smith FA, Blake DR, Rowland FS (2000) Energy and material flow through the urban ecosystem. Annu Rev Energ Env 25:685–740

Demirbaş A (2006) Global renewable energy resources. Energy Sources 28:779–792

Demirbas A (2011) Waste management, waste resource facilities and waste conversion processes. Energy Convers Manag 52:1280–1287

DeVoy JE, Congiusta E, Lundberg DJ, Findeisen S, Bhattacharya S (2021) Post-Consumer textile waste and disposal: differences by socioeconomic, demographic, and retail factors. Waste Manage 136:303–309

Dewan AM, Yamaguchi Y (2009) Land use and land cover change in Greater Dhaka, Bangladesh: using remote sensing to promote sustainable urbanization. Appl Geogr 29:390–401

Dixon N, Jones DRV (2005) Engineering properties of municipal solid waste. Geotext Geomembr 23:205–233

Douml F (2011) Using Landsat data to determine land use/land cover changes in Gmhane, Turkey. Sci Res Essays 6:1249–1255

Edenhofer O, Pichs-Madruga R, Sokona Y, Minx JC, Farahani E, Kadner S (2014) IPCC, 2014: climate change 2014: mitigation of climate change. Contribution of working group III to the fifth assessment report of the intergovernmental panel on climate change. Cambridge University Press, Cambridge

El G, Hosam K, Shalaby Heidi A (2016) Eco and Green cities as new approaches for planning and developing cities in Egypt. Alex Eng J 55:495–503

Emadi A, Lee YJ, Rajashekara K (2008) Power electronics and motor drives in electric, hybrid electric, and plug-in hybrid electric vehicles. IEEE Trans Industr Electron 55:2237–2245

Erell EP, Pearlmutter D, Williamson T (2015) urban microclimate: designing the spaces between buildings. Taylor & Francis, New York. ISBN: 978-1-84407-467-469

Farooq S, Ahmad S (2008) Urban sprawl development around Aligarh city: a study aided by satellite remote sensing and GIS. J Indian Soc Remote Sens 36:77–88

Fenger J (1999) Urban air quality. Atmos Environ 33(29):4877–4900

Fletcher DH, Likongwe PJ, Chiotha SS, Nduwayezu G, Mallick D, Md NU, Jones L (2021) Using demand mapping to assess the benefits of urban green and blue space in cities from four continents. Sci Total Environ 785:147238

Ford A, Barr S, Dawson R, Virgo J, Batty M, Hall J (2019) A multi-scale urban integrated assessment framework for climate change studies: a flooding application. Comput Environ Urban Syst 75:229–243

Fraga S, Rodríguez F, Bravo I, Zapata M, Marañón E (2012) Review of the main ecological features affecting benthic dinoflagellate blooms. Cryptogamie Algologie 33:171–179

Franco MAJQ, Pawar P, Wu X (2021) Green building policies in cities: a comparative assessment and analysis. Energy Build 231:110561

Frenkel A (2004) The potential effect of national growth-management policy on urban sprawl and the depletion of open spaces and farmland. Land Use Policy 21:357–369

Gennadievna-Pahomova E, Tomakov IV, Tomakov VM (2017) The formation of the green building market of the Russian cities. J Appl Eng Sci 15:395–400

Ghorab El, Hosam K, Shalaby Heidi A (2016) Eco and Green cities as new approaches for planning and developing cities in Egypt. Alex Eng J 55:495–503

Ghosh AR, Qureshi MS, Kim JI, Zalduendo J (2014) Surges. J Int Econ 92:266–285
Girardet H (2003) Cities, people planet. In: Vertovec S, Posey DA (eds) Globalization, globalism, environment, and environmentalism: consciousness of connections. Oxford University Press, New York, pp 87–102
Godfrey R, Julien M (2005) Urbanisation and health. Clin Med 5:137
Golait N, Moharil RM, Kulkarni PS (2009) Wind electric power in the world and perspectives of its development in India. Renew Sustain Energy Rev 13:233–247
Goswami P, Baruah J (2008) Simulation of daily variation of suspended particulate matter over Delhi: relative roles of vehicular emission, dust, and domestic appliances. Mon Weather Rev 136:3597–3607
Grimm NB, Faeth SH, Golubiewski NE, Redman CL, Wu J, Bai X, Briggs JM (2008) Global change and the ecology of cities. Science 19:756–760
Guan X, Wei H, Lu S, Dai Q, Su H (2018) Assessment on the urbanization strategy in China: achievements, challenges and reflections. Habitat Int 71:97–109
Gupta JP (2006) Land use planning in India. J Hazard Mater 130:300–306
Haase D, Haase A, Rink D (2014) Conceptualizing the nexus between urban shrinkage and ecosystem services. Landsc Urb Plann 132:159–169
Hanjra MA, Qureshi ME (2010) Global water crisis and future food security in an era of climate change. Food Policy 35(5):365–377
Hassan MM (2017) Monitoring land use/land cover change, urban growth dynamics and landscape pattern analysis in five fastest urbanized cities in Bangladesh. Remote Sens Appl Soc Environ 7:69–83
Hawkins TR, Gausen OM, Strømman AH (2012) Environmental impacts of hybrid and electric vehicles—a review. Int J Life Cycle Assess 17:997–1014
Herath G (2004) Incorporating community objectives in improved wetland management: the use of the analytic hierarchy process. J Environ Manage 70:263–273
Hossain K, Maruthi YA, Das NL, Rawat KP, Sarma KSS (2018) Irradiation of wastewater with electron beam is a key to sustainable smart/green cities: a review. Appl Water Sci 8:1–11
Howarth PJ, Boasson E (1983) Landsat digital enhancements for change detection in urban environments. Remote Sens Environ 13:149–160
Howarth PJ, Wickware GM (1981) Procedures for change detection using Landsat digital data. Int J Remote Sens 2:277–291
Huang W, Huang Y, Lin S, Chen Z, Gao B, Cui S (2018) Changing urban cement metabolism under rapid urbanization–A flow and stock perspective. J Clean Prod 173:197–206
Jassby D, Cath TY, Buisson H (2018) The role of nanotechnology in industrial water treatment. Nat Nanotechnol 13(8):670–672
Jayasooriya V, Ng A, Muthukumaran S, Perera B (2017) Green infrastructure practices for improvement of urban air quality. Urban Urban Green 21:34–47
Jhansi SC, Mishra SK (2013) Wastewater treatment and reuse: sustainability options. Consilience: J Sustain Dev 10:1–15
Ji X, Han M, Ulgiati S (2020) Optimal allocation of direct and embodied arable land associated to urban economy: understanding the options deriving from economic globalization. Land Use Policy 91:104392
Jones GW (2002) Southeast Asian urbanization and the growth of mega-urban regions. J Popul Res 19:119–136
Jouhara H, Khordehgah N, Almahmoud S, Delpech B, Chauhan A, Tassou S (2018) Waste heat recovery technologies and applications. Therm Sci Eng Progr S2451904918300015
Kalnay E, Cai M (2003) Impact of urbanization and land-use change on climate. Nature 423:528–531
Kellenberg D (2012) Trading wastes. J Environ Econ Manag 64:68–87
Kennedy C, Cuddihy J, Engel-Yan J (2007) The changing metabolism of cities. J Ind Ecol 11:43–59
Kesari KK, Soni R, Jamal QMS et al (2021) Wastewater Treatment and reuse: a review of its applications and health implications. Water Air Soil Pollut 232:208

Kokalj F, Samec N (2013) Combustion of municipal solid waste for power production. In: Ng H (ed) Advances in internal combustion engines and fuel technologies. IntechOpen. https://doi.org/10.5772/55497

Kopsidas ON, Giakoumatos SD (2021) Economics of recycling and recovery. Nat Resour 12:73–90

Krishan S (2011) Economic and political weekly vol. 46, no. 26/27. Supplement: Rev Agric 95:87–93

Kumar A, Agrawal A (2020) Recent trends in solid waste management status, challenges, and potential for the future Indian cities—a review. Current Res Environ Sustain 2:100011

Kuriakose F, Kylasam Iyer D (2016) Regulating eco-sensitive zones. Available at SSRN 2817186

Lederer J, Laner D, Fellner J (2014) A framework for the evaluation of anthropogenic resources: the case study of phosphorus stocks in Austria. J Clean Prod 84:368–381

Li J, Zhou K, Xie B, Xiao J (2021) Impact of landscape pattern change on water-related ecosystem services: comprehensive analysis based on heterogeneity perspective. Ecol Ind 133:108372

Liu Y, Fang F, Li Y (2014) Key issues of land use in China and implications for policy making. Land Use Policy 40:6–12

Longato D, Lucertini G, Dalla Fontana M, Musco F (2019) Including urban metabolism principles in decision-making: a methodology for planning waste and resource management. Sustainability 11:7

Lu N, Wei H, Fan W, Xu Z, Wang X, Xing K, Ulgiati S (2018) Multiple influences of land transfer in the integration of Beijing-Tianjin-Hebei region in China. Ecol Ind 90:101–111

Lyu P, Wang P, Yuanyuan L, Yuanqing W (2021) Review of the studies on emission evaluation approaches for operating vehicles. J Traffic Transp Eng 8(4):493–509

Magutu P, Onsongo C (2011) Operationalising municipal solid waste management. In: Kumar S (ed) Integrated waste management. IntechOpen, vol II. https://doi.org/10.5772/16457.

Mahgoub MESM, van der Steen NP, Abu-Zeid K, Vairavamoorthy K (2010) Towards sustainability in urban water: a life cycle analysis of the urban water system of Alexandria City, Egypt. J Clean Prod 18:1100–1106

Maillard P, Santos NAP (2008) A spatial-statistical approach for modeling the effect of non-point source pollution on different water quality parameters in the Velhas river watershed–Brazil. J Environ Manage 86:158–170

Maiti S, Agrawal PK (2005) Environmental degradation in the context of growing urbanization: a focus on the metropolitan cities of India. J Hum Ecol 17:277–287

Manisalidis I, Stavropoulou E, Stavropoulos A, Bezirtzoglou E (2020) Environmental and health impacts of air pollution: a review. Front Public Health 8:14

Mansour S, Al-Belushi M, Al-Awadhi T (2020) Monitoring land use and land cover changes in the mountainous cities of Oman using GIS and CA-Markov modelling techniques. Land Use Policy 91:104414

Meloni E, Palma V (2020) Most recent advances in diesel engine catalytic soot abatement: structured catalysts and alternative approaches. Catalysts 10:745

Misra D, Sreedharan B, Bhuyan S, Sethunathan N (1993) Accelerated degradation of fenitrothion in fenitrothion-or 3 methyl-4-nitrophenol-acclimatized soil suspensions. Chemosphere 27:1529–1538

Monstadt J (2009) Conceptualizing the political ecology of urban infrastructures: insights from technology and urban studies. Environ Plann A 41:1924–1942

Moore M, Gould P, Keary BS (2003) Global urbanization and impact on health. Int J Hyg Environ Health 206:269–278

Morello J, Buzai GD, Baxendale CA, Rodríguez AF, Matteucci SD, Godagnone RE, Casas RR (2000) Urbanization and the consumption of fertile land and other ecological changes: the case of Buenos Aires. Environ Urban 12:119–131

Nakatani J, Hirao M (2011) Multicriteria design of plastic recycling based on quality information and environmental impacts. J Ind Ecol 15:228–244

Nandan A, Tauseef SM, Siddiqui NA (2017) Assessment of ambient air quality parameters in various industries of Uttarakhand, India. In: Materials, energy and environment engineering. Springer, Singapore, pp 279–290
Newman G, Sansom GT, Yu S, Kirsch KR, Li D, Kim Y, Musharrat S (2022) A framework for evaluating the effects of green infrastructure in mitigating pollutant transferal and flood events in Sunnyside, Houston, TX. Sustainability 14:4247
Nieto A, Borrull F, Pocurull E, Marcé RM (2010) Occurrence of pharmaceuticals and hormones in sewage sludge. Environ Toxicol Chem 29:1484–1489
Pan M, Chu LM (2016) Adsorption and degradation of five selected antibiotics in agricultural soil. Sci Tot Environ 545:48–56
Pasricha PK, Gera BS, Shastri S, Maini HK, John T, Ghosh AB, Garg SC (2003) Role of the water vapour greenhouse effect in the forecasting of fog occurrence. Bound-Layer Meteorol 107:469–482
Pathan SK, Sastry SVC, Dhinwa PS, Rao M, Majumdar KL, Sampat Kumar D, Phatak VN (1993) Urban growth trend analysis using GIS techniques—a case study of the Bombay metropolitan region. Int J Remote Sens 14:3169–3179
Pedersen JA, Soliman M, Suffet IH (2005) Human pharmaceuticals, hormones, and personal care product ingredients in runoff from agricultural fields irrigated with treated wastewater. J Agr Food Chem 53:1625–1632
Piña B, Bayona JM, Christou A, Fatta-Kassinos D, Guillon E, Lambropoulou D, Michael C, Polesel F, Sayen S (2020) On the contribution of reclaimed wastewater irrigation to the potential exposure of humans to antibiotics, antibiotic resistant bacteria and antibiotic resistance genes–NEREUS COST Action ES1403 position paper. J Environ Chem Eng 8:102131
Pokhrel D, Viraraghavan T (2004) Diarrhoeal diseases in Nepal vis-à-vis water supply and sanitation status. J Water Health 2:71–81
Ponting C (2007) A new green history of the world: the environment and the collapse of great civilizations. Penguin Books, New York
Quarmby NA, Cushnie JL (1989) Monitoring urban land cover changes at the urban fringe from SPOT HRV imagery in south-east England. Int J Remote Sens 10:953–963
Quarmby S, Santos G, Mathias M (2019) Air quality strategies and technologies: a rapid review of the international evidence. Sustainability 11(10):2757
Ravindra K, Mor S, Kamyotra JS, Kaushik CP (2003) Variation in spatial pattern of criteria air pollutants before and during initial rain of monsoon. Environ Monitor Assess 87:145–153
Ravishankar C, Nautiyal S, Seshaiah M (2018) Social acceptance for reclaimed water use: a case study in Bengaluru. Recycling 1:4
Reddy MK, Suneela M, Sumathi M, Reddy RC (2005) Indoor air quality at Salarjung museum, Hyderabad, India. Environ Monit Assess 105:359–367
Ritter KSL (2002) Sources, pathways, and relative risks of contaminants in surface water and groundwater: a perspective prepared for the Walkerton inquiry. J Toxicol Environ Health Part A 65:1–142
Rizwan S, Nongkynrih B, Gupta SK (2013) Air pollution in Delhi: its magnitude and effects on health. Indian J Commun Med Offic Publ Indian Assoc Prev Soc Med 38(1):4–8
Sanchez Rodriguez R, Ürge-Vorsatz D, Barau AS (2018) Sustainable development goals and climate change adaptation in cities. Nat Clim Chang 8:181–183
Saseanu AS, Gogonea R-M, Ghita SI, Zaharia RŞ (2019) The impact of education and residential environment on long-term waste management behavior in the context of sustainability. Sustainability 11:3775
Saravanan P, Ilangovan P (2010) Identification of urban sprawl pattern for Madurai region using GIS. Int J Geomatics Geosci 1:141
Schneider A, Woodcock CE (2008) Compact, dispersed, fragmented, extensive? A comparison of urban growth in twenty-five global cities using remotely sensed data, pattern metrics and census information. Urb Stud 45:659–692

Senecal PK, Felix L (2019) Diversity in transportation: why a mix of propulsion technologies is the way forward for the future fleet. Results Eng 4:100060

Shah H, Ruparelia J (2022) Comparative studies for the treatment of industrial effluents employing advanced processes: towards enhancement of environmental performance. Discov Water 2:4

Shao Z, Sumari NS, Portnov A, Ujoh F, Musakwa W, Mandela PJ (2020) Urban sprawl and its impact on sustainable urban development: a combination of remote sensing and social media data. Geo-Spat Inform Sci 1–15

Sharma RK, Gulati S, Puri A (2014) Green chemistry solutions to water pollution. In: Ahuja S (ed). Water reclamation and sustainability. Elsevier, pp 57–75

Shekdar AV (2009) Sustainable solid waste management: an integrated approach for Asian countries. Waste Manage 29(4):1438–1448

Shetty S, Reid N (2014) Dealing with decline in old industrial cities in Europe and the United States: problems and policies. Built Env 40:458–474

Simon GL, Bumpus AG, Mann P (2012) Win-win scenarios at the climate–development interface: challenges and opportunities for stove replacement programs through carbon finance. Glob Environ Chang 22:275–287

Singal SK (2007) Review of augmentation of energy needs using renewable energy sources in India. Renew Sustain Energy Rev 11:1607–1615

Singh A (1989) Review article digital change detection techniques using remotely-sensed data. Int J Remote Sens 10:989–1003

Singh S, Kumar M (2006) Heavy metal load of soil, water and vegetables in peri-urban Delhi. Environ Monit Assess 120:79–91

Sirmon DG, Hitt MA (2003) Managing resources: Linking unique resources, management, and wealth creation in family firms. Entrep Theory Pract 27:339–358

Slovic A, Ribiero H (2018) Policy instruments surrounding urban air quality: the cases of São Paulo, New York City and Paris. Environ Sci Policy 81:1–9

Soni HB, Thakur K (2015) Preliminary checklist of marine mollusks from Beyt Dwarka, Gulf of Kutch (eco-sensitive zone), Gujarat, India. Int J Environ 4:243–255

Strassburg BB, Iribarrem A, Beyer HL, Cordeiro CL, Crouzeilles R, Jakovac CC, Visconti P (2020) Global priority areas for ecosystem restoration. Nature 586:724–729

Streets DG, Yarber KF, Woo JH, Carmichael GR (2003) An inventory of gaseous and primary aerosol emissions in Asia in the Year 2000. J Geophys Res 108:8809–8823

Tainter JA (2000) Problem solving: complexity, history. Sustain Popul Environ 22:3–41

Tan PY, Zhang J, Masoudi M, Alemu JB, Edwards PJ, Gret-Regamey A et al. (2020) A conceptual framework to untangle the concept of urban ecosystem services. Landsc Urb Plann 200:103837

Tong Z, Baldauf R, Isakov V, Deshmukh P, Max Zhang K (2016) Roadside vegetation barrier designs to mitigate bear-road air pollution impacts. Sci Total Environ 541:920–927

Turker M, Asik O (2002) Detecting land use changes at the urban fringe from remotely sensed images in Ankara, Turkey. Geocarto Int 17:47–52

Turker M, Derenyi E (2000) GIS assisted change detection using remote sensing. Geocarto Int 15:51–56

Turner BL, Lambin EF, Verburg PH (2021) From land-use/land-cover to land system science. Ambio 50:1291–1294

Ulgiati S, Zucaro A (2019) Challenges in urban metabolism: sustainability and well-being in cities. Front Sustain Cities 1:00001. https://doi.org/10.3389/frsc.2019.00001

UNEP (2002) http://www.unep.or.jp/Ietc/Publicati/spc/State_of_Managementwaste/index.asp

United Nations (2014) World urbanization prospects: the 2014 revision, highlights. Population division. United Nations

Van Leeuwen CJ, Frijns J, Van Wezel A, van de Ven FH (2012) City blueprints: 24 indicators to assess the sustainability of the urban water cycle. Water Resour Manage 26:2177–2197

Vining DR Jr, Plaut T, Bieri K (1977) Urban encroachment on prime agricultural land in the United States. Int Reg Sci Rev 2:143–156

Wang S, Zeng J, Huang Y, Shi C, Zhan P (2018) The effects of urbanization on CO_2 emissions in the Pearl River Delta: a comprehensive assessment and panel data analysis. Appl Energy 228:1693–1706

Warwick C, Guerreiro A, Soares A (2013) Sensing and analysis of soluble phosphates in environmental samples: a review. Biosens Bioelectron 41:1–11

Weber C, Puissant A (2003) Urbanization pressure and modeling of urban growth: example of the Tunis Metropolitan Area. Remote Sens Environ 86:341–352

Winkler SL, Anderson JE, Garza L, Ruona WC, Vogt R, Wallington TJ (2018) Vehicle criteria pollutant (PM, NOx, CO, HCs) emissions: how low should we go? Clim Atmosph Sci 1:26

Wu S, Liang Z, Li S (2019) Relationships between urban development level and urban vegetation states: a global perspective. Urban For Urb Greening 38:215–222

Yadav M, Gupta R, Sharma RK (2019) Green and sustainable pathways for wastewater purification. In: Advances in water purification techniques. Elsevier, pp 355–383

Yang J, Chang Y, Yan P (2015) Ranking the suitability of common urban tree species for controlling PM2.5 pollution. Atmos Pollut Res 6:267–277

Yokoo HF (2010) An economic theory of reuse. Sustain Sci 5:143–150

Zhang Q, Wang C, Shinohara F (2007) Automatic extraction of water body based on EOS/MODIS remotely sensed imagery. In: MIPPR 2007: automatic target recognition and image analysis; and multispectral image acquisition 6786:678642

Zhang H, Hu D, Chen J, Ye X, Wang SX, Hao J, Wang L, Zhang R, Zhisheng A (2011) Particle size distribution and polycyclic aromatic hydrocarbons emissions from agricultural crop residue burning. Environ Sci Technol 45:5477–5482

Zucaro A, Maselli G, Ulgiati S (2022) Insights in urban resource management: a comprehensive understanding of unexplored patterns. Front Sustain Cities 3:807735

Zupancic T, Westmacott C, Bulthuis M (2016) The impact of green space on heat and air pollution in urban communities: a meta-narrative systematic review. David Suzuki Foundation, Vancouver, BC, Canada

Chapter 14
Species Selection in Urban Forestry—Towards Urban Metabolism

Ombir Singh

Abstract With increasing urbanization in the twentieth century, the incorporation of trees into urban settlements has increased and management of trees within the urban area is considered as a distinct discipline of forestry. Urban plantations have many positive effects on urban metabolism like reducing urban runoff, heat islands, energy requirements, effects of climate change and to accelerate urban metabolism. Selection of species with array of considerations, e.g., capacity to sequester more carbon, control of air, noise and water pollution, wastewater treatment through phytoremediation, etc., is an important part of urban forestry. Planting of trees and shrubs in an urbanized area involves the selection of species, planting site, planting activities, maintenance, surgery of trees, etc. Several types of planting sites are unique to urban areas including street lawn, undulating areas, avenue and cluster planting, etc. These sites may require special considerations when selecting a species and choosing a proper planting technique. The choice of species should synchronize with the type of habitation, building patterns, the colour of the country side, the nature of the terrain (rocky, undulating, plains, etc.) and its texture (broken, gently sloped, smooth or with abrupt transitions, etc.). These landforms are the guiding factors to finalize the planting plan so that the trees grow well and synchronize with the city conditions. Species characteristic's includes grandeur of the size, gracefulness of the branching patterns, tree form and stem character, the harmony of line and symmetry of form, density or pattern of foliage in relation to shade, the elegance of foliage, spectacular foliage colour, spectacular floral display colour, attractive fruits, the luxury of fragrance, etc. In addition, trees and other species in the urban environment are subjected to a number of stresses which are very different from those suffered by trees in typical rural conditions. Hence, in selecting the species for planting in urban areas, the basic properties of species like climatic adaptation, disease resistance, large phenotypic plasticity, etc., should be given more weightage during plantation in these areas.

Keywords Climate change · Green space · Landscape · Pollution · Tree species · Urban forestry

O. Singh (✉)
Forest Research Institute, Dehradun, Uttarakhand 248006, India
e-mail: ombirfri@gmail.com

R. Bhadouria et al. (eds.), *Urban Metabolism and Climate Change*,
https://doi.org/10.1007/978-3-031-29422-8_14

14.1 Introduction

Nowadays, nearly 55% of the world's population lives in urban areas, a proportion that is expected to increase to 68% by the year 2050. The projections of UN (2018) showed that the world's population could add another 2.5 billion people to the urban areas by 2050 due to gradual shift in residence of the human population from rural to urban areas and from this increase about 90% will be taken place in Asia and Africa only. However, in developed countries, the majority of the population is living in cities, e.g., more than 80% of USA (Wolf 1998) and 85% of the Australian population are living in and around urban cities (Brack 2002). Due to this urbanization trend, distance between city inhabitants and nature is increasing and urban areas are now becoming the hot spots for environmental changes (Schell and Ulijaszek 1999; Chaudhary and Tewari 2011). Urban greenery/forestry is one of the ways to bridge this gap between people and nature. Urban metabolism studies are necessary to assess the environmental performance of cities and design resource-efficient strategies for urban development in the changing world (Bhadouria et al. 2022). The metabolism of urban areas is enormous and each person needs about 1–10 ha of non-urban area to support the needed resources. The problems of large metabolic throughput, low metabolic efficiency and disordered metabolic processes are a major cause of unhealthy urban systems (Zhang 2013). Natural and human-induced stressors elicit changes in energy metabolism and stress physiology in many populations and urban metabolism provides a metaphorical framework to study the interactions of natural and human systems. Urban trees can help to reduce the ecological footprint of this metabolism and improve ecosystem carrying capacity by delivering an array of ecosystem services such as reducing greenhouse gases through carbon storage, decreasing stormwater runoff through interception and absorption of rainwater and mitigate the urban heat island effect through reductions in surface and air temperatures at a local scale (Chen and Jim 2008; Kathleen et al. 2020).

Urban forests or urban green spaces are the green lungs of any city, which are more known for their non-priced benefits (like pollution control, energy conservation, leisure/recreation, carbon sequestration, etc.) than priced benefits in urban settings. They provide multiple ecosystem services like physical, emotional and psychological benefits to the residents. However, provision of green spaces even in very small amounts and education based on a local environmental perspective is essential in creating a sustainable urban landscape (Verma et al. 2020). These urban forests are also act as sponges by holding water during flood time and helping in the recharging of groundwater by reducing surface runoff (Locatelli 2016). In addition, due to a high leaf area index, forests maintain high evaporation fluxes (transpiration and interception), which can exceed evaporation fluxes over the oceanic surface. Evaporation and transpiration cause elevated moisture content in the vertical column of air over a forest, which in turn, produces elevated condensation well above the canopy due to the adiabatic ascent of moist air. This 'sucks in' air horizontally from elsewhere via the formation of horizontal pressure gradients; and under the right circumstances, can bring moist air from the ocean (Ellison et al. 2017). Moist ocean air enhances condensation above the forest, further strengthening the process and thus stabilizing land moisture supply. Air returns to the ocean in the upper atmosphere after precipitation of moisture over the continent, completing the circulation pattern. Such biotically induced atmospheric circulation sustaining the hydrological cycle on land has been termed as the 'biotic pump' of atmospheric moisture (Makarieva and Gorshkov 2007).

Trees also act as pollutant filtration by trapping dust, dirt and smoke on their leaves and bark and help improve air quality by absorbing excess carbon from the atmosphere, particulate matter and other harmful gases. Urban areas with greater tree and vegetation cover and fewer impervious surfaces tend to exhibit lower temperatures than those blanketed by solid surfaces (Bounoua et al. 2015). Trees in cities can help cool the air between 2 and 8 °C, thus reducing the urban heat island effect. Moreover, they also act as carbon sinks in the equations relevant within the context of global warming. Trees reduce stormwater runoff and can assist with processing wastewater and minimizing air pollution and diseases from sewage water through their use for tree planting. Because of their impermeable surfaces, cities are vulnerable to flooding, but urban parks or trees can reduce runoff through infiltration. Many cities have established and conserved forests for protecting their drinking water resources. The New York City water department decided to allocate 1.5 billion USD to preserve the Catskill and Delware watersheds, rather than spending 6 billion USD to construct a new water treatment system (Shwartz 2000). In central Himalaya, a valley fill located in the watershed of Lake Nainital is shown to contribute significantly to its health by providing filtered water and diluting pollutants (Singh 2002). Urban forests protect soils and moderate harsh urban climates, for instance, by cooling the air, reducing wind speeds and by giving shade. The shade provided by trees may result in building energy saving and lowering air conditioning cost. In arid regions, forest shelterbelts around cities help combat desertification.

In most of our urban cities, the spaces are badly packed with construction and the open areas are normally dull and unattractive. The blue sky is all dulled with pollution haze and to improve this vista, we must therefore set new priorities to create a good landscape with suitable species. The assemblage and display tree species should provide desired effects, hence, judicious selection of species for a particular site is of paramount importance. That is the reason that tree planting and landscaping have been an integral part of the city's master plan. However, urban conditions are very different than the conditions in rural or natural area (Sabo et al. 2003). Complex and harsh urban growing conditions require a careful selection of the right tree species. There is a need for an integrated focus on the identification and selection of cultivars and species for urban areas, based on sound knowledge of site conditions and tree characteristics. As urban tree populations are often dominated by only a few tree species or families, the risks related to pests and diseases are larger. The dramatic impact of Dutch elm disease in North America and Europe is an example of this (Konijnendijk and Randrup 2004). In order to develop a more resistant urban forest resource, it is advisable to provide for sufficient variation among tree species, genus and families. The use of exotic versus native species has also been a topic of debate.

However, the use of native species is a nature-oriented management of green areas and can maintain genetic resources as well.

The establishment of trees in urban environments often requires a different approach, viz., efforts should be made to provide growing conditions below the surface and to have sufficient space for root development. For that, special soil substrates have been developed for application in urban conditions including the development of efficient establishment techniques in terms of planting holes, adequate water supply and protection against various threats (Bhadouria et al. 2016; Tripathi et al. 2018). In addition, several types of planting sites are unique to urban areas including street lawns, undulating areas, avenue and cluster planting, etc. These sites may require special considerations when selecting a species. Collins et al. (2000) suggested a number of reasons why new human-imposed scales for ecological processes are found within urban areas. First, compared with ecosystems in rural areas, urban ecosystems are highly patchy and the spatial patch structure is characterized by a high point-to-point variation and degree of isolation between patches. Second, disturbances such as fire and flooding are suppressed in urban areas, and human-induced disturbances are more prevalent. Third, because of the higher temperatures in urban systems, in temperate climates, there are longer vegetation growth periods. Fourth, ecological successions are altered, suppressed, or truncated in urban green areas and the diversity and structure of communities of plants and animals may show fundamental differences from those of non-urban areas (Pickett et al. 1997; Niemela 1999). Though urban plantings have close links to forestry, but tends to be more multidisciplinary. In recent years, the research community has emerged with growing knowledge and developed new approaches and techniques for the purpose. Additionally, urban forestry is dominated by urban societies and their values, norms and demands and present forestry must respond to these urban perceptions. Hence, multiple-use forestry is being transformed into modern landscape planning in urban areas.

As urban plantations have many positive effects on urban metabolism and to accelerate urban metabolism, selection of species with array of considerations is an important constituent in urban forestry and this chapter discuss range of parameters for selecting suitable tree and other species for planting in urban areas. The objective of this chapter is to explore the species selection process of several stakeholders who are planting or selling trees in an urban context.

14.2 Selection of Tree Species in Relation to Site Conditions

Trees are one of the most important components of urban green space and they play an important role in street aesthetics. So, tree species selection and planting design will strengthen the feeling of identity and distinctiveness of a city. As an integral component of urban green infrastructure, trees selection is crucial to successfully shaping a better urban environment (Jacobs 1993; Li et al. 2011; Conway and Vetch 2015). Urban conditions are very different from natural and rural areas, hence, require a careful selection of the right tree species for planting in these areas (Sabo et al. 2003). Trees can tolerate different types of stress and respond to stress in an array of conditions. Some tree species can tolerate stress such as poor soils and adverse environmental conditions better than other species (Bhadouria et al. 2020; Tripathi et al. 2020). Several important factors of tree species related to growth such as mature size and form, growth rate, branching pattern, foliage characters, bark texture, flowers, fruits and seeds, should be considered when selecting a suitable species to be planted in urban areas.

The choice of species should synchronize with the type of habitation, building patterns, the colour of the country side, nature of the terrain (rocky, undulating, plains, etc.) and its texture (broken, gentle slope, smooth or with abrupt transitions, etc.). These landforms are the guiding factors to finalize the planting plan so that the trees grow well and synchronize with the city's conditions. The assemblage and display of floral trees can achieve very significant and attractive effects on the landscape of crowded cities, slum areas and dry roads. However, these trees should occupy only well-planned positions in the landscape. The physical characteristics of planted trees species must be fast-growing, having straight stems, can be prune as desired, evergreen, wind firmness, deep root system, relatively long-lived, not subject to wind throw or breakage of large branches and comparatively resistance to drought and must be suitable in the local climate (McKinney 2008; Pandey and Luitel 2020). Based on the environmental performance to increase urban metabolism, landforms present and aesthetic value to cities, the followings are the main considerations for the selection of tree species for urban landscape.

14.2.1 Grandeur of the Size

The grandeur of the size of a tree is a synthesis of several components and dimensions. It includes the robustness of stem, massive branching pattern, girth and height of trees and overall build of the trees. Features like buttressing (height and width of plant buttresses), ruggedness of proportion of clean stem, density of foliage, etc., also have a significant value. The grandeur can be best judged when these trees are planted though in single numbers and occupy a dominant space in the landscape. Trees can also be chosen for their spectacular branching pattern, curiously distorted and variously blotched branches, twisted as unusual drooping branches, distinguished foliage or floral composition and shape of the crown.

14.2.2 Branching Pattern

Trees can be selected as per the branching characteristic of a particular tree as per the land form specifications of the site in urban areas (Table 14.1).

The importance of this character can be visualized in the following points:

Table 14.1 Branching pattern and trees

Branching pattern	Trees
Round form	*Ailanthus excelsa, Tectona grandis, Tamarindus indica, Azadirachta indica, Emblica officinalis*
Broad form	*Ficus infectoria, Mangifera indica, Saraca asoca*
Square form	*Alstonia scholaris, Mimusops elengi*
Spreading form	*Albizia procera, Casssia siamea, Plumeria alba*

- Trees with large leaves and high foliage which permit light penetration and good circulation of air should be preferred for planting in parks.
- Round and broad tree forms can be planted on pedestrian paths.
- Species with heavy foliage and thorny habits such as *Acacia nilotica* should not be planted on roadside as the heavy leaf shedding and thorns make them undesirable for the purpose.

14.2.3 Tree Form

Table 14.2 provides the various tree forms with tree names for selecting trees for planting in urban localities as per the landscape demands of a particular site.

Various tree forms can be utilized in urban plantings as follows:

- The broad conical forms are planted to provide colour and texture to the background and form a foreground natural screen.
- Pyramidal forms in groups help to obstruct undesirable views very effectively and also serve a good back ground for a flowering tree in the foreground.

Table 14.2 Tree form and trees

Tree form	Trees
Conical form	*Grevillea robusta, Polyalthia longifolia, Bombax ceiba, Spathodia companulata*
Columnar form	*Mellingtona hortensis, Cedrus deodara, Pinus roxburghii*
Tapering	*Polyalthia longilolia, Araucaria cunninghamii*
Arrow form	*Casuarina equisetifolia*
Weeping form	*Callistemon* spp., *Salix babylonia, Putranjiva roxburghii, Hardwickia binata*
Horizontal form	*Araucaria* spp., *Terminalia arjuna, Chukrasia tabularis, Toona ciliata*
Contorted Forms	*Cassia fistula, Jacaranda mimosifolia, Delonix regia, Butea monosperma*
Multi-trunk forms	*Albizia lebbeck, Acacia nilotica, A. auriculiformis, A. planiformis, Adanasonia digitata*

Table 14.3 Trees with different foliage colours

Foliage colour pattern	Trees
Colour tints (leaf shedding)	Orange—*Bischofia javanica, Bridelia retusa* Greyish red—*Lagerstroemia speciosa* Red purple—*Careya arborea*
Colour tints (leaf sprouts)	Reddish pink—*Saraca asoca* Red—*Schleichara oleosa, Acer oblongum*
Species with variegated leaves	*Ficus elastica*—yellowish spots *Erythrina indica*—yellowish spots *Psidium guajava*—blood red spots

- The evergreens keep on adding green when the deciduous have shed their leaves and become dominant features of beauty in the gardens and avenues.
- Weeping forms are famous for their beauty, gracefulness, femininity, sorrow, and elegance. These forms should be used occasionally to produce the effects at special points such as small roads, relaxing spots and corners of the park. They can also be planted along the moving watersides, stagnant pool of water and in the middle of lawn to create loneliness and sculpturing value and sudden change in the landscape.

14.2.4 Foliage Colour

The foliage of some trees is spectacular in colour and changes in different seasons. These trees provide foliage colour which impart distinct characteristics at a particular time of the season. The variegated foliage also provides a graceful appearance in the landscape. New leaves of many trees show spectacularly tinted foliage in the spring. Table 14.3 provides the tree species with various foliage colours for varied landscapes.

14.2.5 Elegance of Foliage

In large open areas or tracts, large trees at greater distances can be planted. The following species are having elegance in their foliage: *Acrocarpus fraxinifolius, Jacaranda mimosifolia, Duabanga grandiflora, Araucaria cunninghamii, Melia composita.*

Table 14.4 Trees for the harmony of line and symmetry of form

Character	Trees
Harmony of line	*Mimusops elengi, Barringtonia acutangula, Pinus oocarpa, Pinus wallichiana*
Symmetry of form	*Alstonia scholaris, Acacia plantiformis, Michelia champaca, Agathis robusta, Populus nigra*

14.2.6 Density and Pattern of Foliage in Relation to Shade

Trees and shrubs found in urban parks or gardens not only provide shade but also enhance the beauty of the area and at the same time attract avifauna. Shade is an essential ingredient of the recreation because it provides coolness to users in addition to the scenic beauty. The following trees have such characteristics and produce shade to the desired effects and can be planted in different landforms: *Alstonia scholaris, Barringtonia acutangula, Dillenia indica, Chukrasia tabularis, Ficus elastica, Ficus retusa, Mimusops elengi, Madhuca latifolia, Syzygium cumini, Terminalia bellerica, Tamarindus indica.*

14.2.7 Harmony of Line and Symmetry of Form

Trees can break monotony if they are planted to show the symmetry of form and harmony when placed in line at variable distances (Table 14.4).

14.2.8 Stem Bark Characteristics

Many trees can be selected to suite urban forestry designs on the basis of various coloured and shaped stems, patterns of bark, natural ramifications of stems and branches (Table 14.5).

Table 14.5 Trees with different bark characteristics

Character	Trees
Crocodile bark	*Terminalia alata, Cinnamomum camphora*
Ash-coloured bark	*Eucalyptus* spp.
Pinkish grey bark	*Terminalia arjuna*
Yellowish bark	*Albizia procera*

14.2.9 Floral Display

A flowering tree provides a cool feeling to the eyes when they bloom and change the colour of the entire landscape. Brilliancy, colour harmony, fineness, sequence of blooming, bloom density, periodicity of bloom, and the time longevity of bloom are some of the characters, which should be considered in urban landscaping design. Necessary plans can be prepared based on the periodicity of flowering, deciduous habit and colour of flowers of tree species.

14.2.9.1 Trees with Attractive Fruits

If the objective of planting in urban landscapes is to provide shelter for avifauna, attractive fruit trees should be planted. However, trees with a tendency to drop large and heavy fruits (*Durio* spp.) and emit a bad smell (*Sterculia foetida*) should not be planted in urban areas.

14.2.9.2 Luxury of Fragrance

Apart from aesthetic value, the fragrance of trees relieves the man from his tension. The mild delicate fragrance brings pleasure to urbanites. The tree species, able to produce highly aromatic flowers, follow as: *Aegle marmelos, Holarrhena antidysenterica, Michelia champaca, Millingtonia hortensis.*

14.3 Selection of Species to Control Pollution

The major benefits of urban forestry include: minimize the heat through evapotranspiration, provide shading to streets and buildings, improve human comfort, reduce the risk of heatstroke, improve air quality by absorbing pollutants, carbon sequestration, water absorption, noise control, traffic control (Negi 1998; Konijnendijk et al. 2005; Kielbaso 2008; Pearlmutter et al. 2017) as well as promote social harmony inclusiveness (Lamichhane and Thapa 2012). The important considerations for the selection of species for pollution control (Kumar et al. 2013) to be planted in the urban landscape include:

- Plants should be evergreen, large-leaved, rough bark, indigenous, ecologically compatible, low water requirement, minimum care, high absorption of pollutants, resistant pollutants and agro-climatic suitable.
- Different types of leaves have differences in the aspects of their surfaces. Some types of leaves have greater surface rigidity or roughness than other leaves, which may affect their stickiness or particle solubility and are considered better for particle collection. Therefore, certain plant leaves may be more useful for efficient

dust capturing than other plants. The crown area of plants is also depending upon the morphological features of the leaf.
- Selection of easily propagated and readily available, medium growing, ecologically much suitable, pest and disease resistant tree species with less maintenance should be given top priority.

14.3.1 Species to Control Air Pollution

Trees and shrubs having specific traits to improve air quality have been selected for planting in urban areas, viz., different leaf structures and surfaces enable species to capture particulate matter better, coniferous trees are more effective for overall particulate capture, etc. Trees such as Tamarind (*Tamarindus indicus*) having smaller compound leaves are generally more efficient particle collectors than larger leaves. Particle deposition is heaviest at the leaf tip and along the leaf margin. The preliminary survey of dust falls on common roadside trees in Mumbai, carried out by Shetye and Chaphekar (1989) reported that the shape of leaves of Mango (*Mangifera indica*), Ashoka (*Polyalthea longifolia*), *Pongamia pinnata* and Umbrella (*Thespepsia populnea*) trees captured higher amounts of dust as compared to other neighbouring plants. Dochinger (1973) a plant pathologist of USDA Forest Service, Ohio, reported that the filtering effects of evergreen trees are better than those of the deciduous trees. In Singapore, it has been noted that a single row of trees planted with or without shrubs can reduce particulate matter by 25% and each hectare (2.471 acres) of plantation can produce enough oxygen to keep about 45 persons alive (Anon 1981). Columnar and medium-sized trees are preferred. Ingold (1971) reported that the leaves with complex shapes and large circumference areas reported to be collected particles more efficiently. Many trees like Neem (*Azadirchta indica*), Silk cotton (*Bombax ceiba*), Indian laburnum (*Cassia fistula* and *C. siamea*), Gulmohar (*Delonix regia*), Pipal (*Ficus religiosa*), Jacaranda (*Jacaranda mimosifolia*), Indian lilac (*Lagerstroemia indica*), Temple or Pagoda tree (*Plumeria rubra* and *P. alba*), Java plum (*Syzygium cumini*) and several other roadside and street trees have found more suitable to control air pollution in the urban environment in India (Maheshwari 1963; Pokhriyal and Subba Rao 1986). The tree species, which can be planted as air pollution tolerant are: *Ficus benghalensis, Polyalthia longifolia, Eucalyptus citriodora, Mangifera indica, Casuarina equisetifolia, Azadirachta indica, Albizzia lebbeck, Moringa olifera,* etc.

14.3.2 Species to Control Noise Pollution

Noise pollution control is a major challenge in each city around the world. A good relationship between urban planning and different factors such as urban density, urban morphology, urban land use, street distribution, street environment and green spaces may lessen the problem. The density, height and width are critical factors in designing an adequate noise screen plantation. Trees, having thick and fleshy leaves with petioles flexible and the capacity to withstand vibration, are suitable to control noise pollution. Heavier branches and the trunk of the trees also deflect or refract the sound waves. A combination of trees and shrubs together with suitable landforms and design appears to be the best system for combating noise pollution (Anon 2001). In general, more than 65 decibels of noise are produced in urban localities are unhealthy for living beings. The followings are the suitable species of trees and shrubs to be planted in urban areas to reduce noise pollution:

Trees: *Alstonia scholaris, Azadirachta indica, Melia azedarach, Butea monosperma, Grevillea pteridifolia, Grevillea robusta, Tamarindus indica, Terminalia arjuna.*

Shrubs and grasses: *Calotropis gigantea, Inga dulcis, Saccharum munja, Nyctanthus arbortristics, Nerium orodrum, Ipomea* sp.

14.3.3 Species to Control Water Pollution

Urban forests can help in the protection of urban water supply, wastewater treatment systems and stormwater management. Most poor cities face significant wastewater treatment challenges and could integrate stabilization ponds into park systems and reuse wastewater for urban forestry. There is a need to use plants to restore ecological environment. Phytoremediation is a bioremediation process that uses various types of plants to remove, transfer, stabilize and/or destroy contaminants in the soil and groundwater. It is a plant-based approach in which plants are used to extract and remove elemental pollutants or lower their bioavailability in soil. Plants have the ability to absorb ionic compounds in the soil even at low concentrations through their root system. Certain plants can extend their root system into the soil matrix and establish a rhizosphere ecosystem to accumulate heavy metals and modulate their bioavailability, thereby, reclaiming the polluted soil and stabilizing soil fertility. Phytoremediation is widely accepted as a cost-effective environmental restoration technology and is an alternative to engineering procedures that are usually more destructive to the soil. The following plants can to remove pollutants from soil and water and are effective in nitrifying effluents, converting ammonia into nitrates and nitrites resulting in lowering the eutrophication of water bodies and useful in sewage water purification (Ansari et al. 2020).

14.3.3.1 Reeds and Marsh Plants

Water reed plants have two major qualities related to sustainable approach to nature that is: (i) the ability to filter water, and (ii) the capacity to produce more biomass. Marsh plants like *Typha latifolia* (broadleaf cattle), *Phragmites communis* (common reed), (Fig. 14.1) and *Iris pseudacorus* (yellow flag) can proliferate over rhizomes

Fig. 14.1 Common reed in Yamuna Biodiversity Park, Delhi (©Author)

in the water ground (Rhizosphere) and the leaves over the surface of water. Such ability makes these reeds to transport oxygen from the atmosphere to the ground and this oxygen oxidates the phosphorus and nitrogen present in the water/soil of the polluted environment underneath. As a result, these water reeds provide a favourable environment for the growth of special bacteria present between the thick mesh of roots and these bacteria decompose the organic and inorganic toxins into nutrients. The final product of this physiological process of reeds and marsh plants is clean water (Lei et al. 2022).

14.3.3.2 Vetiver Grass

Chrysopogon zizanioides/Vetiveria zizanioides provides ample opportunities for phytoremediation with a bio-economic return. Vetiver is a dynamic and multipurpose perennial grass for carbon farming and essential oil production during phytoremediation and has a large geographical distribution with good adaptive phyto-management. Therefore, the exploitation of the vetiver system (multifunctional application) is a potential and sustainable way for environmental remediation and socio-economic development of nearby localities (Kapoor et al. 2020). Vetiver plays an important role in various ecosystem services (controlling soil erosion, water conservation, essential oil, phytoremediation, etc.) (Lavania and Lavania 2009; Vasavi et al. 2010).

14.3.3.3 Indian Grass

Shorgastrum nutans growing along the roadsides are known for their power to detoxify common agrochemical residues such as well-known pesticides and herbicides related to atrazine and metolachlor.

14.3.3.4 Poplar and Willows

Poplar and Salix species also have water purifying capacities.

14.3.3.5 Aquatic Plants

Hydrilla verticillata, Spirodela polyrrhiza, Bacopa monnierii, Phragmites karka, Scirpus lacustris, Pennywort (*Hydrocotyle umbellata),* Duck weed *(Lemna minor) and* Water velvet *(Azolla pinnata)* are suitable to remove heavy metals from water bodies in urban areas (Vasavi et al. 2010).

14.4 Trees/Urban Forest to Combat Global Warming

Climate change threatens the health and survival of urban trees and the various benefits delivered to urban inhabitants. This risk is more in cities at lower latitudes like New Delhi and Singapore (Rodrigues et al. 2022). Woodlands, forests and forestry have a substantial role in helping to reduce climate change. Trees remove carbon dioxide from the atmosphere and store the carbon in solid form as wood. The harvesting and use of wood from sustainably managed forests transfers the carbon into wood products where it can continue to be stored, often over long periods, in materials such as those used for construction and furniture. With the growing concern of climate change and atmospheric degradation, the use of the urban metabolism model has become a key element in determining and maintaining levels of sustainability and health in cities around the world. As urban forests and trees regulate temperatures (through shade and evaporative cooling) and water (through rain interception and infiltration), they play a role in urban adaptation to climate variability and change. The urban heat island effects, which increase the health impacts of heat waves, are moderated by the green cover, as observed, e.g. in Manchester, UK (Gill et al. 2007). In cities, ecosystem-based adaptation requires a good understanding of landscape ecology and the potential of green infrastructure to improve the well-being of vulnerable communities. Adaptation needs also to be designed at multiple scales, including ecosystem management outside the urban areas and for upper watershed protection. For example, three scales are proposed in Beijing, China, viz., (i) for a green infrastructure at the regional scale (forest belts), (ii) in the city (urban parks and green corridors) and (iii) in neighbourhoods (road and vertical greening) (Li et al. 2005). However, urban ecosystem-based adaptation raises concerns about high opportunity costs of land and possible management constraints, e.g. during droughts, scarce water consumed to maintain trees may be needed for other uses.

Urban forests-like any forest-help mitigate climate change by capturing and storing atmospheric carbon dioxide during photosynthesis, and by influencing energy needs for heating and cooling buildings. Trees typically reduce cooling costs but can increase or decrease winter heating use depending on their location around a building and whether they are evergreen or deciduous. Urban forests enable cities to better adapt to the effect of climate change on temperature patterns and weather events. Forests influence local and global temperatures and the flow of heat. At the local scale, forests can remain much cooler during the daytime due to shade and the role of evaporation and transpiration in reducing sensible heat (Hesslerová et al. 2013).

Cities are generally warmer than their surroundings (typically by about 1–2 °C, though this difference can be as high as 10 °C under certain climactic conditions (Kovats and Akhtar 2008) meaning that average temperature increases caused by global warming are frequently amplified in urban areas. Urban forests help control this 'heat island' effect by providing shade and by reducing urban albedo (the fraction of solar radiation reflected back into the environment), and cooling evapotranspiration (Romero-Lankao and Gratz 2008). Cities are also particularly susceptible to climate-related threats such as storms and flooding. Urban trees can help control runoff from these by catching rain in their canopies and increasing the infiltration rate of deposited precipitation. Reducing stormwater flow reduces stress on urban sewer systems by limiting the risk of hazardous combined sewer overflows (Fazio 2010). Furthermore, well-maintained urban forests help buffer high winds, control erosion and reduce drought in cities (Cullington et al. 2010; Norwak et al. 2013).

In addition, urban forests using Miyabaki method could be useful in creating fascinating complex ecosystem to balance today's soil and climatic conditions. Such forests help in lowering temperatures in concrete heat islands, reduce air and noise pollution, attract birds and insects and also create carbon sinks in these areas. Bhadouria et al. (2017) advocated to have the multi-factorial interaction studies under various resource-disturbance combinations to help current climate change scenario.

14.5 Trees Representing Various Zodiac Signs

Trees in urban forests have a pivotal role in environmental and forestry education in urban areas. Easily accessible trees and woodlands provide a vital facility for both formal and informal learning. Trees in urban forests create awareness of the role of trees in human life which is very much essential for the conservation of tree resources. In astrology, there are 27 nakshatra/stars and 12 zodiac signs and there are 12 trees to represent each one of them. For this, plantations/small gardens can be established in cities for the peoples for awareness and motivation for tree planting. Table 14.6 illustrates the plants and trees associated with different rashi/zodiac sign.

Table 14.6 Zodiac sign plants (As per astrology)

S. No.	Rasi/zodiac sign	Common name	Botanical name
1	Aries (Mesh)	Red Sanders	*Pterocarpus santalinus*
2	Taurus (Vrishabh)	Sapthaparni	*Alstonia scholaris*
3	Gemini (Mithun)	Jackfruit/Kathal	*Artocarpus heterophyllus*
4	Cancer (Kark)	Pallas	*Butea monosperma*
5	Leo (Singh)	Bel	*Aegle marmelos*
6	Virgo (Kanya)	Mango	*Mangifera indica*
7	Libra (Tula)	Bakula	*Mimusops elengi*
8	Scorpio (Vrishchik)	Khair	*Acacia catechu*
9	Sagittarius (Dhanusu)	Pipal	*Ficus religiosa*
10	Capricorn (Makar)	Shisham	*Dalbergia sissoo*
11	Aquarius (Kumbh)	Shami	*Prosopis cineraria*
12	Pisces (Meen)	Indian Banyan/Vat Vriksha	*Ficus benghalensis*

14.6 Other Considerations for Urban Plantings

One of the great joys of trees in our parks, gardens and landscapes is their diversity and the many aesthetic attributes that they provide us with, including flowers, fruit, bark colour and leaf shape and colour. The diversity of tree species represents an opportunity to plant the right tree in the right location, giving it the best chance of a sustainable future, thereby creating resilience within the tree population to climate change and the introduction of exotic pests and pathogens (Hirons and Sjomann 2019). High species diversity is also vital to enhance the resilience of urban forest to abiotic and biotic challenges in future (Alvey 2006).

In addition, the following considerations are also very important to make the urban environment healthy.

- The species for turfing are *Cynodon dactylon, Cythoclinic perpurea, Xanthium strumerium.*
- Species for waterlogged areas are *Eucalyptus* spp., *Terminalia arjuna, Syzygium cumini.*

- Avoid using highly flammable species in plantings near natural areas, viz., pine or eucalyptus trees or chaparral shrubs.
- Select species and cultivars that are less susceptible to pests and pathogens.
- Incorporate a mixture of plant functional types to provide year-round interception and evapotranspiration benefits (e.g. deciduous and coniferous trees, turfgrass).
- Select species that can provide multiple ecosystem services, such as carbon sequestration, stormwater mitigation or pollinator habitat.
- Plant trees near major emissions sources, such as industrial or manufacturing sites, to reduce air pollution. Consider species selection, tree size and leaf area, as well as the position of plantings relative to nearby buildings and other features that can influence wind direction and speed.
- Remove existing invasive species using non-chemical treatments. Care must be taken in urban areas as these are susceptible to the introduction and spread of invasive plants owing to global transport linkages.

14.7 Conclusions and Future Perspectives

The continuously increasing world population in cities necessitates adapting ecologically based urban development to work towards sustainability. Natural and human-induced stressors elicit changes in energy metabolism and stress physiology in many populations and urban metabolism provides a metaphorical framework to study the interactions of natural and human systems. Urban trees can help to reduce the ecological footprint of this metabolism and improve ecosystem carrying capacity by delivering an array of ecosystem services such as reducing greenhouse gases through carbon storage, decreasing stormwater runoff through interception and absorption of rainwater and mitigate the urban heat island effect through reductions in surface and air temperatures at a local scale. Urban sustainability can be improved profoundly by increasing the tree cover in these areas. Trees can tolerate different types of stress and respond to stress in an array of conditions. Some tree species can tolerate stress such as poor soils and adverse environmental conditions better than other species. Several important factors of tree species related to growth such as mature size and form, growth rate, branching pattern, foliage characters, bark texture, flowers, fruits, seeds, etc., should be considered while selecting a species to be planted in urban areas.

Trees also act as pollutant filtration by trapping dust, dirt and smoke on their leaves and bark and help in improving air quality by absorbing excess carbon from the atmosphere, particulate matter and other harmful gases. Synchronization between urban planning and different factors such as urban density, urban morphology, urban land use, street distribution, street environment and green spaces may lessen the noise pollution problem in cities. Urban forests can help in the protection of urban water supply, wastewater treatment systems and stormwater management. Urban forests, like any forest, can help mitigate climate change by capturing and storing atmospheric carbon dioxide during photosynthesis, and by influencing energy needs for heating and cooling buildings. Trees in urban forests create awareness of the role of trees in human life which is very much essential for the conservation of tree resources. Hence, urban forestry has many positive effects on urban metabolism and to plant right species at right place can lessen the various stresses of urban environment and also combat global warming by accelerating urban metabolism.

In addition, there is a need to develop areas of more natural vegetation in cities and establishing urban nature reserves in Indian cities on the patterns of the 1215-ha Rock Creek Park in Washington DC, 1-ha Camley Street Reserve adjacent to Kings Cross Station in London, 164-ha Bukit Timah Reserve in Singapore and the 11-ha Bukit Nanas Reserve in Kuala Lumpur.

References

Alvey AA (2006) Promoting and preserving biodiversity in urban forest. Urb For Urb Greening 5:195–201

Anon (1981) A guide to tree planting. Parks and Recreation Department, Ministry of National Development, Singapore

Anon (2001) Tree plantation strategy. Grand Trunk Road Improvement Project. National Highway Authority of India. 1 Eastern Avenue, Maharani Bagh, New Delhi

Ansari AA, Naeem M, Gill SS, Alzuaibri FM (2020) Phytoremediation of contaminated waters: an ecofriendly technology based aquatic macrophytes application. Egypt J Aquat Res 46:371–376. https://doi.org/10.1016/j.ejar.2020.03002

Bhadouria R, Singh R, Srivastava P, Raghubanshi AS (2016) Understanding the ecology of tree-seedling growth in dry tropical environment: a management perspective. Energy Ecol Environ 1(5):296–309

Bhadouria R, Srivastava P, Singh R, Tripathi S, Singh H, Raghubanshi AS (2017) Tree seedling establishment in dry tropics: an urgent need of interaction studies. Environ Syst Decis 37(1):88–10

Bhadouria R, Srivastava P, Singh S, Singh R, Raghubanshi AS, Singh JS (2018) Effects of Effects of light, nutrient and grass competition on growth of seedlings of four tropical tree species. Indian For 144(1):54–65

Bhadouria R, Srivastava P, Singh R, Tripathi S, Verma P, Raghubanshi AS (2020) Effects of grass competition on tree seedlings growth under different light and nutrient availability conditions in tropical dry forests in India. Ecol Res 35(5):807–818

Bhadouria R, Upadhyay S, Tripathi S, Singh P (2022) Urban ecology and global climate change. Wiley-Blackwell

Bounoua L, Zhang P, Mostovoy G, Thome K, Masek J, Imhoff M, Shepherd M, Quattrochi D, Santanello J, Silva J, Wolfe R, Toure AM (2015) Impact of urbanization on US surface climate. Environ Res Lett 10:84010. https://doi.org/10.1088/1748-9326/10/8/084010

Brack CL (2002) Pollution mitigation and carbon sequestration by an urban forest. Environ Pollut 116(1):195–200

Chaudhry P, Tewari VP (2011) Urban forestry in India: development and research scenario. Interdiscip Environ Rev 12(1):80–93

Chen WY, Jim CY (2008) Assessment and valuation of the ecosystem services provided by urban forests. In: Carreiro MM, Song Y-C, Wu J (eds) Ecology, planning, and management of urban forests: international perspectives. Springer, New York, pp 53–83

Collins JP, Kinzig A, Grimm NB, Fagan WF, Hope D, Wu JG, Borer ET (2000) A new urban ecology. Am Sci 88(5):416–425

Conway TM, Vecht JV (2015) Growing a diverse urban forest: species selection decisions by practitioners planting and supplying trees. Landsc Urban Plan 138(6):1–10. https://doi.org/10.1016/j.landurbplan.2015.01.007

Cullington J, Gye J (2010) Urban forests: a climate adaptation guide. Part of the BC regional adaptation collaborative (RAC). Ministry of Community, Sport and Cultural Development: British Columbia, Canada

Dochinger LS (1973) Miscellaneous Publication No. 1230. USDA, Forest Service, Upper Darby, p 22

Ellison D, Morris CE, Locatelli B, Sheil D, Cohen J, Murdiyarso D, Gutierrez V, Noordwijk MV, Creed IF, Pokorny J, Gaveau D, Spracklen DV, Tobella AB, Sands DC, Muys B, Verbist B, Springgay E, Sugandi Y, Sullivan CA (2017) Trees, forests and water: cool insights for a hot world. Glob Environ Chang 43:51–61

Fazio J (ed) (2010) How trees can retain stormwater runoff. Tree City USA Bulletin No. 55. Arbor Day Foundation, Nebraska City

Gill SE, Handley JF, Ennos AR, Pauleit S (2007) Adapting cities for climate change: the role of the green infrastructure. Built Environ 33:115–133

Hesslerová P, Pokorný J, Brom J, Rejšková-Procházková A (2013) Daily dynamics of radiation surface temperature of different land cover types in a temperate cultural landscape: consequences for the local climate. Ecol Eng 54:145–154. https://doi.org/10.1016/j.ecoleng.2013.01.036

Hiron AD, Sjomann H (2019) Tree species selection for green infrastructure: a guide for specifiers. Trees & Design Action Group, Issue 1.3

Ingold CT (1971) Fungal spores. Clarendon Press, Oxford

Jacobs J (1993) The death and Life of Great American Cities. Random House, New York

Kapoor V, Tripathi S, Devi RS, Srivastava P, Bhadouria R (2020) Ecological economics of an urban settlement: an overview. Urban Ecol 91–110

Kathleen LW, Sharon TL, Jennifer KM, Gregory RAR, van den Matilda B, Adrina CB (2020) Urban trees and human health: a scoping review. Int J Environ Res Public Health 17(12):4371. https://doi.org/10.3390/ijerph17124371

Kielbaso JJ (2008). Management of urban forests in the United States. In: Carreiro MM, Song YC, Wu J (eds) Ecology, planning, and management of urban forests. Springer, New York

Konijnendijk CC, Randrup TB (2004) Urban forestry. Chapter in landscape and planning. Elsevier, pp 472–478

Konijnendij KC, Randrup T, Nilsson K, Schipperijn J (2005) Urban forests and tree reference book. Springer

Kovats S, Akhtar R (2008) Climate, climate change and human health in Asian cities. Environ Urban 20:165–175

Kumar SR, Arumugam T, Anandakumar CR, Balakrishnan S, Rajavel DS (2013) Use of plant species in controlling environmental pollution—a review. Bull Env Pharmacol Life Sci 2(2):52–63

Lamichhane D, Thapa HB (2012) Participatory urban forestry in Nepal, Gaps and ways forward. Urb For Urb Greening 11:105–111. https://doi.org/10.1016/j.ufug.2011.07.008

Lavania UC, Lavania S (2009) Sequestration of atmospheric carbon into subsoil horizons through deep rooted grasses-Vetiver grass model. Curr Sci 97(5):619–619

Lei Y, Carlucci L, Rijnaarts H, Langernhoff A (2022) Phytoremediation of micropollutants by Phragmites australis, Typha angustifolia and Juncus effuses. Int J Phytorem 13:1–7. https://doi.org/10.1080/15226514.2022.2057422

Li F, Wang R, Paulussen J, Liu X (2005) Comprehensive concept planning of urban greening based on ecological principles: a case study in Beijing, China. Landsc Urban Plan 72:325–336

Li YY, Wang XR, Huang CL (2011) Key Street tree species selection in urban areas. African J Agric Res 6(15):3539–3550. https://doi.org/10.5897/AJAR11.461

Locatelli B (2016) Ecosystem services and climate change. In: Routledge handbook of ecosystem services. In: Potschin M, Haines-Young R, Fish R, Turner RK (eds) Routledge handbook of ecosystem services. Routledge, London, pp 481–490. https://www.routledge.com/products/9781138025080. ISBN 978-1-138-02508-0

Maheshwari JK (1963) The flora of Delhi. Council of Scientific and Industrial Research, New Delhi
Makarieva AM, Gorshkov VG (2007) Biotic pump of atmospheric moisture as driver of the hydrological cycle on land. Hydrol Earth Syst Sci 11(2):1013–1033
McKinney ML (2008) Effects of urbanization on species richness: A review of plants and animals. Urban Ecosyst 11:161–176
Negi SS (1998) Urban and recreational forests. International Book Distributors, Dehradun, India
Niemela J (1999) Ecology and urban planning. Biodivers Conserv 8(1):119–131
Nowak DJ, Greenfield EJ, Hoehn R, LaPoint E (2013) Carbon storage and sequestration by trees in urban and community areas of the United States. Environ Pollut 178:229–236
Pandey HP, Luitel DJ (2020) Diversity and species selection in urban forestry: reflection from Maitighar to Tinkune Road of Kathmandu Valley, Nepal. J Environ Sci 6:19–26
Pearlmutter D, Calfapietra C, Samson R, O'brien L, Krajter S, Sanesi G, Delamo RA (2017) The urban forest, cultivating green infrastructure for people and the environment. Springer International Publishing AG
Pickett STA, Burch WR Jr, Dalton SE, Foresman TW, Grove JM, Rowntree R (1997) A conceptual framework for the study of human ecosystems in urban areas. Urban Ecosyst 1(4):185–199
Pokhriyal TC, Subba Rao BK (1986) Role of forests in mitigating air pollution. Indian For 112:573–582
Rodrigues ME, Tjelkar MG, Leneir J, Baumgartner JB, Beument LJ, Nipperess DA, Power SA, Reichard B, Rymer PD, Gallgher BV (2022) Climate change increases global risk to urban forests. Nat Clim 12:950–955
Romero-Lankao P, Gratz DM (2008) Urban areas and climate change: review of current issues and trends. Issues paper for the 2011 global report on human settlements
Sabo A, Benedikz T, Randrup T (2003) Selection of trees for urban forestry in the Nordic Countries. Urb For Urb Greening 2(2):101–114. https://doi.org/10.1078/1618-8667-00027
Schell LM, Ulijaszek SJ (1999) Urbanism, health and human biology in industrialized countries. Cambridge University Press, Cambridge
Shetye RP, Chaphekar SB (1989) Some estimation on dust fall in the city of Bombay, using plants. Progr Ecol 4:61–70
Shwartz M (2000) Stanford report, NewService/Press Release, December 1, 2000
Singh SP (2002) Balancing the approaches of environmental conservation by considering ecosystem services as well as biodiversity. Curr Sci 82(11):1331–1335
Tripathi S, Bhadouria R, Srivastava P, Devi RS, Chaturvedi R, Raghubanshi AS (2020) Effects of light availability on leaf attributes and seedling growth of four tree species in tropical dry forest. Ecol Process 9(1):1–16
Tripathi S, Bhadouria R, Srivastava P, Singh R, Raghubanshi AS (2018) The effects of interacting gradient of irradiance and water on seedlings of five tropical dry forest tree species. Trop Ecol 59(3):489–504
UN (2018) News, May 16, 2018, New York. Monthly Newsletter-UN DESA Voice. Department of Economic and Social Affairs
Vasavi A, Usha R, Swamy PM (2010) Phytoremediation—an overview review. J Ind Pollut Control 26(1):83–88
Verma P, Singh R, Bryant C, Raghubanshi AS (2020) Green space indicators in a social-ecological system: a case study of Varanasi, India. Sustain Cities Soc 60:102261
Wolf K (1998) Urban nature benefits: psycho-social dimensions of people and plants, fact sheet # 1. Centre for Urban Horticulture, Washington
Zhang Y (2013) Urban metabolism: a review of research methodologies. Environ Pollut 18:463–473. https://doi.org/10.1016/j.envpol.2013.03.052

Part IV
Smart Urban Metabolism

Chapter 15
Geospatial Analyses for Urban Metabolism and Climate Change

Sunil Bhaskaran, Asami Minei, Jin Shin, Sanjiv Bhatia, and Andrella Collins

Abstract The amount of CO_2 sequestered by trees in any region may play an important role in reducing the harmful impacts of CO_2 and assisting in designing strategies for addressing global climate change. Given the current growth of global economies, the energy demand is increasing at an alarming rate leading to an increase in CO_2 emissions and a subsequent increase in global temperatures. It has been reported that cities are responsible for more than 70% of greenhouse gas (GHG) emissions, and they have an obligation to the decarbonization of the global economy. Global acknowledgment of the challenges posed by climate change events is reflected by the increasing number of extreme weather events. This has led to an unprecedented demand for novel ideas and methods for addressing the challenges emanating from the increase in CO_2 emissions. Mapping different types of vegetation and the levels of CO_2 sequestered may provide some solutions toward addressing and reducing global temperatures. We present a methodology to map an accurately detailed land cover and forested/grassland areas over the Bronx in New York City. The estimation of CO_2 sequestered by vegetation areas will be derived from multispectral data, digital elevation models (DEMs), and tree sample data through height-derived allometric equations. Since the existing New York City tree database does not record tree age or total height, elevation data from the LiDAR data will be critical in terms of biomass and carbon dioxide estimation. The results may enable city planners and policymakers to develop new strategies for developing sustainable city planning for reducing the adverse impacts of GHGs and the sustenance of urban metabolism.

Keywords Carbon · Climate change · Digital elevation models · Land cover · LiDAR

S. Bhaskaran (✉)
E.E.S, Graduate Center, BCC Geospatial Center of the CUNY CREST Institute, City University of New York (CUNY), New York, USA
e-mail: Sbhaskaran@gc.cuny.edu

A. Minei · J. Shin · A. Collins
BCC Geospatial Center of the CUNY CREST Institute, Bronx, USA

S. Bhatia
Department of Computer Science, University of Missouri, Columbia, USA

R. Bhadouria et al. (eds.), *Urban Metabolism and Climate Change*,
https://doi.org/10.1007/978-3-031-29422-8_15

15.1 Introduction

Global climate change poses several serious risks and challenges to urban communities. The rising global mean temperatures raise sea levels, increasing the number and frequency of extreme weather events (Bhadouria et al. 2022; Qian et al. 2022), which can directly harm or destroy urban communities. New York City is one of the world's biggest urban communities in terms of culture and economics and is not impervious to global climate change. For example, historical catastrophic river and urban flooding were caused by the post-Tropical Depression Ida in 2021 September. This horrific event caused casualties, up to $95 million in damage, and economic loss according to the estimation of Joel N. Myers, the AccuWeather founder, and chief executive, and a devastating impact on the infrastructure such as submerging the subway system in the New York area (Puleo 2021). Another example of global climate change is the rising mean temperatures. The mean annual temperatures in New York City and the surrounding regions are projected to increase by 4.0–7.5 °F by the 2080s (Horton et al. 2010). The impacts may include increased infrastructure maintenance caused by the summertime stress on materials and the increased use of electricity in summer. All those rapid global climate changes and impacts are caused by greenhouse gas (GHG) emissions by human activities. GHGs, including carbon dioxide (CO_2) and methane (CH_4), have caused the temperature on our planet to rise, resulting in an increase in the frequency and intensity of extreme weather events. Cities are responsible for over 70% of global CO_2 emissions (IPCC 2014). As of 2010, more than half of the global population lived in urban areas, and the global urban population is expected to increase up to 70% (United Nations 2019). Therefore, controlling CO_2 emissions in urban areas is an extremely important key to climate change mitigation.

To tackle those challenges, the state of New York has been making efforts to support the energy transition, including renewable energy, building decarbonization, a clean energy economy, a durable and distributed energy system, and lastly, GHG emissions reduction using green space (York 2022). Managing urban outdoor green space is an important key to controlling CO_2. Urban vegetation adds CO_2 to the atmosphere when they die and decompose, increasing the total CO_2 emissions. In contrast, urban vegetation, such as forests in parks, greenery, and street trees, can be functional in sequestering atmospheric CO_2 and storing a carbon-based sugar molecule called glucose and oxygen by photosynthesis, resulting in the reduction of the amount of atmospheric CO_2. Not just sequestering CO_2 from the atmosphere, urban green spaces can also lower surface and air temperatures by offering shade and through evapotranspiration. Research shows that peak summer temperature can be reduced by 1–5 °C through evapotranspiration or in combination with the shading of vegetation (Huang et al. 1990; Kurn et al. 1994). Therefore, monitoring and collecting a record of how much outdoor green space can sequester atmospheric CO_2 in urban settings would provide important insights into global climate change and its impacts over time. However, developing an accurate CO_2 sequestration map has been challenging because accurate calculations require detailed tree species growth characteristics, age, height, and the tree's wood density (Birdsey 1992; DOE US 1998). However, estimating CO_2 sequestered by an individual tree is a labor-intensive and time-consuming process, making it extremely challenging for researchers to measure every single tree in the region of interest.

Remote sensing technologies have increasingly remedied the challenges of the traditional estimation of CO_2 sequestration in a tree. Remotely sensed data are often coupled with field data to train and validate a model to estimate CO_2 sequestration through regression algorithms using machine learning methods. This means machine learning methods allow us to predict continuous outcome values using the sample CO_2 sequestration data and predictor variables from remotely sensed data. Oftentimes, multispectral and light detection and ranging (LiDAR) sensors are used to derive predictors including vegetation indices, texture features, and elevation models (Lu 2006; Zhang and Shao 2021).

The fusion of active LiDAR data and passive multispectral imagery is shown to be useful for accurate tree detection, which can help classify the land cover and estimate sequestered CO_2 by trees. LiDAR data are often creating high-quality digital elevation models (DEMs), and the tree height data derived from the LiDAR data are extremely useful for tree biomass and CO_2 estimations. Multispectral data are often used to derive vegetation indices that measure the condition of vegetation. Research shows that vegetation indices are deeply correlated to biomass and CO_2 estimation.

Among other remote sensing platforms, orbital satellite platforms have been traditionally used to monitor large areas of land cover. This approach provides valuable insight into land cover management, land cover change detection, and vegetation health. Although the satellite platforms can collect large areas of data at once, the trade-off is the relatively low pixel resolution, which is commonly about 30 m. However, with the recent advances in remote sensing technologies, higher-resolution imagery has become more affordable and available to researchers. The higher spatial resolution enables a more detailed analysis of urban areas which are heterogeneously organized with dense buildings, roads, and vegetation laid out in small areas (Ali et al. 2022).

The objective of this study was to foster a methodology for accurately estimating CO_2 sequestered by urban vegetation—open grassland, bushes, and tree cover from a region in Bronx City, New York. The methodology was to process and format different types of datasets and to create twenty-four variables that were to be used in the estimation of CO_2 sequestered. A detailed land cover classification was performed on the 8-band multispectral imagery (WorldView-2) for extracting the forest and grassland areas. Tree cover and grassland areas were extracted separately for estimation of the biomass which was used in two separate allometric equations. The annual CO_2 sequestration rate of grassland was calculated by using allometric equations that ingested different types of vegetation index.

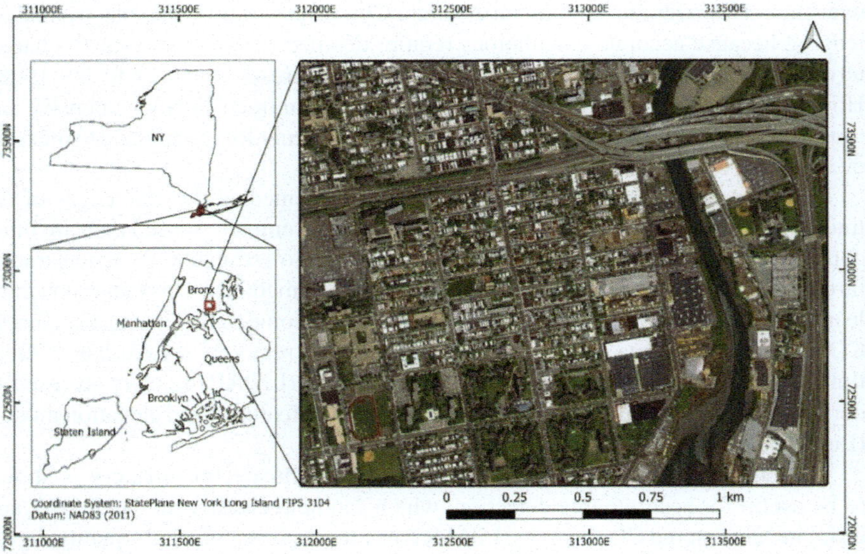

Fig. 15.1 Sketch map of the study site. The 2015 true color WorldView-2 imagery was used to show the study area and its extent

15.2 Study Area

The study region was in the Bronx, New York, USA (40° 49′ 20.3″ N and 73° 50′ 47.7″ W) (Fig. 15.1). This area covers 3.1 km^2 and is classified as a warm/humid continental climate according to the Köppen climate classification. The summers are warm and wet while the winters are freezing, snowy, and windy. Over the year, the temperature typically varies from −2.7 °C to 29.3 °C (27 °F to 85 °F). The area receives mean annual precipitation of 121 cm (47.6 inches). The land cover comprises dense residential sites, highways, trees, grassland, and water bodies including the Westchester River that runs through the study area.

15.3 Datasets

15.3.1 WorldView-2

We used WorldView-2 multiband high-resolution spaceborne imagery acquired on the May 10, 2016, for the study. The WorldView-2 instrument collects images in the panchromatic at a spatial resolution of 0.5 m in the panchromatic (450–800 nms) and 1.8 m in the multispectral bands in the 400–800 nms wavelength (DigitalGlobe).

15.3.2 New York (NYC) Topo Bathymetric Data

In May 2017, Quantum Spatial (QSI) collected topographic, topo bathymetric light detection and ranging (LiDAR) data for New York City (NYC) in the spring and summer of 2017. The LiDAR data were used to derive 1-foot (30 cm) resolution topo bathymetric digital elevation (DEM) and surface models (DSMs). The elevation data are publicly available on the official Web site of New York State (gis.ny.gov). These datasets were used as independent variables for land cover classification and in the extraction of the elevation data for biomass estimation.

15.3.3 2015 Street Tree Census

Volunteers and staff organized by NYC Parks and Recreation and partner organizations managed a detailed street tree census in 2015. The tree data include tree species, diameter, and its locational data, which was used for verification of vegetation areas of land cover classification and in the CO_2 sequestration of sample trees. The street tree data can be downloaded as a geodatabase (NYC Open Data).

15.3.4 Summary of Datasets Used in the Study

Datasets used for this tutorial are listed in Table 15.1 with the name, data type, collection date, purpose, and source. The multispectral high-resolution 8-band (VIS–NIR) WorldView-2 satellite data (top of the atmosphere) were used for delineating the different land cover classes including tree cover and open grass. The New York City 0.30 m by 0.30-m topo bathymetric digital elevation model (DEM) was utilized to estimate the tree height which is a key variable for calculating the above-ground biomass. The New York City tree point datasets were used to perform random validation of the classification results, and the socioeconomic datasets were acquired from the United States Census data repository. The socioeconomic datasets were used to visualize the correlations between CO_2 sequestered versus population density, income levels, and ethnicity. This was done to better understand environmental justice, awareness levels, and other related issues.

15.4 Methodology and Analyses

The methodology involved image processing, feature extraction estimation of sequestered carbon (Fig. 15.2). The LiDAR-generated DEM and DSM were used to calculate the canopy height model (CHM) by using the raster calculator. The

Table 15.1 Datasets used in the analyses

Name	Data type	Collection date	Purpose
WorldView-2	8 multiband (1.8 m) 1 panchromatic (0.46 m)	10/05/2016	Land classification/vegetation extraction
NYC Topo bathymetric DEM/DSM	0.30-m raster	05/03/17 and 07/26/17	Land classification/tree height extraction for biomass calculation
NYC street tree census	Point data	2015	Classification validation
2010 Census data (population, ethnicity, and income)	CSV	2022	See the correlation of the census data to the vegetation area
NYC census boundary	Shape		Census boundary to map data

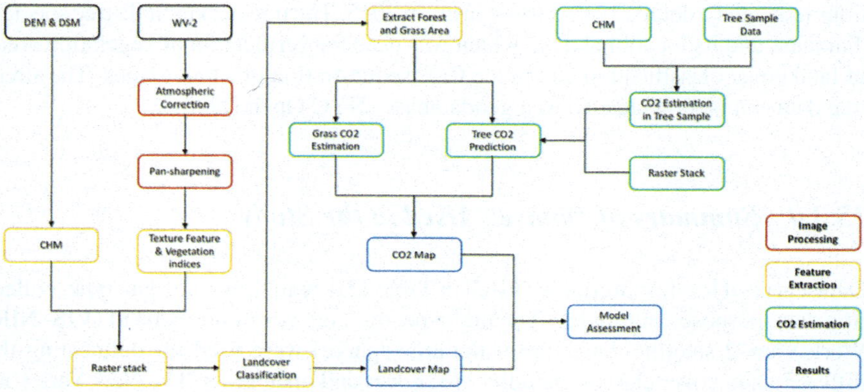

Fig. 15.2 Flowchart of data processing workflow

WV-2 imagery was corrected to generate a surface reflectance product by using an atmospheric correction algorithm that was used to correct the wavelengths from the visible near infrared (VNIR) through the shortwave infrared (SWIR) by determining the atmospheric compensation parameters using the observed spectra without using any ancillary information. We used QUAC, which (Quick Atmospheric Correction (QUAC)) which is an atmospheric correction method for multispectral and hyperspectral imagery that works with the visible and near-infrared through shortwave infrared (VNIR-SWIR) wavelength range. The results of the conversion and final product are shown in Fig. 15.3a.

The image was sharpened by using the higher-resolution panchromatic band to merge with the multispectral lower resolution bands and resampled to the resolution of the higher-resolution data which was the pan band (0.46 m by 0.46 m). Since the objective was to calculate the amount of CO_2 sequestrated in urban areas as

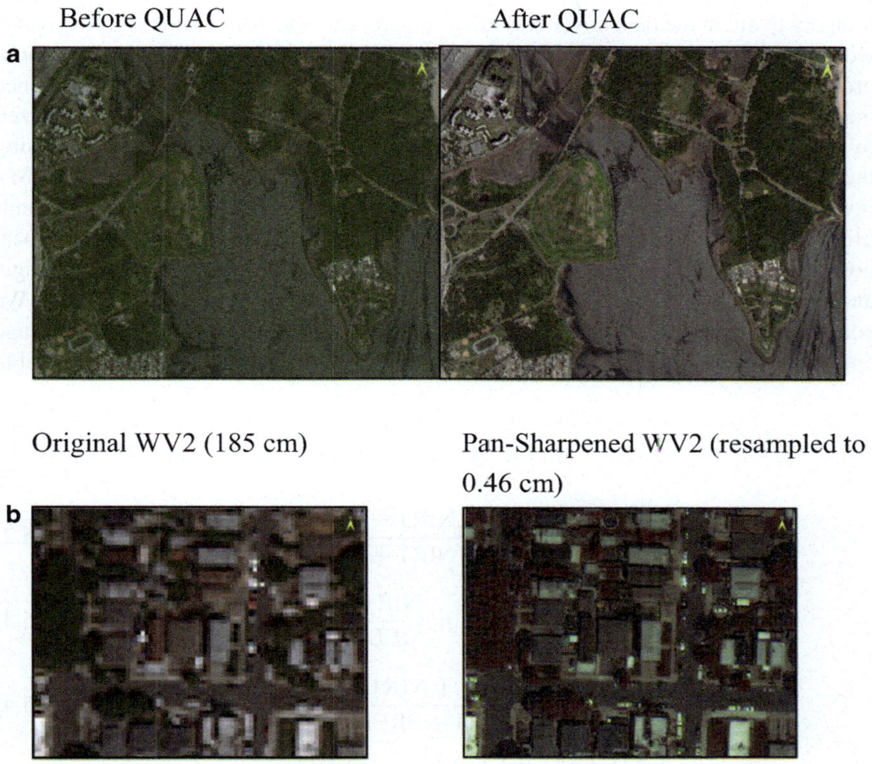

Fig. 15.3 **a** WorldView-2 before and after applying the QUAC correction. **b** Imagery after the Gramm–Schmidt spectral sharpening method was used

accurately as possible, the Gramm–Schmidt spectral sharpening (GSPS) method in ENVI was applied to render a sharp resampled image in Fig. 15.3b.

The DEM and DSM were derived from an airborne LiDAR dataset. The DEM and DSMs were merged with the clipped area to the study area (WV-2), and the units of measurement were standardized to meters by using a raster calculator. The canopy height model (CHM) was generated for each pixel by using DEM and DSM as shown by Eq. 15.1 in Table 15.2. We derived relevant indices from the imagery that were critical for the study including the normalized difference vegetation index (NDVI), ratio and enhanced vegetation indices (RVIs and EVIs). Additionally, the green NDVI or GNDVI, the modified and optimized soil-adjusted vegetation indices (MSAVIs, OSAVIs) were derived from the imagery. These indices and equations (below) were derived from earlier research conducted by (Kaufman 1992; Gitelson et al. 1996; Huete 1988; Aiazzi 2002; Huete et al. 2002; Klonus and Ehlers 2007; Ahmad 2012; Johnson 2014; Xue 2017; Gholineiad and Fatemi 2019; Ghimire et al. 2020). The WV-2 has broad bands which present a challenge to spectrally unmix the pixels; therefore, we computed different types of texture attributes from the imagery to aid in the land cover classification process. The near-infrared (NIR) bands were found to be the most suitable for computing the texture attributes. A second-order statistics texture model was performed by using the gray-level co-occurrence matrices (GLCMs) (Haralick 1979). Since urban environments are characterized by heterogeneity, we were able to delineate between land cover classes by using their texture attributes on the ground. We used the angular second moment (ASM), inverse difference moment (IDM), contrast, correlation matrices to calculate second-order statistics from the image. These matrices allowed us to compute the local homogeneity of pixels, acquire direct measures of local homogeneity of the image, and enabled us to estimate linear dependency of gray levels of nearby pixels. We computed attributes for twenty-four (24) variables that were derived from the image and resampled to a single resolution for further calculations and estimations (Table 15.2).

$$\mathrm{CHM} = \mathrm{DSM} - \mathrm{DEM} \tag{15.1}$$

$$\mathrm{NDVI} = \frac{((\mathrm{NIR1} - \mathrm{RED}))}{((\mathrm{NIR1} + \mathrm{RED}))} \tag{15.2}$$

$$\mathrm{RVI} = \frac{\mathrm{NIR1}}{\mathrm{RED}} \tag{15.3}$$

$$\mathrm{EVI} = 2.5 \times \frac{((\mathrm{NIR1} - \mathrm{RED}))}{((\mathrm{NIR1} + 6\mathrm{RED} + 7\mathrm{Blue}) + 1)} \tag{15.4}$$

$$\mathrm{GNDVI} = \frac{(\mathrm{NIR2} - \mathrm{GREEN})}{(\mathrm{NIR2} + \mathrm{GREEN})} \tag{15.5}$$

$$\mathrm{SAVI} = \frac{(\mathrm{NIR1} - \mathrm{RED})}{(\mathrm{NIR1} + \mathrm{RED} + L)(1 + L)} \quad \text{where } L = 0.5 \tag{15.6}$$

$$\mathrm{MSAVI} = \left(2\mathrm{NIR2} + 1 - \frac{\sqrt{((2\mathrm{NIR2} + 1)^2 - 8(\mathrm{NIR2} - \mathrm{RED}))}}{2} \right) \tag{15.7}$$

$$\mathrm{OSAVI} = (1 + 0.16)\frac{(\mathrm{NIR1} - \mathrm{RED})}{(\mathrm{NIR1} + \mathrm{RED} + 0.16)} \tag{15.8}$$

15.4.1 Land Cover Classification and Vegetation Extraction

An object-based approach was employed to classify the land cover. Object-based approaches were successfully employed in a New York City study using time series of Ikonos multispectral imagery (Bhaskaran et al. 2013). Object-based approaches use

Table 15.2 Equations used for extracting vegetation indices from the WV-2 bands

Band #	Variable	Band #	Variable	Band #	Variable
1	B1	9	CHM	17	VAR
2	B2	10	DEM	18	EVI
3	B3	11	DSM	19	GSAVI
4	B4	12	ASM	20	MSAVI
5	B4	13	CONTR	21	NDVI
6	B6	14	CORR	22	OSAVI
7	B7	15	ENTRE	23	RVI
8	B8	16	IDM	24	SAVI

elements of photointerpretation like shape, size, texture, and association of features to delineate the land covers and are useful to delineate spectrally similar features (Bhaskaran et al. 2010). A detailed land cover classification was performed to extract land cover including urban vegetation areas to estimate the above-ground biomass (AGM). The vegetation areas are further subdivided into grassland and forest/tree cover areas. The ENVI feature extraction (Fx) was used for the land cover classification by segmenting the image into regions of pixels, followed by computing the spectral and spatial attributes for each region to create image objects and classification. A multi-dimensional raster layer was created as input data by stacking multispectral, DEM, and DSM data. The stack layer was processed to create a segmentation image in the feature extraction process, and the attributes for each region were automatically chosen from the stack layer by ENVI to maximize the classification performance. This was followed by creating training datasets on a segment of the image. A support vector land cover classification algorithm was used to map the different classes including grass, forest, road, hard pavement, rooftop, shadow, bare ground, and water (Fig. 15.4a). An example-based method yielded a land cover map consisting of eight land cover classes which were tested for its accuracy against a validation or ground truth dataset. An accuracy assessment of the classification yielded satisfactory results (Fig. 15.4b). The project-specific variables of open grass and tree cover were extracted from the land cover classification (Fig. 15.4c).

15.4.2 Above-Ground Estimation Model

The annual CO_2 sequestered was estimated by using the open grass and tree covers by using the methods described below. The CO_2 estimates were derived from the image through regression algorithms and empirical allometric equations.

a

b

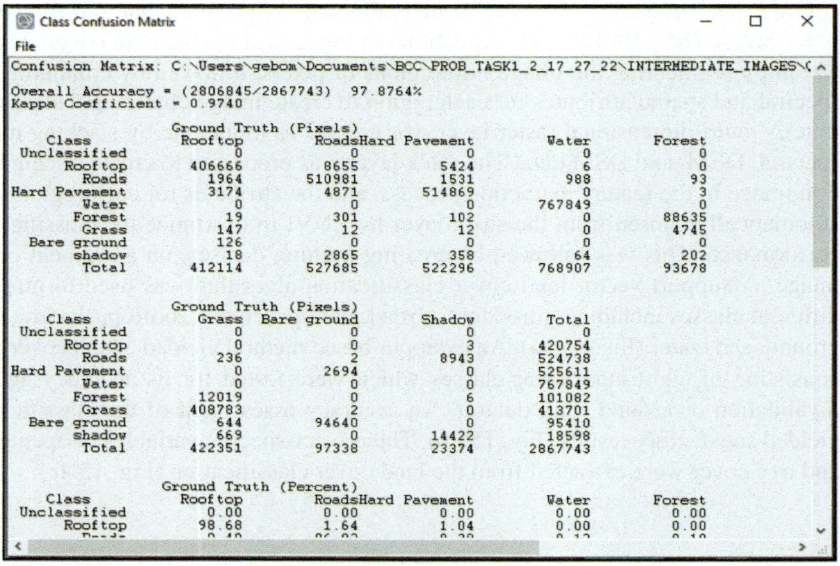

Class Confusion Matrix

File

Confusion Matrix: C:\Users\gebom\Documents\BCC\PROB_TASK1_2_17_27_22\INTERMEDIATE_IMAGES\(

Overall Accuracy = (2806845/2867743) 97.8764%
Kappa Coefficient = 0.9740

Ground Truth (Pixels)

Class	Rooftop	Roads	Hard Pavement	Water	Forest
Unclassified	0	0	0	0	0
Rooftop	406666	8655	5424	6	0
Roads	1964	510981	1531	988	93
Hard Pavement	3174	4871	514869	0	3
Water	0	0	0	767849	0
Forest	19	301	102	0	88635
Grass	147	12	12	0	4745
Bare ground	126	0	0	0	0
shadow	18	2865	358	64	202
Total	412114	527685	522296	768907	93678

Ground Truth (Pixels)

Class	Grass	Bare ground	Shadow	Total
Unclassified	0	0	0	0
Rooftop	0	0	3	420754
Roads	236	2	8943	524738
Hard Pavement	0	2694	0	525611
Water	0	0	0	767849
Forest	12019	0	6	101082
Grass	408783	2	0	413701
Bare ground	644	94640	0	95410
shadow	669	0	14422	18598
Total	422351	97338	23374	2867743

Ground Truth (Percent)

Class	Rooftop	Roads	Hard Pavement	Water	Forest
Unclassified	0.00	0.00	0.00	0.00	0.00
Rooftop	98.68	1.64	1.04	0.00	0.00

Fig. 15.4 **a** Land cover map over a section of the Bronx Borough, New York City. **b** Accuracy assessment report. **c** Open grass and tree cover map—study relevant layers extracted from the land cover map

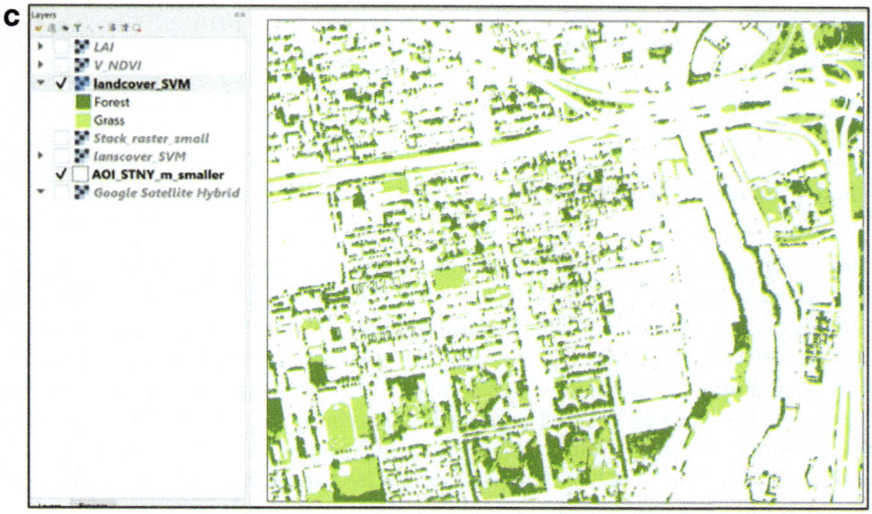

Fig. 15.4 (continued)

15.4.3 Grassland and Low Bush Area

The CO_2 sequestration rates for open grass and low bushes were estimated by calculating the biomass and converting it into the annual CO_2 sequestration rate. For grassland and low bush areas biomass, the EVI was used. An EVI-derived allometric Eq. (15.9) was used for calculating biomass by using an existing and proven algorithm (Meshesha et al. 2020). The actual carbon stored by a tree is the combined value of carbon captured during photosynthesis and sequestered by it. A tree's actual dry weight total volume is used to estimate its average carbon content which is approximately equal to 50% of its actual dry weight volume. Therefore, we used the dry weight to calculate and determine the weight of carbon in the tree [Wcarbon = 0.5 * Wdry weight]. The mass of the wood in the tree consists of the above-ground dry weight minus the moisture in the tree.

To find carbon storage (kg), we multiplied the dry weight by 0.5 (Eq. 15.15). The total amount of carbon dioxide found in a tree is the amount of carbon dioxide stored in the tree. CO_2 is converted from C by multiplying by 3.667 (Quantification methods for tree planting projects). The calculated biomass was converted to CO_2 sequestration by multiplying by 0.5 and 3.667 (Eq. 15.10). The initial annual CO_2 sequestration rate is in tons * hectare^{-1} * year^{-1}. Since the study area is relatively small, we converted the unit to kg m^{-2} year^{-1} by using Eq. 15.11.

$$\text{AGB grass (t/ha.year)} = (11.21(\text{EVI}))^2 + 0.27(\text{EVI}) + 0.038 \tag{15.9}$$

$$CO_2\ \text{(t/ha.year)} = \text{DW} \times 0.5 \times 3.667 \tag{15.10}$$

$$CO_2(\text{kg/m}^2.\text{year}) = \frac{CO_2(\text{t/ha.year}) \times 1016}{10{,}000} \quad (15.11)$$

15.4.4 Forest/Tree Cover

The height of the tree and diameter of the trunk breast height (DBH) were used in an allometric equation to calculate the CO_2 sequestered by tree cover. The NYC Street tree sample data that contain the locational information for each tree, DBH, and the elevation data derived from LiDAR data were used to estimate the CO_2 sequestered for each WorldView image pixel with an assumption that all pixels were representative of a tree cover. The workflow for estimating CO_2 from tree cover is to determine the above-ground green weight (GW) of a tree, the above-ground dry weight (DW) of a tree, the carbon storage in a tree, the carbon dioxide sequestered in a tree, and above-ground green weight (GW).

The above-ground green weight of a tree (GW) in kg for each tree is calculated based on the diameter in breast height (DBH) in cm and tree height (H) in meters. For the trees with a DBH less than 28 cm, Eq. 15.12 was used. For trees with a DBH of more than 28 cm, Eq. 15.13 was used. The equations used here are an average for trees in the Southeast including sweetgum and red oak; they are the best options for this research (Clark and Schroeder 1985) since some of the southeast trees are also found in the study area. The 2015 NYC street tree data have a diameter in breast height (DBH) but not the height of trees. Therefore, the height of a tree was extracted from the 2017 CHM data.

$$\text{GW} = 0.0577 \times (\text{DBH})^2 \times H \quad (15.12)$$

$$\text{GW} = 0.0346 \times (\text{DBH})^2 \times H \quad (15.13)$$

$$\text{DW} = \text{GW} \times 0.5 \quad (15.14)$$

$$C = \text{DW} \times 0.5 \quad (15.15)$$

$$CO_2 = C \times 3.667 \quad (15.16)$$

Once the amount of CO_2 in each tree of the NYC street tree sample was calculated, a regression model was used to estimate the CO_2 stored for the whole forested area. The biomass for the open grassland was estimated by using an EVI. According to earlier research (Meshesha et al. 2020), the EVI-based biomass model is effective in estimating about 92% of the grass biomass, and the NDVI-based biomass may be used to estimate about 87% of the biomass. We used the raster calculator tool to run Eqs. (15.10) and (15.11) and *R* software to mask values for all other classes other than open grass cover. A raster layer statistic was generated to get the CO_2 sequestration metrics (Fig. 15.5a). The distribution of CO_2 sequestered by grassland is shown by a graph and a boxplot (Fig. 15.5b), and the statistics of carbon sequestered by open grass is shown in Table 15.3.

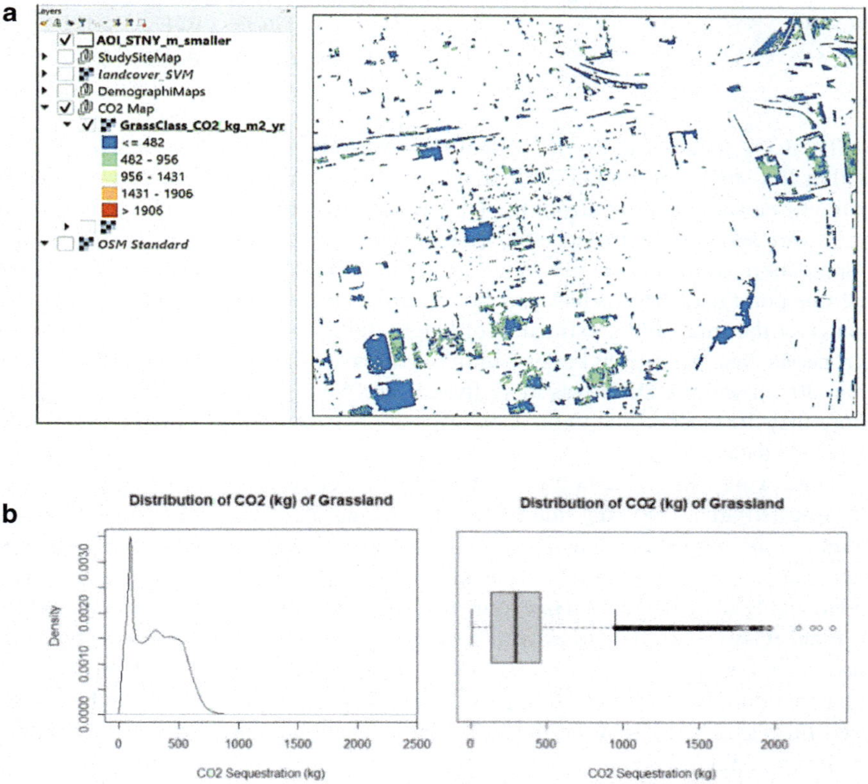

Fig. 15.5 **a** CO_2 sequestered by open grass. **b** Distribution of CO_2 sequestration of grassland represented in smooth density estimation and boxplot

Table 15.3 Statistics for CO_2 sequestered by open grass

Minimum value	6.775922298431396
Maximum value	2380.62548828125
Range	2373.849565982819
Sum	890,492,392.0161834
Mean value	309.6608730724475
Standard deviation	187.1969762715857

Table 15.4 Tree point data extracted from tree cover land class using the NYC tree cover dataset

Tree ID	Username	Latitude	Longitude	City	Height (m)
402,561	NYC Parks Staff (NYCPS)	1,024,413.16	241,655.68	10,472	0.058
402,383	NYCPS	1,024,349.39	231,139.40	10,472	0.055
402,062	NYCPS	1,025,149.48	241,246.30	10,472	1.957
402,285	NYCPS	1,025,803.28	241,904.27	10,462	3.368
402,555	NYCPS	1,024,654.39	241,698.46	10,472	4.785
401,936	NYCPS	1,024,477.57	241,126.33	10,472	5.121
402,559	NYCPS	1,024,486.02	241,669.60	10,472	0.012

The CO_2 sequestered by individual trees was computed by extracting all trees within the study site biomass for tree cover. The NYC tree dataset was used to extract tree sample points from the land cover class (tree) to which the heights of trees were added. An attribute table showing the sample, datasets, and extracted tree point data from the land cover was created (Table 15.4). There were some errors in the tree point data. This is because the tree points were not collected directly in the center of the tree, so the extraction of height did not work well. To exclude those erroneous data, we selected trees within the forest class where tree height was more than 30 cm and the DBH was more than 1 cm. We also excluded trees with DBH more than 100 inches due to the lack of positional accuracy of the tree location on the NYC dataset.

There were 793 tree samples out of 2581 that met the criteria of: (a) being within the forested area, (b) being taller than 30 cm, and (c) having a diameter of breast height of more than 1 inch and less than 100 inches. The units were converted from inches to centimeters using a field calculator. The following equations where DBH is in cm and H is the height of a tree in meters were used to calculate the Green Weight in Field (GW) in kg. For trees with a diameter of < 28 cm: We used the equation the $\text{GW} = 0.0577 \times (\text{DBH})^2 \times H$, and for trees with a diameter > 28 cm: We used the equation $\text{GW} = 0.0346 \times (\text{DBH})^2 \times H$. For calculating DW, C, and CO_2, we used Eqs. 15.17–15.19 in the field calculator. The cumulative attributes generated are given in Table 15.5.

$$\text{DW} = \text{GW} \times 0.5 \tag{15.17}$$

$$C = \text{DW} \times 0.5 \tag{15.18}$$

$$CO_2 = C \times 3.667 \tag{15.19}$$

A supervised regression model estimated the CO_2 sequestered from the study area. The response variables of CO_2 estimate of trees in WV-2 bands, vegetation indices, and texture attributes (Table 15.6) were used to derive gross estimates (kgs) of the CO_2 sequestered by the tree cover (Fig. 15.6). We achieved this by converting

Table 15.5 Calculation of different CO_2-related attributes

Species	ID	HGT_m	LC	DBH_cm	Carbon	GW	DW	CO_2
London planetree	99,289	0.39	5	5.08	0.145	0.58	0.290	0.532
Golden raintree	10,378	0.64	5	5.08	0.238	0.95	0.475	0.873
Kentucky coffeetree	500,746	0.75	5	5.08	0.280	1.12	0.560	1.027
Japanese Zelkova	504,838	0.34	5	22.86	2.550	10.20	5.100	9.351
Honey Locust	402,110	3.10	5	7.62	2.598	10.39	5.195	9.527
London planetree	501,706	10.22	5	53.34	251.595	1006.38	503.190	922.599
Littleleaf linden	496,792	14.03	5	45.72	253.680	1014.72	507.360	930.245
Golden raintree	405,491	2.34	5	35.56	25.570	102.28	51.140	93.765
London planetree	503,241	11.44	5	50.80	255.418	1021.67	510.835	936.618
London planetree	100,382	10.41	5	53.34	256.395	1025.58	512.790	940.200
Gingko	405,072	8.69	5	58.42	256.538	1026.15	513.075	940.725
Northern Red Oak	506,487	9.51	5	55.88	256.860	1027.44	513.720	941.906

the individual CO_2 estimates to a raster format. All 24 predictor values were aligned to the newly created raster layer and stacked together.

The R software was used to develop codes for deriving the regression model. We developed a methodology to estimate the amount of CO_2 sequestered by the forest

Table 15.6 Variables used in the study

Band #	Variable	Band #	Variable	Band #	Variable
1	CO_2	10	CHM	18	VAR
2	B1	11	DEM	19	EVI
3	B2	12	DSM	20	GSAVI
4	B3	13	ASM	21	MSAVI
5	B4	14	CONTR	22	NDVI
6	B4	15	CORR	23	OSAVI
7	B6	16	ENTRE	24	RVI
8	B7	17	IDM	25	SAVI
9	B8			26	Land cover classification

Fig. 15.6 Estimated CO_2 stored in each tree (kgs)

area in the study area using a regression model. We prepared 24 variables (predictor values) metrics related to biomass as dependent variables, and CO_2 sequestered sample data were used as an independent variable.

15.4.5 Biophysical Metrics

A total of 24 biophysical metrics related to biomass and CO_2 sequestration were extracted from the WV-2 multispectral and topo bathymetric data. Based on the preliminary literature research (Lu 2006; Zhang and Shao 2021) and computer capabilities, eight wave bands, seven vegetation indices, six texture features, and three elevation models were selected. Multiple bands of the WorldView-2 imagery were used in the study to derive vegetation indices including NDVI (Eq. (15.20), RVI (Eq. (15.21), EVI (Eq. (15.22)), GNDVI (Eq. (15.23)), SAVI (Eq. (15.24)), modified soil-adjusted vegetation index (MSAVI) (Eq. (15.25)), and optimized soil-adjusted vegetation index (OSAVI) (Eq. (15.26)).

$$\text{NDVI} = \frac{((\text{NIR1} - \text{RED}))}{((\text{NIR1} + \text{RED}))} \tag{15.20}$$

$$\text{RVI} = \frac{\text{NIR1}}{\text{RED}} \tag{15.21}$$

$$\mathrm{EVI} = 2.5 \times \frac{((\mathrm{NIR1}-\mathrm{RED}))}{((\mathrm{NIR1} + 6\mathrm{RED} + 7\mathrm{Blue}) + 1)} \tag{15.22}$$

$$\mathrm{GNDVI} = \frac{(\mathrm{NIR2} - \mathrm{GREEN})}{(\mathrm{NIR2} + \mathrm{GREEN})} \tag{15.23}$$

$$\mathrm{SAVI} = \frac{(\mathrm{NIR1} - \mathrm{RED})}{(\mathrm{NIR1} + \mathrm{RED} + \mathrm{L})(1 + L)} \quad \text{where } L = 0.5 \tag{15.24}$$

$$\mathrm{MSAVI} = \frac{\left(2\mathrm{NIR}_2 + 1 - \sqrt{((2\mathrm{NIR}_2 + 1)^2 - 8(\mathrm{NIR}_2 - \mathrm{RED}))}\right)}{2} \tag{15.25}$$

$$\mathrm{OSAVI} = \frac{(1 + 0.16)(\mathrm{NIR1} - \mathrm{RED})}{(\mathrm{NIR1} + \mathrm{RED} + 0.16)} \tag{15.26}$$

The texture features were derived by the WV-2 NIR1 band using the algorithms developed by Haralick et al. (1973). The NIR1 band was chosen because it reflected the vegetation information the most compared to other bands of WV-2. The canopy height model (CHM) values were derived (Eq. 15.27) by computing the difference between the digital elevation model (DEM) and digital surface models (DSMs). The DEM and DSM were acquired from the New York City open-source data repository (gis.ny.gov).

$$\mathrm{CHM} = \mathrm{DSM} - \mathrm{DEM} \tag{15.27}$$

15.4.6 CO_2 Sequestration Amount

The amount of CO_2 sequestered by street trees was estimated by using the NYC TreesCount! and CHM data. TreesCount! is the third citizen participatory inventory of street trees in New York City, and the data include street tree location, species, and diameter at breast height (DBH). First, we selected trees that fell within the area of the forested class based on the land cover classification that was previously created. Height-derived allometric equations were used to calculate the CO_2 sequestered by the selected DBH and the height of a tree, which was derived from the TreesCount! 2015 and CHM, respectively. We used the attribute field calculator in QGIS to plug in the equations to determine the amount of CO_2 sequestered for individual trees.

The four steps in estimating the amount of CO_2 sequestered by a tree were as follows: (a) to determine the above-ground green weight (GW) of a tree, (b) the above-ground dry weight (DW) of a tree, (c) the carbon storage in a tree, and (d) the carbon dioxide sequestered in a tree.

The first step was to determine the above-ground green weight of a tree (AGW) in kg. The green weight of a tree is an estimate of the mass of the tree when it is alive. The diameter at breast height (DBH) in centimeters and tree height (H) in meters was used in the estimation of AGW. For trees with a DBH of less than 28 cm, Eq. (15.28) was used. For a tree with a DBH of more than 28 cm, Eq. (15.29) was used. Those equations used in the study are an average for trees in the Southeast including sweetgum and red oak (Clark and Schroeder 1985), included in the study area.

The second task was to determine the above-ground dry weight (ADW) of a tree in kg. The next step was to determine the stored carbon in a tree. The CO_2 sequestered by a tree in kg of a tree was calculated by multiplying the carbon storage by 3.667 (Eq. 15.32).

$$\mathrm{GW}_{\mathrm{DBH}<28\ \mathrm{cm}} = 0.0577 \times (\mathrm{DBH})^2 \times H \tag{15.28}$$

$$\mathrm{GW}_{\mathrm{DBH}>28\ \mathrm{cm}} = 0.0346 \times (\mathrm{DBH})^2 \times H \tag{15.29}$$

$$\mathrm{DW} = \mathrm{GW} \times 0.5 \tag{15.30}$$

$$C = \mathrm{DW} \times 0.5 \tag{15.31}$$

$$\mathrm{CO}_2 = C \times 3.667 \tag{15.32}$$

The CO_2 sequestered by the tree sample points was calculated, and the point vector data were converted into a raster layer for creating a stacked raster of all dependent and independent variables. All the variables were resampled to the smallest spatial resolution of elevation data of 30 cm, and all pixels were aligned to perform a pixel-based regression. A forest regression algorithm created a predictive model. The model was optimized and assessed by using the R^2 (coefficient of determination) and root-mean-square error (RMSE).

15.4.7 CO_2 Estimation of Grass Area

Grasslands and shrubs can sequester some atmospheric carbon each year. However, the quantity of sequestration is less than that of trees due to their small size and short lifespan. Because of these characteristics, equations to calculate biomass and CO_2 sequestration are often derived from an empirical allometric equation using the relationship between vegetation indices and CO_2 sequestration. Therefore, we use a modified allometric equation (Eq. 15.33) developed by Meshesha et al. (2020) to estimate the annual CO_2 sequestration rate of grassland.

$$\mathrm{CO}_2\left(\mathrm{kg/m^2.year}\right) = \left((11.21(\mathrm{EVI}))^2 + 0.27(\mathrm{EVI}) + 0.038\right) \times \frac{1.8335}{10} \tag{15.33}$$

15.5 Results and Discussion

15.5.1 Land Cover Classification

Land cover classification was performed using SVM algorithms by using the Environment for Visualizing Images (ENVI) software (Fig. 15.7). The overall classification accuracy and kappa coefficients are 97.88% and 0.97, respectively. Those performance matrices indicate that the classification map is highly accurate based on the validation dataset. Table 15.7 summarizes the resulting class distribution in the study area. The vegetation areas (forest and grass classes combined) occupy 23.04% (0.12 km^2), the artificial objects (rooftop, road, and hard pavement) add up to 67.99% (2.1 km^2), and other natural properties (water and soil) include 6.7% (0.2 km^2) of the study area. The area covered by grassland and trees is small (23.04) compared to the other classes which were used to determine CO_2 sequestration for the study.

Fig. 15.7 SVM land cover classification for the study area

Table 15.7 Distribution of classification results

Class	%	Area (km^2)
Forest	14.44	0.45
Grass	8.60	0.27
Water	4.79	0.15
Soil	1.94	0.06
Rooftops	18.64	0.58
Roads	32.52	1.01
Hard Pavement	16.83	0.52
Shadow	2.24	0.07
Total	100	3.10

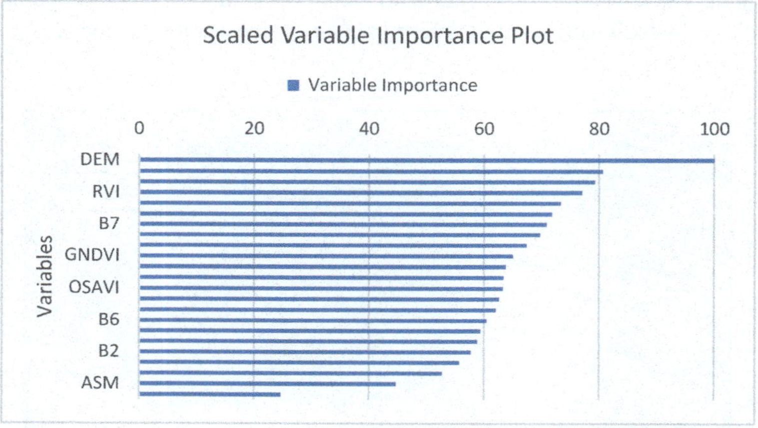

Fig. 15.8 Scaled variable importance plot derived by the random forest regression model

15.5.2 *Regression Model*

The regression model to estimate the CO_2 sequestered amount of the whole study area was performed, and the results showed high R^2 and RMSE ($R^2 = 0.99$, RMSE = 0.92 kg m^{-2}). Figure 15.8 shows the scaled variable importance plots, which indicate that the top variables are all elevation models and some vegetation indices.

15.5.3 *Annual CO_2 Sequestration of Grassland*

Grass sequesters CO_2 with trees, and therefore, we used the WV-2 dataset to calculate annual sequestration rates over grasslands as well (Fig. 15.9a) and summarized (Fig. 15.9b). The mean CO_2 sequestration rate is 0.30 kg/m^2/yr. Ground truthing

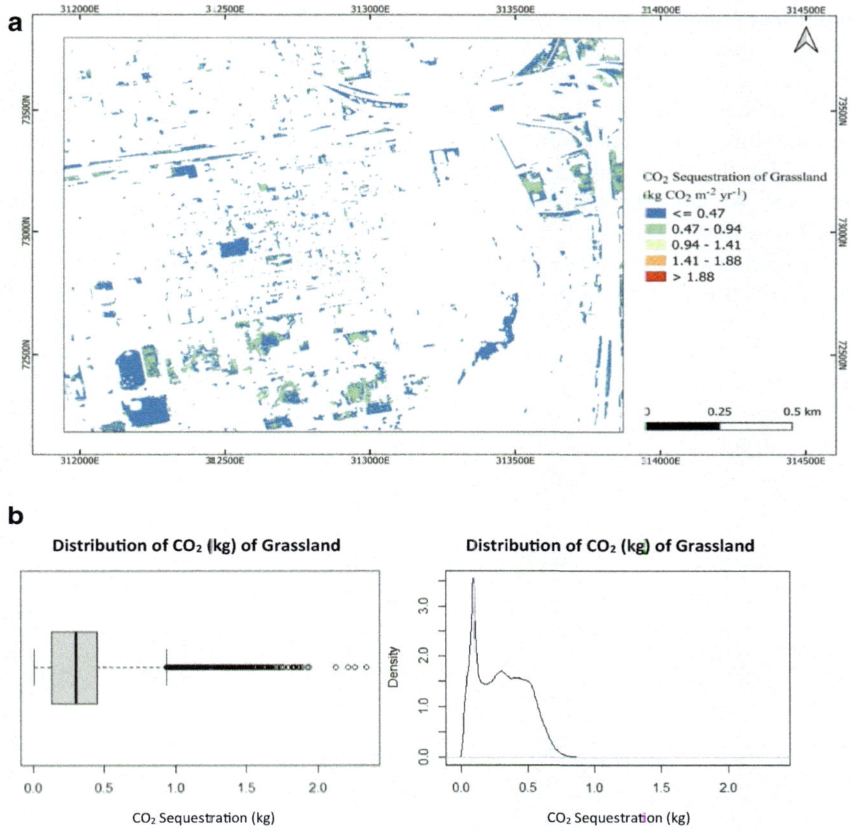

Fig. 15.9 **a** Annual CO_2 sequestration of grassland map. **b** Distribution of CO_2 sequestration in boxplot and smooth density histogram

could not be conducted in this study to estimate the accuracy of the sequestration rates due to COVID-19 for the time being.

Total CO_2 Sequestration Storage of Forest

The amount of CO_2 sequestered was calculated and shown in Fig. 15.10a. In general, the area has low sequestration levels which clearly shows that the tree type and grass cover may not be ideal for CO_2 sequestration. The distribution of the CO_2 sequestered values is summarized in a smooth density histogram and boxplot in Fig. 15.10b. The regression algorithm we used assumed that each pixel was populated and dominated by a tree which was not the case. Therefore, the mean of CO_2 sequestration (576 kg m^{-2}) is invariably overestimated because the regression algorithms assume each pixel has a tree while several pixels cover a single tree. Also, there seem to be some outliers, which increase the mean of CO_2 storage higher. Future research would focus on CO_2 sequestered by individual trees by using appropriate resolution multispectral datasets or by using masking techniques to avoid areas without tree cover. This would enable a more accurate calculation of CO_2 sequestered. The age of the tree is also a key parameter that was not used in this study because of the unavailability of data. For future studies, we will conduct detailed field investigations for logging the DNA and age of trees that would lead to better CO_2 sequestration estimates. We will also explore object-based approaches to detect and extract individual tree types. The object-based approach yielded successful results in a study where Ikonos high-resolution datasets were used to delineate individual rooftops in New York City (Bhaskaran et al. 2013).

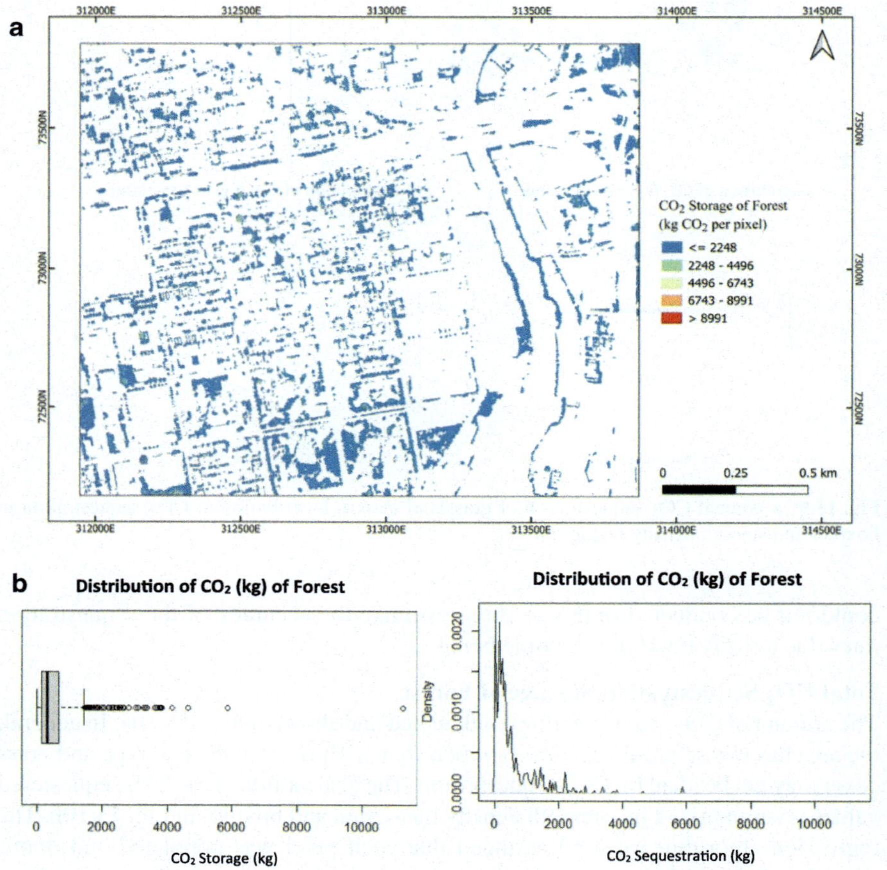

Fig. 15.10 **a** CO_2 sequestered by tree map. **b** Distribution of predicted CO_2 sequestration of the forest area in boxplot and smooth density histogram

15.5.4 *Demographic Data and the CO_2 Sequestration*

We integrated demographic characteristics of the population in the Bronx with the rates of CO_2 sequestered to visualize the spatial patterns of CO_2 sequestration in the region. In general, most regions are characterized by low sequestration rates even though there are variations within that observation. Regions with an ethnically diverse population with low income are correlated with low sequestration rates. The Asian community has higher income levels compared to Afro-Americans, but both races are characterized by low sequestration rates. These analyses may assist in developing policies to plan specific species of trees that will increase the sequestration rates and play an important albeit small role in achieving the net-zero emissions targeted by NYSERDA. For example, the awareness levels of people in this area may need to be elevated so that they support the planting of more trees that sequester carbon at higher rates. The map will also assist in designing policies to address equity issues around climate mitigation strategies. Most of the Afro-American communities have lower CO_2 sequestration rates through trees and grass cover Urban development must integrate strategies to mitigate the effects of carbon emissions by using informed decisions. Composite maps of socioeconomic and carbon sequestered were derived to show the distribution between carbon sequestered, population density (Fig. 15.11a), median household income (Fig. 15.11b), and ethnicity (Fig. 15.11c).

Future Studies

We will explore the eddy covariance technique (Aubinet et al. 2012) in our future study, which has been developed and successfully applied for net ecosystem exchange (NEE) measurements under numerous conditions over an array of vegetation types. The eddy covariance system offers continuous flux measurements of CO_2 that will then be used to derive an empirical relationship for NEE under various environmental conditions that allow the modeling of CO_2 exchange of different vegetation types. We can possibly reduce the gap between actual field data by using satellite data and field measurements for different vegetation types and land management practices. We will also conduct detailed field investigations to determine the accuracy of the analyses and acquire data related to the age of the tree from reliable sources.

15.6 Conclusions

Changing lifestyle characterized by increasing use of resources and infrastructure may impact the environment causing a reduction of critical resources and damage to the environment. Climate change induced by human activities on a global scale can alter weather and climate patterns causing an increase in temperature and extreme weather events. Urban environments are heavily reliant on available but limited infrastructure and resources which facilitates their sustenance and challenges their growth and progress. Extreme weather events can raise many questions about urban

a

b

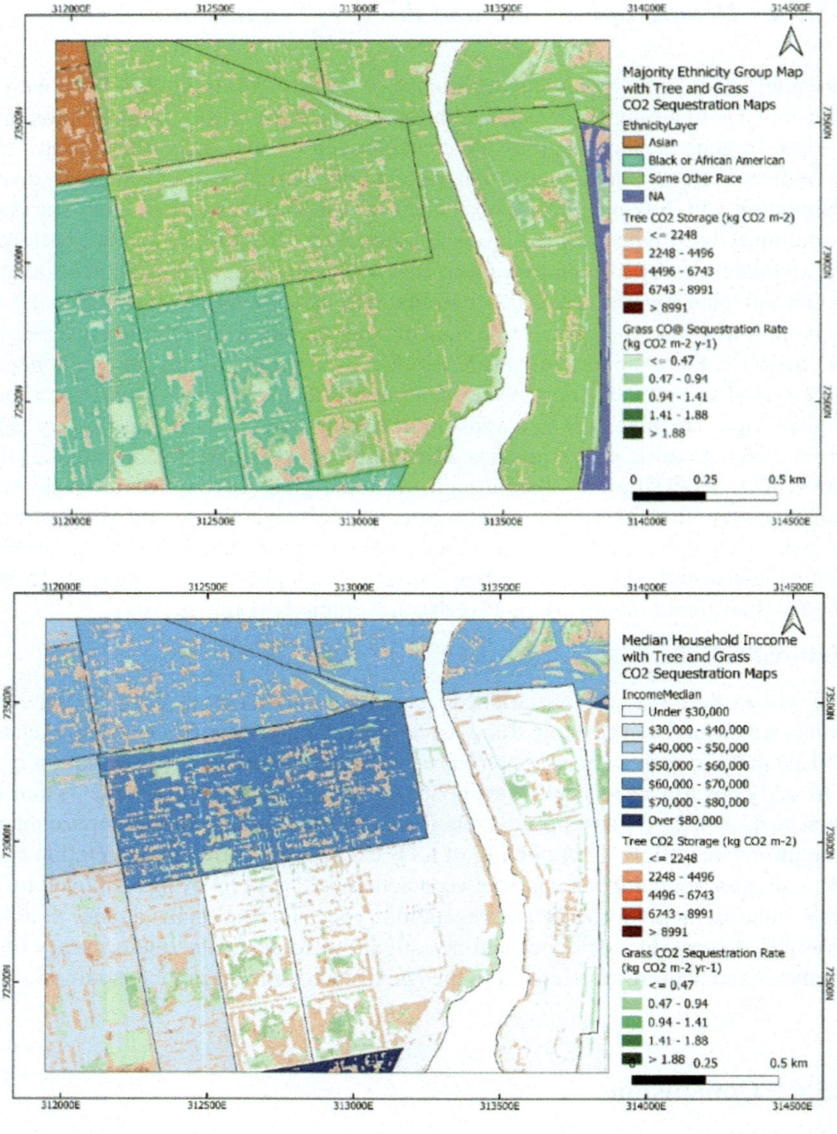

Fig. 15.11 **a** Majority ethnicity group map with tree and grass CO_2 sequestration. **b** Median household income map with tree and grass CO_2 sequestration. **c** Population density map with tree and grass CO_2 sequestration

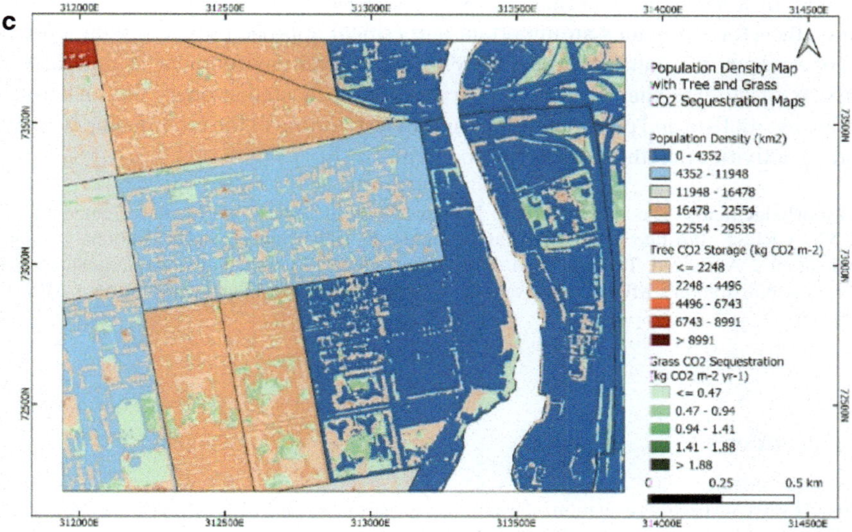

Fig. 15.11 (continued)

resilience, sustenance, and the environmental quality that needs to be better understood and addressed in a timely manner. Carbon sequestration by urban trees signifies a natural carbon sink in urban areas. Through this process, the CO_2 released into the atmosphere can be reduced, depending on land use and vegetation distribution in the urban region. Large urban trees are great filters for urban pollutants and fine particulates. Urban vegetation plays a significant role in maintaining a healthy environment and providing clean air and an esthetic value that powers the sustenance of urban environments. The CO_2 sequestered by trees offsets the costs incurred by cities to cool down homes and businesses during summer. Therefore, they provide a natural engine that assists in maintaining the ecological balance and fuels urban metabolism in cities. In the State of New York, particularly New York City, huge investments and funds are diverted to retrofit and electrify existing urban infrastructure and buildings to achieve net-zero carbon emissions. While these initiatives can make changes and impact urban metabolism positively, their extrapolation to other cities around the world will be contingent on the availability of funding and technology to drive these initiatives. Furthermore, successful implementation of retrofitting projects demands extensive holistic knowledge about the design of buildings and the demographic characteristics of residents in any city. A more viable strategy to address issues related to urban metabolism as demonstrated in this study must be promoted at a larger and coordinated global scale to sustain urban metabolism and ecology. The natural ability of trees to sequester CO_2 and create the environmental balance to sustain urban environments will play an important role in urban metabolism and benefit urban conglomerations in the long run. However, the lack of a global database on the type of tree species and their ability to sequester CO_2 is still not available. The study, therefore, provides an important and critical approach to addressing climate change issues and enhances the ability of the natural environment to mitigate the impacts of climate change. It demonstrates a multifaceted strategy for sustainable urban metabolism and promotes a superior understanding of the relationship between human activities and the natural environment.

Acknowledgements The research was conducted at the BCC Geospatial Center of the CUNY CREST Institute (BGCCCI) and funded by the United States National Science Foundation's Advanced Technological Education (NSF-ATE) and National Aeronautics and Space Administration's-MUREP Innovations in Space Technology Curriculum (NASA-MISTC) programs.

References

Ahmad F (2012) Spectral vegetation indices performance evaluated for Cholistan Desert. J Geograph Reg Plan 5(6). https://doi.org/10.5897/jgrp11.098

Aiazzi B, Alparone L, Baronti S, Garzelli A (2002) Context-driven fusion of high spatial and spectral resolution images based on oversampled multiresolution analysis. IEEE Trans Geosci Remote Sens 40(10):2300–2312. https://doi.org/10.1109/tgrs.2002.803623

Ali M, Hasnat GNT (2022) Urban landscape change detection using GIS and RS. In: Bhadouria R, Upadhyay S, Tripathi S, Singh P (eds) Urban Ecology and Global Climate Change. Wiley-Blackwell, pp 104–120. https://doi.org/10.1002/9781119807216.ch6

Aubinet M, Vesala T, Papale D (eds) (2012) Eddy covariance: a practical guide to measurement and data analysis. Springer Science & Business Media

Bhadouria R, Upadhyay S, Tripathi S, Singh P (eds) (2022) Urban ecology and global climate change. Wiley-Blackwell

Bhaskaran S, Nez E, Jimenez K, Bhatia SK (2013) Rule-based classification of high-resolution imagery over urban areas in New York City. Geocarto Int 28(6):527–545. https://doi.org/10.1080/10106049.2012.726278

Bhaskaran S, Paramananda S, Ramnarayan M (2010) Per-pixel and object-oriented classification methods for mapping urban features using IKONOS satellite data. Appl Geogr 30(4):650–665. https://doi.org/10.1016/j.apgeog.2010.01.009

Birdsey RA (1992) Carbon storage and accumulation in United States forest ecosystems, vol 59. U.S. Department of Agriculture, Forest Service

Cheng W-C, Chang J-C, Chang C-P, Su Y, Tu T-M (2008) A fixed-threshold approach to generate high-resolution vegetation maps for ikonos imagery. Sensors 8(7):4308–4317. https://doi.org/10.3390/s8074308

Clark A, Schroeder JG (1985) Weight, volume, and physical properties of major hardwood species in the southern Appalachian mountains. Research Paper SE-253, United States Department of Agriculture Forest Service

DOE US (1998) Method for calculating carbon sequestration by trees in urban and suburban settings. Voluntary Reporting of Greenhouse Gases, US Department of Energy, Energy Information Administration

DigitalGlobe. Satellite maps, satellite imagery, GIS data | Satellite Imaging Corp (n.d.) Retrieved April 4, 2022, from https://www.satimagingcorp.com/

Ghimire P, Lei D, Juan N (2020) Effect of image fusion on vegetation index quality—a comparative study from gaofen-1, gaofen-2, gaofen-4, landsat-8 oli and Modis imagery. Remote Sens 12(10):1550. https://doi.org/10.3390/rs12101550

Gholinejad S, Fatemi SB (2019) Optimum indices for vegetation cover change detection in the Zayandeh-Rud River Basin: a fusion approach. Int J Image Data Fusion 10(3):199–216. https://doi.org/10.1080/19479832.2019.1601642

Gis.ny.gov. NYC Topobathymetric Data. http://gis.ny.gov/elevation/NYC-topobathymetric-DEM.htm

Gitelson AA, Kaufman YJ, Merzlyak MN (1996) Use of a green channel in remote sensing of global vegetation from Eos-Modis. Remote Sens Environ 58(3):289–298. https://doi.org/10.1016/s0034-4257(96)00072-7

Haralick RM (1979) Statistical and structural approaches to texture. Proc IEEE 67(5):786–804. https://doi.org/10.1109/proc.1979.11328

Haralick RM, Shanmugam K, Dinstein IH (1973) Textural features for image classification. IEEE Trans Syst Man Cybern SMC-3(6):610–621. https://doi.org/10.1109/tsmc.1973.4309314

Horton R, Rosenzweig C, Gornitz V, Bader D, O'Grady M (2010) Climate risk information: climate change scenarios and implications for NYC infrastructure New York City panel on climate change. Ann N Y Acad Sci 1196(1):147–228. https://doi.org/10.1111/j.1749-6632.2010.05323.x

Huang YJ, Akbari H, Taha H (1990) The wind-shielding and shading effects of trees on residential heating and cooling requirements. ASHRAE Proc 96 (pt. 1)

Huete AR (1988) A soil-adjusted vegetation index (SAVI). Remote Sens Environ 25(3):295–309. https://doi.org/10.1016/0034-4257(88)90106-x

Huete A, Didan K, Miura T, Rodriguez EP, Gao X, Ferreira LG (2002) Overview of the radiometric and biophysical performance of the Modis vegetation indices. Remote Sens Environ 83(1–2):195–213. https://doi.org/10.1016/s0034-4257(02)00096-2

IPCC Climate Change (2014) Mitigation of climate change. Contribution of Working Group III to the Fifth Assessment Report of the Intergovernmental Panel on Climate Change 1454:147

Johnson B (2014) Effects of pansharpening on vegetation indices. ISPRS Int J Geo Inf 3(2):507–522. https://doi.org/10.3390/ijgi3020507

Kaufman YJ, Tanre D (1992) Atmospherically resistant vegetation index (ARVI) for Eos-Modis. IEEE Trans Geosci Remote Sens 30(2):261–270. https://doi.org/10.1109/36.134076

Klonus S, Ehlers M (2007) Image fusion using the Ehlers spectral characteristics preservation algorithm. Giscience Remote Sens 44(2):93–116. https://doi.org/10.2747/1548-1603.44.2.93

Kurn DM, Bretz SE, Huang B, Akbari H (1994) The potential for reducing urban air temperatures and energy consumption through vegetative cooling (No. LBL-35320). Lawrence Berkeley Lab, CA (United States)

Lu D (2006) The potential and challenge of remote sensing-based biomass estimation. Int J Remote Sens 27(7):1297–1328. https://doi.org/10.1080/01431160500486732

Meshesha DT, Ahmed MM, Abdi DY, Haregeweyn N (2020) Prediction of grass biomass from satellite imagery in Somali regional state, eastern Ethiopia. Heliyon 6(10):e05272. https://doi.org/10.1016/j.heliyon.2020.e05272

NYC Open Data (n.d.) 2015 street tree census—tree data. https://data.cityofnewyork.us/Environment/2015-Street-Tree-Census-Tree-Data/pi5s-9p35

Puleo M (2021) Death toll rises after 'most significant flash-flooding' in NYC history. AccuWeather. Retrieved August 22, 2022, from https://www.accuweather.com/en/hurricane/deadly-flooding-in-nyc-idas-heavy-rains-deluge/1011223

Qian Y et al (2022) Urbanization impact on regional climate and extreme weather: current understanding, uncertainties, and future research directions. Adv Atmos Sci 42. https://doi.org/10.1007/s00376-021-1371-9

QUick Atmospheric Correction (QUAC). L3 Harris Geospatial (n.d.) Retrieved April 4, 2022, from https://www.l3harrisgeospatial.com/docs/quac.html

State of New York (2022) Toward a clean energy future: a strategic outlook 2022–2025. In: N. Y. S. E. R. a. D. Authority (ed)

United Nations (2019) World urbanization prospects: the 2018 revision. Online Edition

Xue J, Su B (2017) Significant remote sensing vegetation indices: a review of developments and applications. J Sens 2017:1–17. https://doi.org/10.1155/2017/1353691
Zhang Y, Shao Z (2021) Assessing of urban vegetation biomass in combination with LiDAR and high-resolution remote sensing images. Int J Remote Sens 42(3):964–985. https://doi.org/10.1080/01431161.2020.1820618fbh

Chapter 16
Smart Urban Metabolism: A Big-Data and Machine Learning Perspective

Ruchira Ghosh and Dipankar Sengupta

Abstract Smart urban metabolism is a contemporary conception of urban metabolism which includes modern-day technologies dealing with the complex challenges of growing smart cities. Traditionally, urban metabolism deals with the influx-efflux of energy and flow of materials through urban space. However, with the growing needs of smart cities, these flow patterns are transiting as a complex network and are subject to interdisciplinary understanding. Furthermore, data availability is a major challenge faced by city planners due to the lack of data inventories and appropriate data management solutions to handle massive datasets, arising from these complex flow patterns. This is ensuing to inefficient adaptation of urban metabolism approaches, especially in developing economies. Thus, the situation remains grave when it comes to resource management of a smart city, and how urban areas may additionally deal with intricate issues like climate change when they are striving to understand their own material and energy cycling. In this chapter, we therefore, discuss how technologies like machine learning can equip urban metabolism, for its transition to "Smart Urban Metabolism." The chapter presents use of technologies like big-data and machine learning, as effective methodologies to channelize and manage heterogeneous multidimensional datasets, adoption of practices, developing self-learning machine learning models, and gain novel insights via predictive analytics, in "Smart Urban Metabolism." Precisely, for urban planners, the "Smart Urban Metabolism" can potentially be an effective approach for identifying complex issues in the flow patterns of energy and material in an urban space. This approach is a step toward sustainable city development.

Keywords Big-data analytics · Machine learning · Smart cities · Smart urban metabolism · Sustainable development · Urban metabolism

R. Ghosh (✉)
Centre for Sustainable Technologies, Belfast School of Architecture and the Built Environment, Ulster University, Jordanstown BT37 0QB, UK
e-mail: r.ghosh@ulster.ac.uk

D. Sengupta
Health Data Sciences Research Group, Centre for Optimal Health, School of Life Sciences, College of Liberal Arts and Science, University of Westminster, London W1W 6UW, UK

R. Bhadouria et al. (eds.), *Urban Metabolism and Climate Change*,
https://doi.org/10.1007/978-3-031-29422-8_16

16.1 Introduction

Urban metabolism is a mandatory concept for growing cities to understand the complex network of urban needs and their resource allocation. However, with the expansion of cities, augmented urban needs, urban mining, and resource management comes huge multifaceted data to manage, along with its dynamism. The initial concept of urban metabolism can be traced back to the nineteenth century (Wolman 1965; Odum 1996). In the current era, it is best defined to be, "*a concept in which the city is using the biological notion referring to the internal processes by which living organisms maintain a continuous exchange of matter and energy with their environment to enable operation, growth, and reproduction*" (Céspedes Restrepo and Morales-Pinzón 2018). Globally, the cities cover < 2% of the Earth's surface, however, consume ~78% of the energy (food, construction minerals, metals, etc.) including the energy consumption demands to generate these resources (Ulgiati and Zucaro 2019). Moreover, cities offer citizens with new opportunities for business, social security, education, and health services, that requires vast resource flows, within and outside the city boundaries. This exchange of energy and materials, to satisfy human settlement and sustenance, gets most often translated into environmental stress posing impacts as micro as locally in the city and as huge as at global scale, due to lack of resource channelization. The majority of the world's economic activities are concentrated in urban areas, generating 80% of the global gross domestic product (GDP), which demands a greater part of this energy consumption to support these economic activities (Ferrão and Fernández 2013; UN-Habitat 2022).

With every fold increase in the population, migration, economic growth, and social changes, the resource metabolism pattern/trend will change, and therefore, with changing urban configurations, urban metabolism alone is not sufficient to understand and address the challenges of today's cities or urban areas. Thus, cities need smart planning and a smart system leading to "Smart Urban Metabolism" (SUM) to grow smart in its functioning. SUM is a contemporary conception of urban metabolism which includes modern-day technologies like big-data machine learning, etc., that aids in dealing with the complex data challenges of growing smart cities. Although urbanization and globalization have been accelerated by technological advancements, however, to attain urban sustainability (UN Sustainable Developmental Goal (SDG) 11—sustainable cities and communities) (United Nations 2015), we also need to consider environmental, social, and economical challenges as precedence. This will help to assess the complexity of urban metabolic processes/systems/services for sustainable human settlement. Therefore, digitalization of urban material-energy flow patterns has become a priority for the development of urban settlements; and consequently, the concept of SUM will play a key role in achieving the objective (Caragliu et al. 2009; Dameri 2013; De Jong et al. 2015; Yu and Xu 2018).

The concept of SUM was proposed to deal with the constraints of urban metabolism (Bibri and Krogstie 2020). Furthermore, this advanced concept integrates both urban smart configuration and urban metabolism, emphasizing both the aspects in a holistic manner (Kitchin et al. 2015; Vinod Kumar and Dahiya 2017;

Macke et al. 2019). In recent years, the assessment of some city's (like Copenhagen, Singapore, Melbourne, etc.) utility services have been equipped with smart solutions, for example, real-time monitoring stations supported by big-data and participatory planning scheme—to augment urban service provisioning (i.e., with respect to water, waste, energy, emissions management, etc.) (Bettencourt 2014; Yigitcanlar et al. 2019). Within such a framework, local authorities, urban utility service providers, academics and research centers, manufacturers and companies, NGOs, and so on are required to work in synergy and actively (Paskaleva 2009; Longa 2011; Glazebrook and Newman 2018). Therefore, integrating natural and artificial intelligence into a unified and coherent multi-facet structure has become a priority and essential for policymakers globally. And thus, the urban data collected from real-time monitoring stations and analyzing them using big-data, machine learning, etc., would benefit in establishing an improved understanding of the city's key functions and performances (Bibri and Krogstie 2020).

Sustainable urban metabolism (SUM) study is a hybrid methodology and multi-dimensional in approach that involves economic and social perspectives of cities, environmental challenges, and technological options to deal with them together in harmony to develop smart and sustainable cities (Goal 11 of the United Nations 2030 Agenda for Sustainable Development encompasses targets to "make cities and human settlements safe, resilient and sustainable") (United Nations 2015). This distinctive character makes SUM a strategic tool to assist urban policymakers, resource managers, and city planners. Aiding urban metabolism with big-data technologies and machine learning (Fig. 16.1) would therefore enable to adopt a data-driven methodology and build up knowledge from the systems investigation in an iterative way.

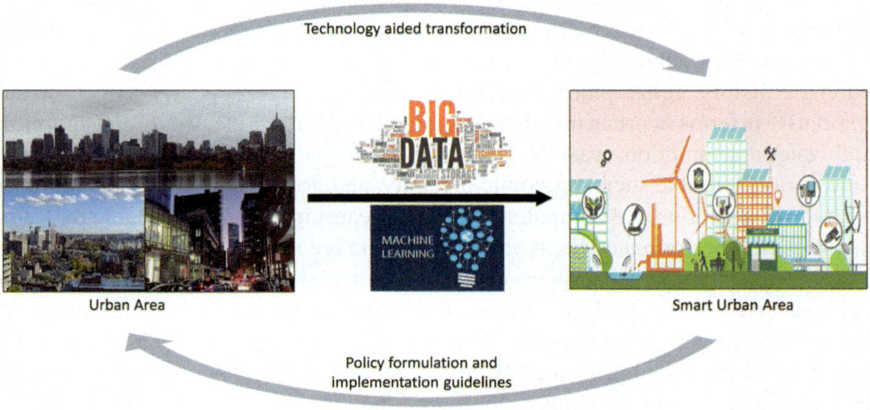

Fig. 16.1 Modern technology like big-data technologies and machine learning can aid landscaping urban metabolism into smart urban metabolism, and thus, assisting the city planners in policy formulation considering the future demands

Data are everywhere and growing in terms of volume, velocity, and variety, which cannot be handled with traditional data processing software, and therefore, is called as the big-data where the volume, velocity, and variety are referred to as the 3Vs of big-data. Estimates suggest, that currently, 2.5 quintillion bytes of data are generated globally every day, which will rise to 463 exabytes by 2025 (Vuleta 2021). And, therefore, big-data technologies have a huge role to play in managing as well as equipping us with the right tools to analyze these datasets for purposes like mining non-trivial patterns, classification, prediction or forecasting, etc. In SUM too, for transforming an "urban area" into a "smart urban area," big-data and the associated technologies hold the key to planning, policy formulation, implementation, and governance (Kandt and Batty 2021). As it encapsulates the computing power to capture and process the real-time data (from components of an urban system like transportation, water consumption, electricity consumption, etc.), and store it historically, also provides with power to detect patterns or make predictions in real time, augmenting the quick decision-making capabilities.

On the other hand, machine learning (ML) provides the extension leading to establishing the analytics component of big-data technologies. ML has been at the forefront since the 1960s, with an aim of designing machines (i.e., computer programs) that can mimic human intelligence (Samuel 1959). The objective set for its development was to attain the computational capabilities for extracting patterns from datasets, making inferences, and using them for decision-making, which is difficult to be achieved from the prevailing statistical methods (Sengupta 2021). In the 1990s, Tom Mitchell defined, ML as "*a computer program learning from experience 'E' with respect to some class of tasks 'T' and performance measure 'P,' if its performance at tasks in 'T' as measured by 'P,' improves with experience 'E'*" (Mitchell 1997). However, functionally in simplistic terms, ML is said to be enabling a computer program (compounding the powers of computer science, mathematics, and statistics) to learn from the data, for example, a computer program recommending products to a customer based on shopping preference along with additional parameters like purchase history, geographical location, ethnic origin, age, gender, occupation, etc. Similarly, in terms of urban metabolism, prediction of future electricity consumption, and water consumption, waste generation pattern can be acquired based on household parameters like socio-economic diversity, and goods' consumption pattern. In both these situations, the computer program is learning from the available data, and its prediction performance improves as more data are made available to it.

Consequently, with the growing data in different realms, as well as the introduction of big-data technologies, the evolution of ML, and advancements in computing techniques (like, as parallel or distributed computing), their applicability has augmented across domains, ranging from art, health care, humanities, social science, to more philosophical and ethical studies. Similarly, they are also changing the way cities are evolving, bringing in historical and real-time data as well as multiple time scales to be considered, raising the prospects of making them smart and sustainable. And therefore, understanding the conventional urban metabolism with technologies like big-data technologies and ML to attain smartness in a city's functioning, provides a greater opportunity for urban functions like the flow pattern of energy and materials

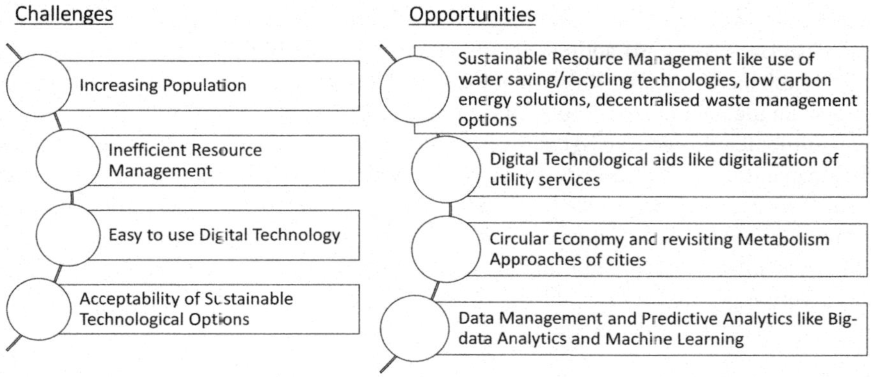

Fig. 16.2 Challenges and opportunities for smart urban metabolism (SUM)

to be assessed through a data-driven lens. Maintaining the urban ecosystem services is vital to sustainable urban development, emphasizing urban system resilience and ensuring public health and well-being. Thus, SUM is the answer to today's urban complexities by building the capacities of urban planners to sustainably design and plan for smart city infrastructure and services.

16.2 Barriers and Opportunities of Smart Urban Metabolism (SUM)

SUM faces several challenges to be adopted by the urban planners (Fig. 16.2); however, at the same time, the complex system in an urban area provides with opportunity to be adapted for future city planning.

16.3 Population and Smart Urban Metabolism Challenges

Better lifestyles and increased job opportunities are driving the population toward cities. And therefore, more and more migration is resulting in augmented resource demand followed by a fold increase in the waste production (with every unit increase in the resource demand and consumption, the waste production escalates, establishing a direct relationship between migration and waste production). This is usually an outcome of a lack of adequate knowledge about the pattern a particular city shows with respect to its relationship between the population influx rate and urban utilities demand and consecutively waste emerging out from the metabolism of these resources (Demaria and Schindler 2016; Facchini et al. 2017; Ipsen et al. 2019). However, the lack of comprehension of urban planners is mainly because they are not well equipped to recognize and analyze the fluctuating urban complexities (energy and material flows). To address this, technologies like big-data, machine learning, etc., which have the capability to handle huge datasets, assess and predict are required to give an idea about a particular or several aspects of a city (Shahrokni et al. 2015). The implementation of SUM cannot achieve success unless the city planners enhance their capacity to accommodate the migration and its associated aspects into the urban utility service planning and provisioning (Lyons et al. 2018). This is vital because people are at the core of driving a city's fate by determining the metabolic flow of materials and energy and guiding the GDP growth, which has a direct impact on the city's future progression.

16.4 Resource Management

Urban metabolism is gaining popularity among urban policymakers, managers, and planners in one hand. However, because of the large range of indicators in this framework, on the other hand, often is cumbersome for them to choose the right ones for developing, monitoring, and evaluating the metabolic pathways. For this, the standardized framework of the urban indicators can provide a greater opportunity to assist the urban planners. ISO 37120 defines urban indicators as quantitative, qualitative, or descriptive measures that provide pattern and trends for a complex system (International Organization for Standardization 2018). Like, the city indicators can be used to estimate the forte and drawbacks of any city (Purnomo and Prabowo 2016). These performance indicators can assist in recognizing critical areas demanding consideration and also the ones performing good (McCool and Stankey 2004). And they can be combined to set a ranking system, that could be used as a tool to measure the competitiveness of any city system, communicate its marketing strategy, and set up an interactive and responsive city administration (Yu and Xu 2018). A few of these key indicators comes from: (a) The "*Key Performance Indicators for Sustainable Digital Multiservice Cities*" ETSI framework identifies 73 city indicators, (b) ISO 37122 provides a complete set of indicators and methodologies to support policymakers, and (c) ISO 37123 on resilient cities is a tool to monitor progress toward a resilient city (Attmsdmc 2017; International Organization for Standardization 2019a, b). Furthermore, the studies on Smart Cities and Communities add to the knowledge in the context of smart urban metabolism (International Telecommunication Union 2016a). Also, the UNECE-ITU "*Smart Sustainable indicators*" framework (Table 16.1) established by the UN in association with the International Telecommunication Union, and others evaluate the urban smartness and the metabolism pathways of urban energy and material flow to comply with the Sustainable Development Goals (United Nations Economic Commission for Europe (UNECE) 2015).

Information communication technology (ICT) indicators for Smart Cities are prerequisite in planning system by cities to adopt. With huge city expansion to support fluctuating socioeconomic variabilities, the information system needs to be revisited,

Table 16.1 List of resources for smart urban metabolism indicators

Urban standards	Functioning areas	Sectors targeted	Indicators	Urban standards
ISO/DIS 37123	Sustainable development indicators for resilient cities	Economic, Social, Environment	73	ISO/DIS 37123
ISO 37122:2019	Sustainable development indicators for smart cities	Economic, Social, Environment	80	ISO 37122:2019
ISO 37120:2018	Sustainable development indicators for city services and quality of life	Economic, Social, Environment	104	ISO 37120:2018
UNECE - ITU	Sustainable indicators for Smart Cities	Economic, Social, Environment	72	UNECE - ITU
ETSI TS 103 463	Indicators for Smart sustainable cities	Economic, Social, Environment	76	ETSI TS 103 463
ITU-T Y.4903/L.1603	Indicators for smart cities to assess the achievement of SDGs	Economic, Social, Environment	52	ITU-T Y.4903/L.1603
ITU-T Y.4902/L.1602	ICT sustainability impacts indicators in Smart Cities	Economic, Social, Environment	30	ITU-T Y.4902/L.1602
ITU-T Y.4901/L.1601	ICT indicators for Smart Cities	Economic, Social, Environment	48	ITU-T Y.4901/L.1601

Sources United Nations Economic Commission for Europe (UNECE) (2015), International Telecommunication Union (2016a, b, c, d), International Organization for Standardization (2018, 2019a, b)

and the performance indicators designed to give a perfect overview for city planners to adapt (International Telecommunication Union 2016a, b). These indicators primarily include information about smart water meters for water supply monitoring and smart electricity meters to track electricity supply and demand-response penetration for electricity customers, access to household sanitation, solid waste collection, wastewater sewer connections, Internet access at households, public transportation network, traffic monitoring, etc. (International Telecommunication Union 2016a, b) (ITU 2016a, b). However, the corresponding data captured in response to these indicators are massive and will further augment in the future with adding population and changing urban structure. Thus, here comes the vital role of technologies like big-data and machine learning to analyze the huge datasets and keep accessibility of the key indicator values and/or patterns handy for urban planners to plan and execute urban utility services.

The advancement in technologies like big-data and machine learning will be beneficial in increasing our understanding of how eminently smart city operates and how can an emerging smart city adapt to sustain in long run. Literature shows the relationship between these technologies, however, smart decision-making is considered indirect and is usually been facilitated through social innovation (Mahmoud et al. 2022). Although machine learning and/or big data have not been used much in the public sector, or urban utility service provisioning but have a promising aspect to investigate technical, managerial, and policy challenges faced by cities today (Kankanhalli et al. 2019). Megacities require a resilient administrative configuration to expediate synergistic working between the society and government to make better policy implementation and ease in smart decision-making (Torfing et al. 2012; Conway 2020). It necessitates integrating internal governance structures and establishing a public–private partnerships with external organizations (Meijer and Bolívar 2015). Analytics and/or data-driven governance will not only equip cities to create smart services but would also be able to provide smart data inventory through sensors and synthesizing data for urban safety governance (Meijer and Thaens 2018). Like, during the COVID-19 pandemic, the entire world did inventory and synthesized data to fight against the Coronavirus (SARS-CoV2). One such successful enactment was put up by the South Korean government, which utilized data-driven technology for encouraging proactive information exchange and implementation of safety protocols among its citizens (Park et al. 2020).

Moreover, the collective smart governance focusing on "Evolution hubs" by nurturing mutual information flow links between information centers, and research institutions can help in improving the social, economic, and ecological performances of smart cities. For example, Amsterdam Smart City, which has a distinctive bonding between municipal administration, academics and research institutes, start-up investors, entrepreneurs, private businesses, ordinary citizens, and other relative stakeholders knitted through a common thread of information flow. The data-driven technology plays a key role within this system in maintaining the "*urban sustenance lab*," by smart data collecting, maintaining, storing, and synthesizing as per demand for future planning (Mora and Bolici 2017). One of the vital steps which Amsterdam adopts is engaging communities through the "Smart Citizen" program. It encourages the residents to participate as data representatives, demonstrating their knowledge of sustainability issues, whereas the data-driven technologies are the "Data Hub Manager." This is one of the best examples exhibiting a collective contribution of humans with technology and ecology toward a city's advancement as a smart city.

16.5 Big-Data Technology and Smart Urban Metabolism (SUM)

Big-data technology may be best defined as the software application(s) for extracting, processing, storing, and analyzing, massive datasets (structured, semi-structured, or

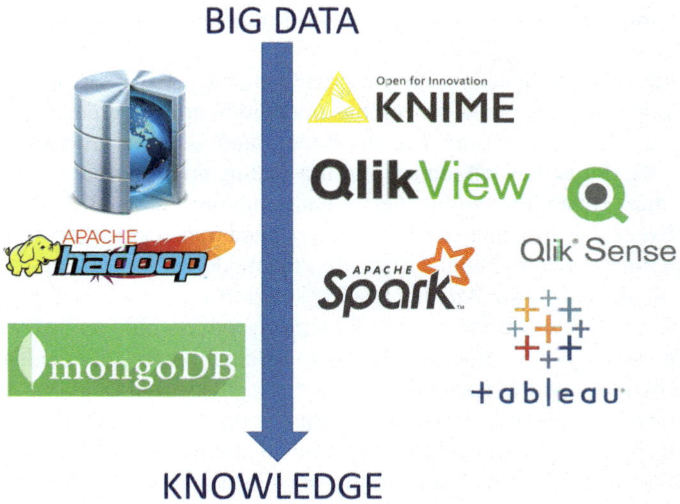

Fig. 16.3 State-of-art big-data technologies—Apache Hadoop and Mongo DB are leading solutions for storage purposes; KNIME and QlikView are among the best for data processing, while Qlik Sense provides cloud-based artificial intelligence (AI) integration platform; Apache Spark and Tableau software's aid in analytics and visualization

unstructured) which cannot be handled with the conventional software's, to obtain knowledge. The topmost solutions frequently used for storage, processing, and analysis (using machine learning) of big-data are (Fig. 16.3)—Apache Hadoop and Spark, KNIME, MongoDB, QlikView, RapidMiner, and Tableau, which have been discussed in the next sub-section.

16.6 Big-Data Solutions—Storage, Processing, and Analysis

Apache Hadoop (https://hadoop.apache.org/) is open-source software, that provides a framework for large-scale distributed computing (most reliable storage and analysis) and has been implemented by companies like Facebook, LinkedIn, IBM, Microsoft, etc., to handle the massive data (Nandimath et al. 2013). In addition to, Spark (https://spark.apache.org/) is one more noteworthy solution from Apache, which was developed considering the benefits of MapReduce (Dean and Ghemawat 2008). It provides an engine supporting data science and machine learning on single-node machines or clusters or on the cloud environment (Zaharia et al. 2016). MongoDB (https://www.mongodb.com/) is another open-source platform that facilitates the management of unstructured or semi-structured or volatile data (i.e., changes frequently) and has been adopted by companies like eBay, MetLife, Google, etc. It is a NoSQL document-oriented database with an easy-to-set-up environment (Bradshaw et al. 2019).

To address the challenges with the processing of a large set of data into any of the storage solutions, data processing operations are required, i.e., commonly referred to as ETL (Extraction, Transformation, and Loading processes) programming. KNIME (https://www.knime.com/) is a leading open-source software solution to provide these capabilities within the big-data structure (Jara et al. 2015). It provides an interactive framework that supports: merging and transforming data → modeling and visualizing → deploying and managing the model → interacting and adapting the model in the real world. Another leading solution available in the market for this objective is QlikView (https://www.qlik.com/us/products/qlikview). It is based on an associative analytics engine and centered on data modeling, that aids in deriving relationships between the data (Troyansky et al. 2015). Qlik has further boosted modern cloud-based analytics with Qlik Sense (https://www.qlik.com/us/products/qlik-sense), which augments and enhances human intuition with AI-powered insights (Troyansky et al. 2015). This in turn would boost the transition from passive to active analytics for real-time collaboration and action.

For analysis, besides Apache Spark, the most prominent software solutions for handling big-data are—RapidMiner and Tableau. RapidMiner (https://rapidminer.com/) is a powerful data tool for building predictive models with the support of machine learning (including deep learning) (Hofmann et al. 2016) and also supports the integration of the Apache Hadoop framework for storage purposes. Tableau (https://www.tableau.com/) is a leading data-driven business intelligence platform for forecasting, decision-making, and strategy implementation (Hoelscher and Mortimer 2018). The major USP of this software is an easy-to-use interface for transforming the raw data into knowledge, without the requirement of any prior programming experience. Additionally, it includes commending in-built options supporting enhanced data visualization in the form of interactive dashboards (Nair et al. 2016).

16.7 Role of Big-Data Solutions in Smart Urban Metabolism (SUM)

In terms of SUM, core essence of smart cities is the integration among utility systems (like electricity, solid waste, water, wastewater, etc.) and capturing of the data being generated. Big-data technologies are therefore becoming a necessity for the efficient functioning of smart as well as sustainable conurbations. The benefit of this would be the granularity (household → street → Particular Area → Urban Area) of the vast urban information that can be further analyzed via techniques like machine learning. Thus, big data equipped SUM can propel opportunities to address local as well as global urbanization challenges, and robust policy formulation to attain the SDG's.

Big-data can play a key role in transforming the way cities are planned considering multiple time scales (Bibri and Krogstie 2020). A ~98% increase in recycling of food waste by South Korea is one of the landmark examples, which highlights how the use of a data-driven technology can make a greater impact. In Seoul alone, the implementation of "Smart Bins," equipped with scales and radio frequency identification which could weigh the food waste as disposed, data in turn mapped to the citizen resident cards, have reduced food waste in the city by 47,000 tons (The World Economic Forum 2019). Another such example is the "city of sensors"—Singapore (Poon 2017). In 2014, ~1000 sensors were installed to capture data as part of the government's initiative to transform Singapore to be the world's first smart country under the "E3A" plan (Everyone, Everything, Everywhere, All the Time). The aim was to connect data from all aspects of urban life (traffic, infrastructure, etc.), with the data integrated among different departmental systems feeding into a central platform (Poon 2017). The benefit of this was evident during the Covid surge in 2020, as Singapore was one of the first countries to develop a tracing app safeguarding public health (Lee and Lee 2020).

So Bigdata Research Infrastructure, a pan-European initiative, provides services for obtaining, analyzing, and visualizing massive datasets for the ethically safe deployment of big-data analytics (Grossi et al. 2018). Furthermore, the emergence of a big-data platform like "Strategic Intelligence," which brings in the latest global research and analysis from the leading research organizations, exploring and monitoring more than 100 global issues with tactical perceptions, is paving the way forward in this direction (The World Economic Forum). Symoto, an ongoing project of the Dutch firm—Except Integrated Sustainability, is another such encouraging prospective (Except Integrated Sustainability B.V.). It aims to build a software application that would enable the development, simulation, and monitoring of large-scale material and energy cycles, industrial symbiosis, and the circular economy. Thus, similar examples of system modelling and strategic decision making can be replicated worldwide to empower SUM and attain sustainable development globally.

To summarize, a few of the key benefits which big-data technologies can offer for SUM:

1. Federated and centralized data storage—for gathering time-centric data recorded by independent utilities, and then merging them at a granular level (each individual/customer) as a centralized storage system.
2. Predictive analysis—for smart planning for basic utilities like water, electricity, traffic, etc.
3. Temporal analytics—for future planning with every fold increase in population, the fold-change (increase/decrease) in demand for basic utilities.
4. Mapping resources—considering the historical data, current demands, and future trends.
5. Planning—devising energy-efficiency programs that aid in improving urban health and security.
6. Cost optimization—in quantitative terms while planning and designing new infrastructure, toward smart and sustainable utility planning.

7. Citizen engagement—promoting awareness, creativity, and innovation among the general populations to be the drivers of the change, as well as encouraging service providers with open data platforms.
8. Aid policymakers—understand the present and future need for policy support in smart city infrastructure and enforce policies accordingly.

16.8 Machine Learning (ML) and Its Role in the Smart Urban Metabolism (SUM)

Machine learning (ML) primarily comprises learning algorithms (computer programs), which are used to analyze datasets (Fig. 16.4). It is applied to the input data, for predicting the corresponding output values or identifying patterns, or for grouping the data based on closest properties (for example, Euclidean distance), with an acceptable performance value which can be improvised based on experience.

The "supervised learning" algorithms (for example, classification, regression, etc.) are used for the datasets in which the class labels (i.e., labeled output) are known, based on which a learning function is obtained by approximating the function for mapping set of input features with the output (Caruana and Niculescu-Mizil 2006). While the "unsupervised learning" algorithms (for example, clustering, association rule mining, etc.) are used for datasets which do not have any class labels (i.e., no labeled outputs), and therefore, the aim of the learning function is to infer the organic construct (like grouping or precedence of occurrences) within a set of closest properties (Ghahramani 2004). In the third category of "semi-supervised" algorithms (for example, reinforcement learning, grey-box modeling), the learning function is trained on a dataset with few class labels and then uses the knowledge to label the unlabeled data points of the dataset (Sinha 2014).

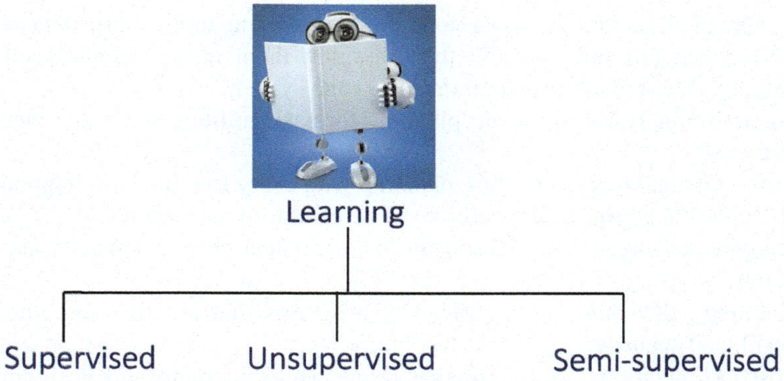

Fig. 16.4 There are three types of learning algorithms based on which ML can be categorized as: supervised, unsupervised, and semi-supervised

16.9 Building a Machine Learning (ML) Model for SUM

For building a robust ML model within a SUM environment (Fig. 16.5), the foremost step is to capture the data and store it historically in a big-data storage structure like Apache Spark or MongoDB (as discussed in the previous section). As the data sources are federated in nature (Architecture, Electricity, Traffic, Water, etc.), they can be stored in distinct data systems and/or maybe unified to be stored as a centralized system. In either case, the first step toward building an ML model would be preprocessing the data (i.e., preparing the data). This is an important step, as to ensure data are without any noise or bias.

Preprocessing the data may involve all or either of the following:

1. Missing Values: This is one of the commonest problems with datasets and needs to be handled, as it can cause bias and/or reduce the overall performance of the model. Typically, two approaches are used to handle this problem—removing records (instances and/or features) with missing values and the imputation method (Manly and Wells 2015; Hegde et al. 2019). There are different imputation methods (examples—mean/median/mode, interpolation, nearest neighbor, maximum likelihood estimate, etc.) that can be used to replace the missing values (Newgard and Lewis 2015). Another approach can be using a learning algorithm (like Naïve Bayes, k-nearest neighbor) that supports datasets with missing values.
2. Data Imbalance: This refers to the problem of class labels not represented equally in the dataset, which can introduce the problem of the model overfitting its predictive capabilities. Most often, this is a common problem, as datasets do not have an equal number of instances for each class label (Thabtah et al. 2020). This can be handled during sampling of instances or while model evaluation using

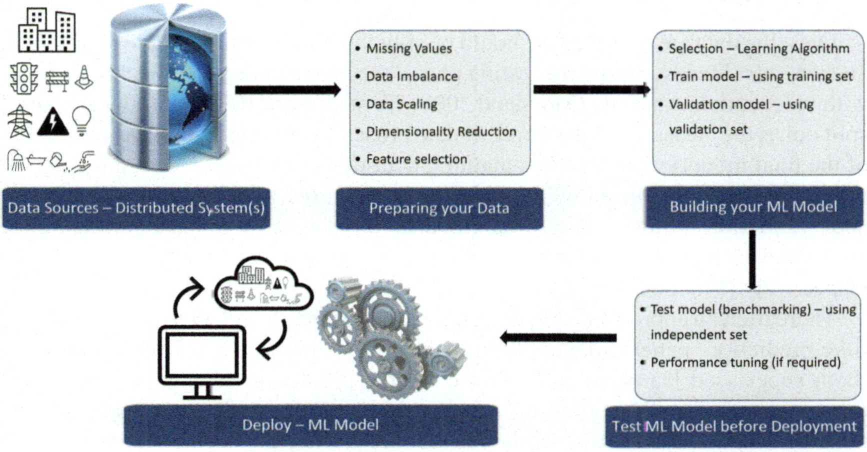

Fig. 16.5 Schematic representation for building a ML model, considering federated data sources in an urban area

performance metrics (like balanced accuracy, Kappa value, receiver operating characteristic (ROC) curve) which consider imbalances of classes in the data or via techniques like grey-box modeling (Grau et al. 2016).

3. Data Scaling: In some cases, the features in the dataset may occur in different ranges or scales (for example, traffic movement may be measured in km/hour, whereas the electricity consumption in KWH) and therefore are not ideal to build a model (using algorithms like regression based, support vector machine (SVM), etc.) as this may impact its overall performance. In such scenarios, scaling methods are used to generalize the data points and minimize the difference among these features. The following methods are commonly used and recommended for data scaling purposes—MaxAbs, MinMax, normalization, quantile transformation, robust scaling, and standardization (Ahsan et al. 2021).
4. Feature Selection: Big-data imply a large number of features to be considered in the data to build a model, which would trigger the complexity of learning to grow exponentially and impact the overall performance of the model. The feature selection step helps to reduce the number of features, i.e., creates a subset of relevant features to build a model from the original set of features. There is supervised and unsupervised feature selection measure that can be applied as a filter (selected before the ML algorithm) or wrapper approach (selected as part of the ML algorithm) (Marsland 2009; Cai et al. 2018).
5. Dimensionality Reduction: aims a similar focus as feature selection, however, the technique also helps to reduce the dimensional space of selected features (like generation of new synthetic features from a combination of the original features and then removing the least significant ones). This technique is also often used for data visualization for complex datasets having many features. Correlation, linear discriminant analysis (LDA), matrix factorization, and principal component analysis (PCA) are some of the commonest methods used for the aforesaid process (Marsland 2009; Reddy et al. 2020).

Once the data are prepared, it should be split into three parts—training, validation, and test sets (ideally, during the testing phase, the model should be benchmarked also with independent datasets) (Marsland 2009). To start with, 10% of this data should be split out as the "test set" (this sample of data is used to provide an unbiased evaluation of the final model) whereas the remaining dataset should be split-out into either 2:1 or k-cross folds, for "training" (sample of the data used to train the model function) and "validation" (this sample of data is used to provide an unbiased evaluation of the model fit from the training set while tuning the model hyperparameters) sets, respectively (Mitchell 1997; Marsland 2009).

Thereafter, an appropriate learning algorithm needs to be selected based on the task (like prediction, pattern identification, etc.). For example, neural network or SVM (both supervised learning) can be used for the prediction of energy consumption (Shapi et al. 2021). The "training" set should be used to train the model, followed by its "validation" with the validation set. Thereafter, the model should be tested upon the "test" set and/or benchmarked against independent datasets. Consequently, the optimized ML model can be implemented in the existing SUM ecosystem, for monitoring real-time activities or decision-making purposes, or the outcomes may be used by urban city managers (or the government) for planning, formulation, and implementation of new guidelines.

16.10 Example of Machine Learning (ML) in Smart Urban Metabolism (SUM)

We have made a basic understanding of how to build an ML model for the SUM ecosystem. Next, we discuss with an example how an ML model designed over a historical dataset can influence in transforming a city into a smart city.

Water is one of the key urban utilities, and its responsible consumption is of utmost importance for smart cities in addressing sustainability challenges, like corresponding energy consumption (Ghosh et al. 2016), which in turn may regulate the health service demands. In this work, Sengupta and Ghosh focused on looking at patterns between residential water consumption, wastewater discharge, and their impact on electricity consumption, with New Delhi (megacity) as a case study (Sengupta and Ghosh 2022). They analyzed the available data for the period 2001–2011 (Kennedy et al. 2015) using ML techniques enabling learning from the data. The predictor models were designed on two best-performing algorithms—logistic regression (generalized linear model) and neural network (a state-of-the-art technique for deep learning applications), using the MLR R-package (Rumelhart et al. 1986; Bischl et al. 2016; Tolles and Meurer 2016).

As the ROC curves (Fig. 16.6) and comparison of the performance measures (Table 16.2) between the two models demonstrate, model-A (based on logistic regression) outperformed model-B (based on the neural network). Using model-A, further analysis (Fig. 16.7) indicated, that there has been a two-fold increase in the electricity requirement with respect to water consumption and wastewater discharge. Such a model, when combined with socio-economic drivers and geographical details, has the potential to be used on a temporal scale to predict fold-change in the demands (water consumption, wastewater discharge, and corresponding electricity consumption). Therefore, this can be a potential interest of application for the smart city planners, as these predictions can be used or further optimized, while designing the water and wastewater management infrastructures, toward a smart and sustainable utility planning.

Thus, such ML models can help focus on enhancing and maintaining urban health by smart planning of basic utility services like water and electricity. The findings may also target mitigating augmentation in residential electricity consumption caused due to coping strategies to deal with poor water and wastewater services. Such studies may also benefit policymakers to understand the present and future need for policy support in smart city infrastructure and enforce policies accordingly.

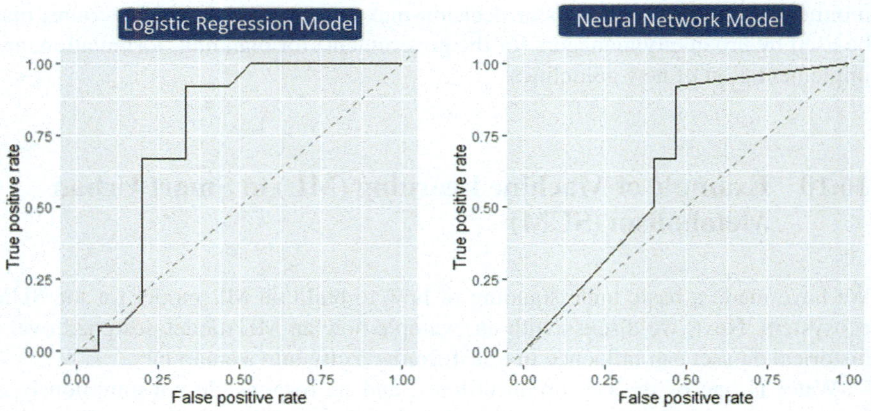

Fig. 16.6 ROC (receiver operating characteristic curve) comparing the performance of the logistic regression and neural network models

Table 16.2 Evaluation parameters for the two ML models

Model	Sensitivity (TPR)	Specificity (TNR)	False positive rate (FPR)	False negative rate (FNR)	Balanced accuracy	MCC
Logistic regression	0.92	0.53	0.46	0.08	0.73	0.47
Neural network	0.67	0.60	0.40	0.33	0.63	0.27

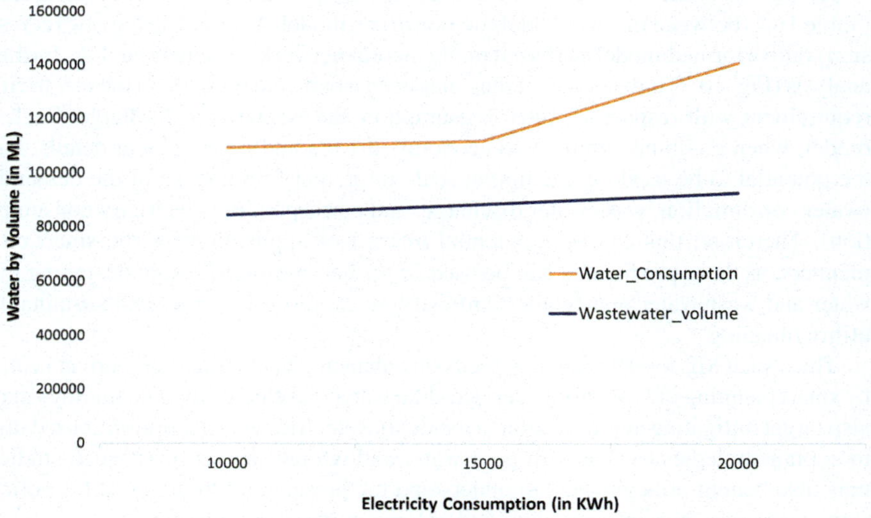

Fig. 16.7 Delhi—electricity consumption versus water consumption and wastewater volume (2001–2011)

16.11 Conclusions

The technologies discussed in this chapter are not the solitary solution to address the challenges in urban areas. However, they are certainly among the strongest tools to increase the efficacy of robust policy implementation and its effective maintenance by actively involving stakeholders through technology amalgamated with social innovation. In conclusion, the combined approach of urban metabolism with advanced technologies like big-data and machine learning is the way forward that will be principally beneficial in understanding the real-time city functions, as they can help urban city planners or governors to make smart decisions, support cities' socioeconomic factors, as well as smart implementation and accomplishment of policies.

References

Ahsan MM, Mahmud MAP, Saha PK et al (2021) Effect of data scaling methods on machine learning algorithms and model performance. Technologies 9:52. https://doi.org/10.3390/TECHNOLOGIES9030052

Attmsdmc (2017) TS 103,463—V1.1.1—access, terminals, transmission and multiplexing (ATTM); key performance indicators for sustainable digital multiservice cities

Bettencourt LMA (2014) The uses of big data in cities. Big Data 2:12–22. https://doi.org/10.1089/BIG.2013.0042

Bibri SE, Krogstie J (2020) Environmentally data-driven smart sustainable cities: applied innovative solutions for energy efficiency, pollution reduction, and urban metabolism. Energy Inform 3:1–59. https://doi.org/10.1186/S42162-020-00130-8/TABLES/7

Bischl B, Lang M, Kotthoff L et al (2016) mlr: machine learning in R. J Mach Learn Res 17:1–5

Bradshaw S, Brazil E, Chodorow K (2019) MongoDB: the definitive guide. O'Reilly Media, Inc.

Cai J, Luo J, Wang S, Yang S (2018) Feature selection in machine learning: a new perspective. Neurocomputing 300:70–79. https://doi.org/10.1016/J.NEUCOM.2017.11.077

Caragliu A, Del CB, Nijkamp P (2009) Smart cities in Europe. In: Proceedings of the 3rd central European conference in regional science, pp 49–59

Caruana R, Niculescu-Mizil A (2006) An empirical comparison of supervised learning algorithms. In: ACM international conference proceeding series

Céspedes Restrepo JD, Morales-Pinzón T (2018) Urban metabolism and sustainability: precedents, genesis and research perspectives. Resour Conserv Recycl 131:216–224. https://doi.org/10.1016/J.RESCONREC.2017.12.023

Conway S (2020) Interactive governance: advancing the paradigm. Administration 68:63–68. https://doi.org/10.2478/ADMIN-2020-0017

Dameri RP (2013) Searching for smart city definition: a comprehensive proposal. Int J Comput Technol 11:2544–2551. https://doi.org/10.24297/IJCT.V11I5.1142

De Jong M, Joss S, Schraven D et al (2015) Sustainable-smart-resilient-low carbon-eco-knowledge cities; Making sense of a multitude of concepts promoting sustainable urbanization. J Clean Prod 109:25–38. https://doi.org/10.1016/J.JCLEPRO.2015.02.004

Dean J, Ghemawat S (2008) MapReduce. Commun ACM 51:107–113. https://doi.org/10.1145/1327452.1327492

Demaria F, Schindler S (2016) Contesting Urban metabolism: struggles over waste-to-energy in Delhi, India. Antipode 48:293–313. https://doi.org/10.1111/ANTI.12191

Except Integrated Sustainability B.V. Except Integrated Sustainability B.V. https://except.eco/. Accessed 29 July 2022

Facchini A, Kennedy C, Stewart I, Mele R (2017) The energy metabolism of megacities. Appl Energy 186:86–95. https://doi.org/10.1016/J.APENERGY.2016.09.025

Ferrão P, Fernández JE (2013) Resource consumption of cities. Sustain Urban Metab 23–42

Ghahramani Z (2004) Unsupervised learning. Lecture notes computer sciences (including Subseries lecture notes artificial intelligence lecture notes bioinformatics). https://doi.org/10.1007/978-3-540-28650-9_5

Ghosh R, Kansal A, Aghi S (2016) Implications of end-user behaviour in response to deficiencies in water supply for electricity consumption—a case study of Delhi. J Hydrol 536:400–408. https://doi.org/10.1016/J.JHYDROL.2016.03.012

Glazebrook G, Newman P (2018) The city of the future. Urban Plan 3:1–20. https://doi.org/10.17645/UP.V3I2.1247

Grau I, Sengupta D, Matilde M et al (2016) Grey-box model: an ensemble approach for addressing semi-supervised classification problems. In: Benelearn 2016: Belgian-Dutch conference on machine learning

Grossi V, Rapisarda B, Giannotti F, Pedreschi D (2018) Data science at SoBigData: the European research infrastructure for social mining and big data analytics. Int J Data Sci Anal 6:205–216. https://doi.org/10.1007/S41060-018-0126-X/FIGURES/6

Hegde H, Shimpi N, Panny A et al (2019) MICE versus PPCA: missing data imputation in healthcare. Inform Med Unlocked 17:100275. https://doi.org/10.1016/J.IMU.2019.100275

Hoelscher J, Mortimer A (2018) Using tableau to visualize data and drive decision-making. J Account Educ 44:49–59. https://doi.org/10.1016/J.JACCEDU.2018.05.002

Hofmann M, Klinkenberg R, Safari an OMC (2016) RapidMiner. Chapman and Hall/CRC, p 525

International Organization for Standardization (2018) ISO—ISO 37120:2018—sustainable cities and communities—indicators for city services and quality of life. In: ISO, Geneva. https://www.iso.org/standard/68498.html. Accessed 26 July 2022

International Organization for Standardization (2019a) ISO—ISO 37122:2019—sustainable cities and communities—indicators for smart cities. In: ISO, Geneva. https://www.iso.org/standard/69050.html. Accessed 26 July 2022

International Organization for Standardization (2019b) ISO—ISO 37123:2019—sustainable cities and communities—indicators for resilient cities. In: ISO, Geneva. https://www.iso.org/standard/70428.html. Accessed 26 July 2022

International Telecommunication Union (2016a) L.1601: Key performance indicators related to the use of information and communication technology in smart sustainable cities. In: Telecommun Stand Sect. https://www.itu.int/rec/T-REC-L.1601-201606-I. Accessed 26 July 2022

International Telecommunication Union (2016b) L.1602: key performance indicators related to the sustainability impacts of information and communication technology in smart sustainable cities. In: Telecommun Stand Sect. https://www.itu.int/rec/T-REC-L.1602-201606-I. Accessed 26 July 2022

International Telecommunication Union (2016c) Y.4903: key performance indicators for smart sustainable cities to assess the achievement of sustainable development goals. In: Telecommun Stand Sect. https://www.itu.int/rec/T-REC-Y.4903/en. Accessed 26 July 2022

International Telecommunication Union (2016d) ITU-T recommendation database. In: Telecommun Stand Sect. https://www.itu.int/ITU-T/recommendations/rec.aspx?rec=12884&lang=en. Accessed 26 July 2022

Ipsen KL, Zimmermann RK, Nielsen PS, Birkved M (2019) Environmental assessment of smart city solutions using a coupled urban metabolism—life cycle impact assessment approach. Int J Life Cycle Assess 24:1239–1253. https://doi.org/10.1007/S11367-018-1453-9/FIGURES/8

Jara AJ, Genoud D, Bocchi Y (2015) Big data for smart cities with KNIME a real experience in the SmartSantander testbed. Softw Pract Exp 45:1145–1160. https://doi.org/10.1002/SPE.2274

Kandt J, Batty M (2021) Smart cities, big data and urban policy: towards urban analytics for the long run. Cities 109:102992. https://doi.org/10.1016/J.CITIES.2020.102992

Kankanhalli A, Charalabidis Y, Mellouli S (2019) IoT and AI for smart government: a research agenda. Gov Inf Q 36:304–309. https://doi.org/10.1016/J.GIQ.2019.02.003

Kennedy CA, Stewart I, Facchini A et al (2015) Energy and material flows of megacities. Proc Natl Acad Sci U S A 112:5985–5990. https://doi.org/10.1073/PNAS.1504315112/SUPPL_FILE/PNAS.1504315112.SD01.XLSX

Kitchin R, Lauriault TP, McArdle G (2015) Knowing and governing cities through urban indicators, city benchmarking and real-time dashboards. Reg Stud Reg Sci 2:6–28. https://doi.org/10.1080/21681376.2014.983149/SUPPL_FILE/RSRS_A_983149_SM0001.JPG

Lee T, Lee H (2020) Tracing surveillance and auto-regulation in Singapore: 'smart' responses to COVID-19. Media Int Aust 177:47–60. https://doi.org/10.1177/1329878X20949545

Longa RD (2011) Urban models and public-private partnership. Springer, Berlin Heidelberg

Lyons G, Mokhtarian P, Dijst M, Böcker L (2018) The dynamics of urban metabolism in the face of digitalization and changing lifestyles: understanding and influencing our cities. Resour Conserv Recycl 132:246–257. https://doi.org/10.1016/J.RESCONREC.2017.07.032

Macke J, Rubim Sarate JA, de Atayde Moschen S (2019) Smart sustainable cities evaluation and sense of community. J Clean Prod 239. https://doi.org/10.1016/J.JCLEPRO.2019.118103

Mahmoud M, Perez-Ortiz M, Asad Bokhari SA, Myeong S (2022) Use of artificial intelligence in smart cities for smart decision-making: a social innovation perspective. Sustainability 14:620. https://doi.org/10.3390/SU14020620

Manly CA, Wells RS (2015) Reporting the use of multiple imputation for missing data in higher education research. Res High Educ 56:397–409. https://doi.org/10.1007/S11162-014-9344-9/TABLES/1

Marsland S (2009) Machine learning: an algorithmic perspective, 2nd edn. CRC Press

McCool SF, Stankey GH (2004) Indicators of sustainability: challenges and opportunities at the interface of science and policy. Environ Manage 33:294–305. https://doi.org/10.1007/S00267-003-0084-4

Meijer A, Bolívar MPR (2015) Governing the smart city: a review of the literature on smart urban governance. Int Rev Admin Sci 82:392–408. https://doi.org/10.1177/0020852314564308

Meijer A, Thaens M (2018) Quantified street: smart governance of urban safety. Inf Polity 23:29–41. https://doi.org/10.3233/IP-170422

Mitchell TM (1997) Machine learning. McGraw-Hill Science/Engineering/Math

Mora L, Bolici R (2017) How to become a smart city: learning from Amsterdam. Smart Sustain Plan Cities Reg 251–266. https://doi.org/10.1007/978-3-319-44899-2_15/COVER

Nair L, Nair L, Shetty S, Shetty S (2016) Interactive visual analytics on big data: tableau vs D3.js. J e-Learning Knowl Soc 12

Nandimath J, Banerjee E, Patil A et al (2013) Big data analysis using Apache Hadoop. Proc 2013 IEEE 14th Int Conf Inf Reuse Integr IEEE IRI 700–703. https://doi.org/10.1109/IRI.2013.6642536

Newgard CD, Lewis RJ (2015) Missing data: how to best account for what is not known. JAMA 314:940–941. https://doi.org/10.1001/JAMA.2015.10516

Odum HT (Howard T (1996) Environmental accounting: EMERGY and environmental decision making 370

Park Y, Choe Y, Park O et al (2020) Contact tracing during coronavirus disease outbreak, South Korea, 2020. Emerg Infect Dis 26:2465. https://doi.org/10.3201/EID2610.201315

Paskaleva KA (2009) Enabling the smart city: the progress of city e-governance in Europe. Int J Innov Reg Dev 1:405. https://doi.org/10.1504/IJIRD.2009.022730

Poon L (2017) Singapore, City of Sensors. https://www.bloomberg.com/news/articles/2017-04-21/privacy-takes-a-backseat-in-singapore-s-vision-of-a-smart-nation. Accessed 29 July 2022

Purnomo F, Prabowo H (2016) Smart city indicators: a systematic literature review. J Telecommun Electron Comput Eng 8:161–164

Reddy GT, Reddy MPK, Lakshmanna K et al (2020) Analysis of dimensionality reduction techniques on big data. IEEE Access 8:54776–54788. https://doi.org/10.1109/ACCESS.2020.2980942

Rumelhart DE, Hinton GE, Williams RJ (1986) Learning representations by back-propagating errors. Nature 323:533–536. https://doi.org/10.1038/323533a0

Samuel AL (1959) Some studies in machine learning using the game of checkers. IBM J Res Dev
Sengupta D (2021) Machine learning in precision medicine. Mach Learn Big Data, IoT Med Inform 405–419. https://doi.org/10.1016/B978-0-12-821777-1.00013-6
Sengupta D, Ghosh R (2022) Artificial intelligence for addressing smart cities poor urban health. Second Int Conf Water, Megacities Glob Chang
Shahrokni H, Lazarevic D, Brandt N (2015) Smart urban metabolism: towards a real-time understanding of the energy and material flows of a city and its citizens. J Urb Technol 22:65–86. https://doi.org/10.1080/10630732.2014.954899
Shapi MKM, Ramli NA, Awalin LJ (2021) Energy consumption prediction by using machine learning for smart building: case study in Malaysia. Dev Built Environ 5:100037. https://doi.org/10.1016/J.DIBE.2020.100037
Sinha K (2014) Semi-supervised learning. In: Data classification: algorithms and applications
Thabtah F, Hammoud S, Kamalov F, Gonsalves A (2020) Data imbalance in classification: experimental evaluation. Inf Sci (ny) 513:429–441. https://doi.org/10.1016/J.INS.2019.11.004
The World Economic Forum (2019) South Korea once recycled 2% of its food waste. Now it recycles 95% | World Economic Forum. In: WeForum. https://www.weforum.org/agenda/2019/04/south-korea-recycling-food-waste/. Accessed 27 July 2022
The World Economic Forum Strategic Intelligence. https://intelligence.weforum.org/. Accessed 29 July 2022
Tolles J, Meurer WJ (2016) Logistic regression: relating patient characteristics to outcomes. JAMA 316:533–534. https://doi.org/10.1001/JAMA.2016.7653
Torfing J, Peters BG, Pierre J, Sørensen E (2012) Interactive governance: advancing the paradigm. Interact Gov Adv Paradig 1–288. https://doi.org/10.1093/ACPROF:OSO/9780199596751.001.0001
Troyansky O, Gibson T, Leichtweis C (2015) QlikView your business : an expert guide to business discovery with QlikView and Qlik Sense
Ulgiati S, Zucaro A (2019) Challenges in urban metabolism: sustainability and well-being in cities. Front Sustain Cities 1:1. https://doi.org/10.3389/FRSC.2019.00001
UN-Habitat (2022) World Cities Report 2022: envisaging the future of cities | UN-Habitat. https://unhabitat.org/world-cities-report-2022-envisaging-the-future-of-cities. Accessed 28 July 2022
United Nations (2015) THE 17 GOALS | Sustainable Development. In: United Nations—Dep Econ Soc Aff. https://sdgs.un.org/goals. Accessed 25 July 2022
United Nations Economic Commission for Europe (UNECE) (2015) UNECE endorses smart sustainable cities indicators to improve performance of cities | UNECE. In: UNECE. https://unece.org/housing-and-land-management/press/unece-endorses-smart-sustainable-cities-indicators-improve. Accessed 26 July 2022
Vinod Kumar TM, Dahiya B (2017) Smart economy in smart cities 3–76. https://doi.org/10.1007/978-981-10-1610-3_1
Vuleta B (2021) How much data is created every day? In: Seedscientific. https://seedscientific.com/how-much-data-is-created-every-day/. Accessed 29 Jul 2022
Wolman A (1965) The metabolism of cities. Sci Am 213:179–190. https://doi.org/10.1038/SCIENTIFICAMERICAN0965-178
Yigitcanlar T, Kamruzzaman M, Foth M et al (2019) Can cities become smart without being sustainable? A systematic review of the literature. Sustain Cities Soc 45:348–365. https://doi.org/10.1016/J.SCS.2018.11.033
Yu W, Xu C (2018) Developing smart cities in China: an empirical analysis. Int J Public Adm Digit Age 5:76-91. https://doi.org/10.4018/IJPADA.2018070106
Zaharia M, Xin RS, Wendell P et al (2016) Apache spark: a unified engine for big data processing. Commun ACM 59:56–65. https://doi.org/10.1145/2934664

Part V
Policy Interventions on Urban Metabolism

Chapter 17
Policy Initiatives on Urban Metabolism in Ghana (2002–2021)

Gladys Nkrumah, Gordon Kofi Sarfo-Adu, and Henry Kwabena Kokofu

Abstract The Government of Ghana has a major role to play in responding effectively to the problems and challenges resulting from poor urban planning and urban metabolism. This has indeed become urgent in the face of the current rate of urbanization in Ghana. The 2021 population census confirms that the population of Ghanaians living in urban areas and big towns is about 57.35%. In 2019, it was 56.71%; whereas in 2018, 2017, and 2016 were, respectively, 56.06%, 55.41%, and 54.75%. This has implications for urban planning and urban metabolism, for which the government has the responsibility to initiate policies to enhance sustainability particularly in the management of urban waste and other related matters. Some policies initiated by the government in 2002 have attempted to provide plans that will sustainably respond to the growing social, economic, and environmental challenges associated with urbanization. For example, the National Housing Policy is dedicated to encouraging and supporting community initiatives that equip residents to be economically empowered, develop and protect the environment, and engender social equity. This chapter attempts to examine some of these policies and make proposals that may enhance the social, economic, and environmental forces that affect urban metabolism in Ghana. In all, eight policies that deal with issues bordering on housing, dangerous and e-waste control and management, landfill, management of healthcare waste, and urban development and planning were reviewed to identify their potential implications on the factors affecting urban metabolism in the country. The functionality of municipal systems in Ghana and many developing economies is weak with limited capacity and autonomy in all the key pillars of sustainable development. This chapter however

G. Nkrumah (✉)
Department of Management and Human Resource, Ghana Communication Technology University, Accra, Ghana
e-mail: gnkrumah@gctu.edu.gh

G. K. Sarfo-Adu
Forestry Commission, Accra, Ghana
e-mail: gsarfoadu.hq@fcghana.org

H. K. Kokofu
Environmental Protection Agency, Accra, Ghana
e-mail: henry.kokofu@epa.gov.gh

R. Bhadouria et al. (eds.), *Urban Metabolism and Climate Change*,
https://doi.org/10.1007/978-3-031-29422-8_17

illustrates attempts being made by the Government of Ghana through policy initiatives that attempt to address these problems and enhance urban metabolism. The policies clearly consider cities in Ghana as organisms with different parts: housing, infrastructure and services, institutions, living environment, and people and that each part should be regulated and managed to ensure the effective functionality of the whole system.

Keywords Environmental deterioration · E-waste · Ghana · Renewable energy · Urban metabolism · Urban policy

Abbreviations

NGO	Non-governmental organizations
CBO	Community-based organizations
UM	Urban metabolism
MLGRD	Ministry of Local Government and Rural Development
HCW	Healthcare waste
GOG	Government of Ghana
NUP	National Urban Policy
UNDP	United Nations Development Programme
UNEP	United Nations Environment Programme
MDGs	Millennium Development Goals
PRSPs	Poverty Reduction Strategy Papers
GEF	Global Environment Policy
EM	Environmental mainstreaming
PBDEs	Phthalates and polybrominated diphenyl ethers
E&E	Electrical and electronic
BAT	Best available techniques
LI	Legal instrument
SECO	Swiss State Secretariat for Economic Affairs
EPA	Environmental Protection Agency
MOH	Ministry of Health
GLG	Ghana landfill guidelines
MWRWH	Ministry of Water Resources, Works, and Housing
GNA	Ghana News Agency
NP	Nonylphenol
PCBs	Polychlorinated biphenyls
TPPs	Triphenyl phosphate
TGESEWM	Technical Guidelines on Environmentally Sound E-waste Management
GNHWMPG	Ghana National Healthcare Waste Management Policy Guidelines

17.1 Introduction

Urban development planning and policies in Ghana have closely been tailored to respond to the challenges associated with basic human needs including water, sanitation and sewerage facilities, electricity, and telecommunication which lag behind urban growth. Coupled with this, there has been inadequate control of urban growth and urban management. Some of these factors have the potential to create problems that can negatively impact on the maintenance of a healthy urban environment. In recent years, apart from urban development policies, other initiatives have been taken by the government to comprehensively respond to the factors that impinge on urban metabolism in Ghana (Palczynski and Scotia 2020). This study seeks to examine selected public policy initiatives in the past two decades (2002–2021) that attempt to enhance the social, economic, and environmental forces that affect urban metabolism in Ghana. Prominent among policies selected for this study include the National Housing Policy (2012), Hazardous and Electronic Waste Control and Management Act [Act 917 (2016), Legislative Instrument 2250 (2018), and Technical Guidelines on Environmentally Sound E-waste Management of 2018]. The others are the Ghana Landfill Guidelines (2002), Ghana National Healthcare Waste Management Policy Guidelines (2020), Renewable Energy Act (2011), and National Housing Policy (2015). Urban areas are known to be centers of consumption of materials and energy with resultant consequences from depletion of natural resources as well as the discharge of wastes and pollution which impact the environment in diverse ways. This is an urgent issue that demands attention from government and relevant stakeholders in the form of policy initiatives. Indeed, public policy seeks to respond to societal problems and challenges that tend to retard the socio-economic development of a country. The aforementioned policy frameworks in diverse ways seek to respond to the problems and challenges which have potential implications for the forces that affect urban metabolism in the country. Table 17.1 and Fig. 17.1 seek to establish the relationship between the policy initiatives and urban metabolism in Ghana.

The chapter is in four parts: part one is the introduction, while part two examines the concept of metabolism in order to put the chapter in the right perspective. Part three is dedicated to selected policy initiatives with potential implications for urban metabolism, and lastly, the conclusion, which also proposes a couple of policy recommendations. The significance of this chapter is its contribution to the literature since only limited studies have been done in African cities despite expected large future populations (Wang 2022); also, there is “little insight into how urban metabolism is implemented in the Global South” (ICLEI 2021).

Table 17.1 Policy initiatives and their contribution to urban metabolism

Policy/legal framework	Potential effects on UM
National urban policy	• Elimination of waste • Sound living and working environment
Hazardous and electronic waste control and management act (Act 917)	• Control and manage hazardous and e-waste environmentally sound manner
Legislative instrument 2250	• Hazardous substances be stored to prevent dispersal of hazardous materials to the environment
Technical guidelines on environmentally sound e-waste management	• Prevent releases of gases, solid particles, etc., to the environment
Ghana landfill guidelines	• Landfill sites should have appropriate gas, leachate, etc., systems to prevent releases of pollutants
Ghana national healthcare waste management and guidelines policy	• Safe management of healthcare waste • Strengthen collaboration between stakeholders
Renewable energy act (Act 832)	• Adequate supply of renewable energy • Build indigenous capacity for renewable energy sources
National housing policy	• Provide safe, secure, decent housing • Housing designed and built to sustainable building

Source Authors' construct 2022

17.2 Urban Metabolism

Urban metabolism (UM), as a concept has received its fair share in the literature, perhaps partly because of its interdisciplinary nature with a scope covering fields as diverse as urban, political, industrial, and political-industrial ecologies, respectively (Newell et al. 2017). According to Derrible et al. (2021), UM is, indeed, a recognized framework that aims at quantifying resource inflows and outflows and resource accumulation (including energy and materials) in urban areas. Currie and Musango (2017) also see UM as a complex web of processes and activities uniquely associated with the city and its suburbs and surrounding communities that are meant to meet the needs of the residents. Other studies have been conducted from the same perspective and discussed in the literature, with significant examples including Zhang (2019), Beloin-Saint-Pierre et al. (2017), Cui (2018), and Cui et al. (2019). In addition, a variety of frameworks at the city level have been examined and adopted for implementation (Beloin-Saint-Pierre et al. 2017) with many case studies examined in various regions of the world (Beloin-Saint-Pierre 2017; Musango et al. 2017). Finally, studies have demonstrated the relationship between the sustainability of the city and the efficiency of urban metabolism (Kissinger and Stossel 2019).

Cities are expected to become efficient in terms of the generation/extraction of resources as well as the consumption of valuable resources to sustain the system of

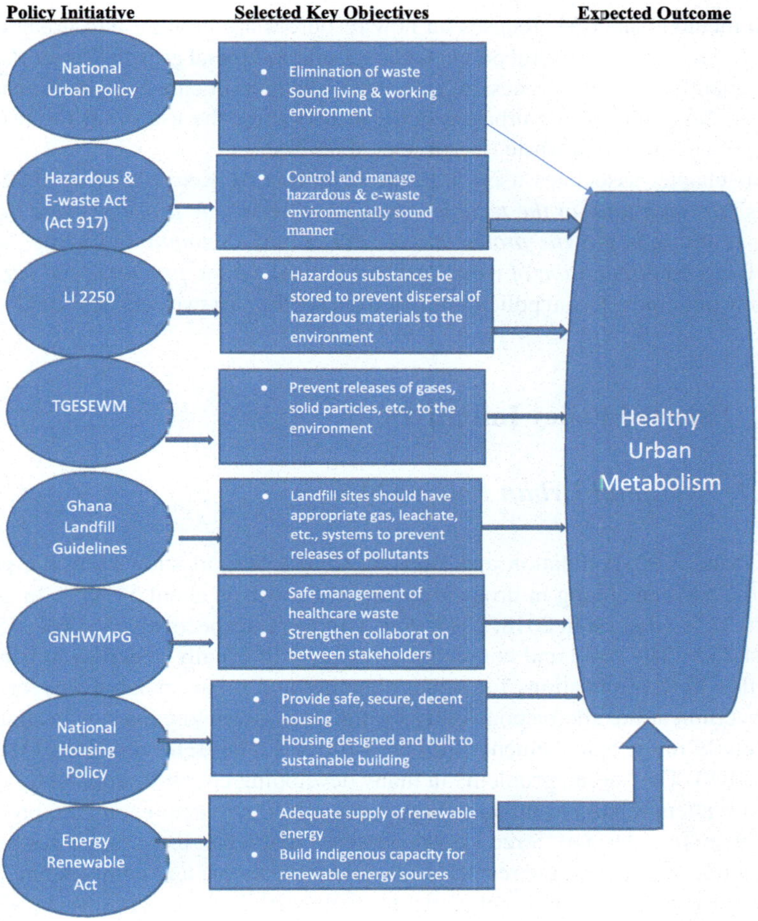

Fig. 17.1 Relationship between policy frameworks and urban metabolism. *Source* Authors' construct 2022

consumption. In Ghana, cities are not only centers of consumption but also serve as places of resource extraction, particularly sand-winning activities and construction of both mechanized and manual bole-holes for the provision of water for domestic and gardening and other related activities. The exponential growth, both spatially and economically, has inevitably increased the demand for more resource extraction and consumption. Domestic organic waste is turned into manure to nourish backyard gardens and in some cases large commercial farms. Thus, cities in Ghana, like their counterparts elsewhere cannot be described as isolated systems; they are in reality networks, flows, and connections. UM explains whether a city is under-performing as it (UM) assesses its ability to utilize available resources. The rationale behind UM is partly to reduce, reuse, repair, and recycle resources as much as possible.

Urban metabolism (UM) focuses on how resources are brought into urban centers and utilized to produce useful output for economic and social growth (Tan et al. 2019). Consequently, the effectiveness of resource utilization in cities can be measured to evaluate the ability of the urban systems to maximize the use of all the resources available with minimal waste (Derrible et al. 2021).

This chapter associates itself with the definition by Kennedy et al. (2007) that UM is "*the sum total of the technical and socio-economic processes that occur in cities, resulting in growth, production of energy, and elimination of waste*" with its main focus on *elimination of waste* and to a limited extent, *production of energy* by focusing on some relevant policies initiated during the past two decades (2002–2021).

17.3 Selected Policy Initiatives

17.3.1 National Urban Policy (2012)

The Ghana 2021 Population and Housing Census Report shows that the population of Ghanaians living in urban areas and big towns is about 57.35%. In 2019, it was 56.71%, whereas in 2018, 2017, and 2016 were, respectively, 56.06%, 55.41%, and 54.75%. The rapid and consistent urban growth usually generates urban problems that have implications for urban planning and urban metabolism, for which the government has the responsibility to initiate policies to enhance sustainability, particularly in the management of urban waste and other related matters (Voskamp et al. 2018). The "urban problem" in many developing countries apparently centers around matters related to urban policy, population, and socio-economic processes. Some three decades ago, Stren (1994) observed that when these processes begin to work at different levels, the results are invariably dynamic and complex and able to provide perspectives that are effective to distinguish between approved and unapproved spaces, formal and informal settlements, and on-going commercial activities. Combes et al. (2022) and Brooks and Byron (2019) have emphasized the implications of this in terms of urban management, which involves diligence in the formulation of policies and strategies; and efficient mobilization and use of resources to facilitate the development, maintenance, and provision of services to the population.

Ghana is no exception to this observation. Like other developing countries, Ghana is confronted with its urban problems that affect urban growth and development, which are highlighted in the National Urban Policy Framework (MLGRD 2012). Among these challenges are "*overconcentration of growth and development in a few cities; weak urban economy; land-use disorder and uncontrolled urban sprawl; increasing environmental deterioration; inadequate urban infrastructure and services; increasing urban insecurity; and urban poverty, slums, and squatter settlements*". The others are "*weak urban governance and institutional coordination; weak information, education, and communication strategy; inadequate urban investment and financing; weak urban transportation planning and traffic management;*

delimitation of urban areas of jurisdiction, lack of integrated planning across jurisdictional boundaries; weak rural–urban linkages; and limited data and information on urban centers" (MLGRD 2012, pp. 15–19, see also Bhadouria et al. 2022). The policy emphasizes the increasing environmental deterioration, which "*arises from conflicting land uses; unsatisfactory collection, disposal, and treatment of waste; choked drains and frequent flooding; coastal erosion and denudation; ineffective management of quarrying of mineral aggregates; ineffective land-use management and environmental protection; and the attitudinal indiscipline of the growing urban population*" (MLGRD 2012).

In response to the "urban problem" in Ghana and a recognition of the broader national agenda, the National Urban Policy (NUP) was developed in 2012. NUP aims at promoting a "*sustainable, spatially integrated and orderly development of urban settlements with adequate housing, infrastructure and services, efficient institutions, and a sound living and working environment for all people to support the rapid socio-economic development of Ghana*". With specific reference to the elimination of waste, a major principle of NUP is to "*mainstream environmental concerns into urban development*" and to promote a healthy living and working environment to enhance and support the development of the country. Mainstreaming environmental concerns into urban development contributes positively toward the enhancement of urban metabolism. As highlighted by the UNDP and UNEP (2020), environmental mainstreaming (EM) "*involves the integration of poverty-environment linkages into development planning processes and their outputs, including Poverty Reduction Strategy Papers (PRSPs) and Millennium Development Goals (MDGs) strategies*". Within the context of urban metabolism, EM "*would influence urban plans, budget processes, sector strategies, and local level implementation to integrate the contribution of environmental management to ensure improved livelihoods, enhanced economic security, and income opportunities for the poor*" (UNDP and UNEP 2020). This will ultimately lead to the establishment of enduring institutional processes with the city to encourage the wider stakeholder community and actors to respond to both entrenched and emerging environmental challenges, including the reduction and management of waste (UNDP and UNEP 2020).

The Global Environment Facility (GEF) rehashes the position of UNDP and UNEP in these words:

> The basic reason why environmental mainstreaming is important is that economic and social development and the environment are fundamentally interdependent—the way we manage the economy and political and social institutions has critical impacts on the environment, while environmental quality and sustainability, in turn, are vital for the performance of the economy and social well-being. As such the task of environmental integration and mainstreaming is at the forefront of development planning and policy formulation (Environment Inside 2007).

Cities in Ghana are caught between the need to balance economic development activities and human development programs. The NUP clearly gives directives for cities to reduce their metabolic footprint without sacrificing what Thomson and Newman (2018) call, their "*livability*". It identifies cities as organisms with different parts: housing, infrastructure and services, institutions, living environment, and people that should be harnessed for urban sustainability. In this context, it is hoped that the policy will be effectively implemented to create the necessary environment for economic development that will strengthen social development-oriented policies without disregarding legal and institutional requirements.

17.3.2 Addressing the E-Waste Menace

There has been an exponential urban growth and associated with it are increasing importation of electronic waste (e-waste) and its effect on cities which raises questions about sustainability and carrying capacity of the urban ecosystem (Kapoor et al. 2020). The "*rapid evolution of electronic technology, coupled with rapid product obsolescence, has compounded the e-waste problem*" (Otsuka et al. 2012). Indeed, some studies estimate that e-waste produced globally increases at about "*three times faster than the growth of overall municipal solid waste*" (Schluep et al. 2009). Research shows that on average about 65% of it is potentially exported to some countries in Latin America, Asia, and Africa including Ghana (Orisakwe and Frazzoli 2010; Environmental Investigation Agency 2011). It has been stressed that Ghana has an unenviable record of being a major destination country for discarded electrical and electronic equipment from the developed world, including China (Azuka 2009; Daum et al. 2017). The situation is worsened by the continuous "*importation of cheap and used electrical devices, electronics, and equipment that have reduced lifespans, with an estimated 15% shown to be non-functional*" (Akon-Yamga et al. 2021).

The seemingly intractable importation of cheap and used electrical equipment and ineffective management of e-waste threaten not only the sustainability of the environment but also human health (Prakash et al. 2010), as it contains dangerous substances with the likelihood of being released if not managed appropriately (Lundstedt 2011; Tysdenova and Bengtsson 2011). According to Azuka (2009), the toxic chemicals that exist in e-waste include "*a wide range of heavy metals, such as cadmium (Cd), lead (Pb), mercury (Hg), arsenic (As), and nickel (Ni), and also persistent organic compounds, such as brominated flame retardants (BFRs) and phthalates. Other chemicals that appear in e-waste include polychlorinated biphenyls (PCBs), nonylphenol (NP), and triphenyl phosphate (TPPs), among others*". A study by Greenpeace International (2008) at Agbogbloshie, a suburb of Accra, the capital of Ghana, revealed some samples containing Cd, Hg, and Pb in quantities that were considered "*especially toxic to aquatic life*". The presence of these harmful chemicals within areas where e-waste is recycled in Ghana is further confirmed in other studies (Brigden et al. 2008; Otsuka et al. 2012; Caravanos et al. 2013). In these studies, soil and ash samples from Agbogbloshie revealed unacceptable levels of toxic metals including "*zinc, lead, and copper, and organic chemicals, such as phthalates and polybrominated diphenyl ethers*" (PBDEs).

The management of e-waste demands policy initiatives that among others should be able to control and eliminate hazardous substances and should be grounded on sustainability principles. It has been hypothesized and confirmed by Cucchiella et al. (2015) and Akon-Yamga et al. (2021) that effective management of e-waste positively correlates with sound environmental, social, and economic benefits. A study by Oteng-Ababio (2014) asserts that the trade in e-waste could partly be explained in terms of the absence of a regulatory framework for its management and disposal in Ghana. The Government of Ghana recognized this policy gap and enacted the Hazardous and Electronic Waste Control and Management Act (Act 917, 2016), Legislative Instrument 2250, and Technical Guidelines to tackle the menace. The passage of the "Hazardous and Electronic Waste Control and Management Act" (Act 917), "Hazardous, Electronic and Other Wastes (Classification), Control and Management Regulations Legislative Instrument" (LI) 2250 in 2016, and "Technical Guidelines on Environmentally Sound E-Waste Management for Collectors, Collection Centers, Transporters, Treatment Facilities and Final Disposal in Ghana" in 2018 is, therefore, a bold attempt to fill the vacuum in the regulatory arena.

The "Hazardous and Electronic Waste Control and Management Act" (Act 917) of 2016 (hereafter referred to as Act 917) aims to "*control and manage the disposal of hazardous waste, electrical and electronic waste in an environmentally sustainable manner*". It seeks to domesticate the Basel Convention and aspects of the Stockholm Convention relating to PCBs in Ghana. In terms of enhancing the elimination of waste, Act 917 specifically seeks to prohibit the import, export, and transportation of hazardous and other wastes in and within Ghana and further directs manufacturers, and distributors of electrical and electronic (E&E) equipment to take back used equipment for recycling in an environmentally sound manner. In addition, a fund has been established to finance the management of E&E waste to reduce its adverse impact on human health and the environment. The disbursement of the fund is mainly for recycling facilities for E&E waste. The passage of Act 917, Legal Instrument (LI) 2250, and the Technical Guidelines makes it mandatory for informal workers in e-waste management to register with relevant state agencies and operate within the legal framework (Kumi et al. 2019).

Before the enactment of Act 917 and the "Hazardous and Electronic Waste Control and Management Regulations" (LI 2250) in Ghana, resource recovery from e-waste was almost exclusively informally organized. It served as a major source of livelihood for many of them but had serious deleterious consequences on the urban environment. Materials that were not needed by the workers were burnt or dumped into the environment, with no consideration of the indirect costs at both the local and global levels. Such actions gave the informal sector a "*competitive advantage vis-à-vis recycling enterprises in the formal sector*" (Hemkhaus-Adelphi 2020). The formal sector was mindful and cared about environmentally sound and socially responsible recycling and applied occupational health and safety standards and ensured for proper treatment and elimination of hazardous substances, which involved financial and other costs.

With the passage of the new legal framework, not only have the producers and private importers registered with Ghana's Environmental Protection Agency (EPA) they also pay an "Advance Eco Levy for electronic goods imported"

(Hemkhaus-Adelphi 2020), which is used to facilitate the implementation, monitoring and enforcement of the policy and "support the formalization of informal actors". According to Hemkhaus-Adelphi (2020), previous projects to "formalize the informal sector" mainly focused on Agbogbloshie which "received considerable public attention as the world's presumably largest dumping ground for e-waste" (Hemkhaus-Adelphi 2020). Further action has now been taken within the new regulatory framework in other urban areas to scale up the involvement of informal workers to "the formal e-waste value chain without jeopardizing their livelihoods" but minimizing and controlling the volume of waste in their operations (Zhang et al. 2018; Hemkhaus-Adelphi 2020).

The Technical Guidelines published in 2018 was funded by the Swiss State Secretariat for Economic Affairs (SECO) and targeted five stakeholders (collectors, collection centers, transporters, treatment facilities, final disposal), according to a study by the Environmental Protection Agency and Sustainably Recycling Industries (2018). For materials management, the guidelines direct that a collector "*shall manage e-waste in a way that prevents releases of gases, liquids or solid particles from any e-waste, or component, to the environment*." A key responsibility of a collector to help eliminate waste is to ensure that "whole or fractions containing hazardous substances shall be stored in a manner that prevents dispersal of hazardous materials to the environment." Furthermore, "a registered collection center shall manage e-waste in a way that prevents the release of gases, liquids, or solid particles from polluting the environment."

Like other organisms, cities are living and always evolving. The way in which resources move through them determines their sustainability, viability, and health within the larger regional and global natural systems (Eberlein 2018). Ghana government's intervention in addressing the e-waste menace and other urban challenges is to regulate resources that flow into urban areas, how they are being used, and what happens with the output, the "city as an organism" does not need to sustain itself to ensure that residents will live balanced and healthy lives.

17.3.3 Ghana Landfill Guidelines (2002)

The Ghana Landfill Guidelines (GLG) (EPA 2002) give directives for the provision of site infrastructure and control mechanisms to mitigate the likelihood of impacts on the environment and human well-being of both existing and closed waste disposal sites on adjoining communities (Keelson 2014). Before the publication of the GLG, landfill sites in Ghana were not regulated in terms of minimum standards for technology, air emission and monitoring standards, and location standards. The GLG guides "*landfill operators, developers, planning authorities and regulatory bodies on the site selection, development, design, construction, operation, closure and post-closure management of municipal solid waste, and commercial and industrial general waste landfill facilities*". The significance of the initiative is that most waste disposal sites are located close to residential facilities and therefore pose danger to human health and general well-being. Indeed, Keelson (2014) has confirmed that landfill sites in the Greater Accra Region did not have "*appropriate gas, leachate, groundwater, or surface water management systems leading to uncontrolled releases of pollutants to the air, soil, and water media*" before GLG was introduced and implemented. Its introduction and implementation obviously would mitigate the effects of wastes hitherto emitted into urban centers and their adjoining communities (Pearce and Chertow 2017).

17.3.4 Ghana National Healthcare Waste Management and Guidelines Policy (2020)

In 2020, the Ghana National Healthcare Waste Management and Guidelines Policy was published by the Ministry of Health (MOH) with a vision to "*manage healthcare waste safely*" and prevent "*its adverse impact on healthcare workers, clients, the public, and the environment*". The main objective of the initiative is to "*provide direction for effective, efficient, and safe management of healthcare waste (HCW) through the adoption of the best available techniques (BAT) and best environmental practices to prevent injuries, infections, and other hazards; protect and promote public health and the environment for sustainable development*" (GNA 2020; MOH 2020). As its contribution to the elimination of waste in health facilities, the policy attempts to strengthen collaborative endeavors between the health sector and other stakeholders and initiate institutional and legal structures to enforce the requirements for the management and disposal of HCW. It further ensures "adherence to proper methods, and infrastructural and technological development of healthcare waste management and mobilizes resources for healthcare waste management at all levels" (MOH 2020).

17.3.5 Renewable Energy Act, 2011 (Act 832)

The Renewable Energy Act (Act 832) was passed in 2011 to "*provide for the development, management, utilization, sustainability, and adequate supply of renewable energy for the generation of heat and power and related matters*" (GOG 2011). Its objective encompasses, among others, "*the provision of a framework to support the development and utilization of renewable energy sources and an enabling environment to attract investment in renewable energy sources.*" It further seeks to "promote the use of renewable energy; improve access to electricity through the use of renewable energy sources; build indigenous capacity in technology for renewable energy sources," and regulate the ·production and supply of wood fuel and biofuel.

Increased reliance on wood fuel coupled with inadequate alternative energy sources in urban areas in Ghana has contributed to deforestation as more trees are felled to ensure adequate wood fuel supply. Its usage does cause accidental fires in homes and marketplaces, while charcoal has the potential to cause deaths from monoxide poisoning. Health problems associated with biofuels and gasoline in urban centers include but are not limited to diseases of the heart, breathing difficulties, and/or premature death. The overall effect of the Renewable Energy Act, on cities and big towns, when fully implemented will lead to the minimization of dangerous atmospheric chemicals and substances that heavily impair the health and well-being of people, especially children and the elderly with underlying health factors. A significant advantage of moving away from fossil fuels toward the use of renewable energy is hugely justified as a result of the former's negative impacts on the environment in general and urban environment in particular. Indeed, fossil fuels do increase environment acidity which can result in serious deleterious and unpredictable impacts on the environment. They emit major greenhouse gases, contributing to unacceptable levels of global warming.

In Ghana and other countries in Africa, the storage and distribution of flammable petroleum products have on countless occasions resulted in explosions leading to fatalities and destruction of property. These explosions do not only cause damage and put the lives of employees and firefighters at risk, but also generate "*noxious gases that are liable to be released into the urban environment*" (Inspire Clean Energy 2020). For example, in 2017, there was an explosion of a gas filling station at Atomic Junction, a suburb of Accra, with at least seven (7) deaths, 192 reported injuries, and destruction of properties. Cities in Ghana, especially Accra and Kumasi, and other regional capitals experience acute traffic jams which affect not only man hours and productivity but cause air pollution and smog (EPA 2019). According to Inspire Clean Energy (2020), "*when sunlight reacts with nitrogen oxide and other volatile organic compounds in the atmosphere, smog is caused.*" A critical emission from vehicles is nitrogen oxide. It has been explained that "*changing energy plans even on a personal level will help give the world's cities cleaner air*" (Inspire Clean Energy 2020). The problem of global warming as a result of fossil fuels has been extensively discussed in the literature and need not be overemphasized in this discourse. Ghana's initiative in the passage of the Renewable Energy Act is a bold attempt to minimize and reverse the effects of global warming and maintain appropriate temperatures in urban centers.

17.3.6 National Housing Policy, 2015

Housing shortages are common in cities in Ghana with many urban residents living in substandard houses, squatter settlements, and slums (Ghana Statistical Service 2021). These lead to unavoidable urban decay in view of deficits, and in some cases non-availability of basic services including appropriate housing units, electricity, sanitation facilities, and many others. Indeed, the generation of revenue and its corresponding expenditure patterns has both contributed to this service provision conundrum. Ironically, the problem defies any reasonable solution and rather worsens since spending in many instances is not directed toward appropriate services but ends up serving the elite and their active collaborators. This places considerable burden on the financial resources of cities which as well displaces development in other jurisdictions.

In response to these challenges in the country's urban housing situation, on behalf of the Government of Ghana, the Ministry of Water Resources, Works, and Housing (MWRWH) published a National Housing Policy in 2015, which envisages a country in which "*everyone can access safe, secure, decent, and affordable housing either owned or rented*" (MWRWH 2015). This vision feeds into the desire to "*leverage the housing sector to facilitate industrialization, local economic development, job creation, environmental sustainability, and social development*" (MWRWH 2015). To promote socio-economic and environmental sustainability, the policy attempts to "*ensure that housing is designed and built to sustainable building principles leading to the creation of green communities while housing programs are made more accessible to the poor*." It is clear that urban Ghana is growing both in size and in population and this could be attributed to fertility and migration (Ghana Statistical Service 2021). The latter could be explained by the growth of commerce/trade and job opportunities in both the formal and informal sectors. An analysis of the National Housing Policy reveals consistencies in the policy and its probable effect on real situations to show its political, economic, and environmental feasibility.

17.4 Conclusion and Future Perspectives

This chapter sets out to examine selected public policy initiatives that seek to enhance the social, economic, and environmental forces that affect UM in Ghana since 2002. The overall picture of urban governance in Ghana and many developing economies relatively shows systemic crises. The urban system, is without doubt, struggling to cope with the demanding needs of a rapidly growing population in the midst of abject poverty and other socio-economic challenges. In fact, their very sustainability is gravely fragile. Their functionality as institutions is weak with limited capacity and autonomy in all the key pillars of sustainable development. Luckily, this study illustrates attempts being made by the Government of Ghana toward the "urban problem" by initiating policies and legal frameworks with the potential to address these problems and enhance urban metabolism.

Selected policies and legal frameworks examined included the National Urban Policy (2012); Renewable Energy Act, 2011 (Act 832); Ghana Landfill Guidelines (2002); "Hazardous, Electronic and Other Wastes (Classification), Control and Management Regulations" (2016); Legislative Instrument (LI) 2250 (2016); and Hazardous and Electronic Waste Control and Management Act, 917 (2016). The rest are National Housing Policy (2015); "Guidelines on Environmentally Sound E-Waste Management for Collectors, Collection Centers, Transporters, Treatment Facilities and Final Disposal in Ghana" (2018); and Ghana National Healthcare Waste Management and Guidelines Policy (2020). In the words of ICLEI (2021), the combined message delivered by these policies demonstrates that "*cities are made of flows: flows of water, energy, food, materials, people, information, money, and power*", which intertwine and weave together as "*the urban fabric, to give a city its unique character*". This organism (urban fabric) extracts resources from the environment and metabolizes them for a holistic and all-embracing qualitative growth and well-being before sending them back into the environment. Such complex and continuous interaction and activities call for moderation which can be guaranteed by the enactment and application of appropriate rules and regulations as illustrated in Fig. 17.1. It is from this perspective that the eight policy initiatives should be viewed and recognized.

The experience in Ghana and other African countries illustrates that in spite of the marginalization and institutional and structural weaknesses of non-state actors, many activities undertaken in the urban setting take place within this sector. When compartmentalized into separate entities, non-governmental organizations (NGOs), community-based organizations (CBOs), civil society, and the informal sector, it shows the existence of many players in the governance arrangement in urban settings. In going forward, municipal authorities in Ghana should embrace and encourage these groups in the provision of services, especially solid waste collection and disposal and e-waste management. As urban authorities encourage these groups in service provision, monitoring and assessment must be considered paramount by the implementing agencies. With specific reference to e-waste, serious efforts should be made to roll out a massive education campaign on Act 917, LI 2250, and the Technical Guidelines. This is extremely important in view of the threats posed by e-waste on humans as well as the environment at large if not handled and processed appropriately. This is not by any means suggesting that awareness should not be created concerning the other initiatives.

References

Akon-Yamga G, Chux U, Daniels A, Quaye W, Ting BM, Asante AA (2021) Transformative innovation policy approach to e-waste management in Ghana: perspectives of actors on transformative changes. Sci Public Policy 48(3):387–397. https://doi.org/10.1093/scipol/scab005

Azuka AI (2009) Influx of used electronics into Africa: a perilous trend. Law Environ Develop J 5:90

Beloin-Saint-Pierre D, Rugani B, Lasvaux S, Mailhac A, Popovici E, Sibiude G, Benetto E, Schiopu N (2017) A review of urban metabolism studies to identify key methodological choices for future harmonization and implementation. J Clean Prod 163:S223–S240

Bhadouria R, Upadhyay S, Tripathi S, Singh P (eds) (2022) Urban ecology and global climate change. Wiley-Blackwell

Brigden K, Labunska I, Santillo D, Johnston P (2008) Chemical contamination at e-waste recycling and disposal sites in Accra and Koforidua, Ghana. Greenpeace Res Lab Tech Note 2008:1–23

Brooks L, Byron L (2019) Vestiges of transit: urban persistence at a microscale. Rev Econ Stat 101(3):385–399

Caravanos J, Clarke EE, Osei CS, Amoyaw-Osei Y (2013) Exploratory health assessment of chemical exposures at e-waste recycling and scrap yard facility in Ghana. J Health Pollut 3(4):11–22

Combes PP, Gobillon L, Zylberberg Y (2022) Urban economics in a historical perspective: recovering data with machine learning. Reg Sci Urban Econ 94:103711

Cucchiella F, D'Adamo I, Lenny-Koh SC, Rosa P (2015) Recycling of WEEEs: an economic assessment of present and future e-waste streams. Renew Sustain Energy Rev 51:263–272

Cui X (2018) How can cities support sustainability: A bibliometric analysis of urban metabolism. Ecol Indic 93:704–717

Cui X, Wang X, Feng Y (2019) Examining urban metabolism: a material flow perspective on cities and their sustainability. J Clean Prod 214:767–781

Currie PK, Musango JK (2017) African urbanization: Assimilating urban metabolism into sustainability discourse and practice. J Ind Ecol 21:1262–1276

Daum K, Stoler J, Grant RJ (2017) Toward a more sustainable trajectory for e-waste policy: a review of a decade of e-waste research in Accra. Ghana, Int J Environ Res Public Health 14(2):135

Derrible S, Cheah L, Arora M, Yeow LW (2021) Urban metabolism. In: Shi W, Goodchild MF, Batty M, Kwan MP, Zhang A (eds) Urban Informatics. The Urban Book Series. Springer, Singapore. pp 5–114. https://doi.org/10.1007/978-981-15-8983-6_7

Eberlein S (2018) Urban metabolism: a real world model for visualizing and co-creating healthy cities. https://www.thenatureofcities.com/2018/07/24/urban-metabolism-real-world-model-visualizing-co-creating-healthy-cities/

Environment Inside (2007) Why do we need to 'mainstream the environment? https://www.environmentalmainstreaming.org/Environment%20Inside/Chapter%201/chapter1-2.html#sdfootnote1sym, Retrieved 24–05–2022

Environmental Investigation Agency (2011) System failure: the UK's harmful trade in electronic waste. http://www.greencustoms.org/docs/EIA_E-waste_report_0511_WEB.pdf

Environmental Protection Agency & Sustainably Recycling Industries (2018) Technical guidelines on environmentally sound e-waste management for collectors, collection centers, transporters, treatment facilities and final disposal in Ghana, Accra, Ghana: EPA & SRI

EPA (2019) Vehicular traffic in urban Ghana: cause for environmental and health concerns. Government Printer, Accra

EPA (2002) Ghana landfill guidelines. Government Printer EPA, Accra

Ghana News Agency (2020) Ghana launches National healthcare waste management and guidelines policy. https://newsghana.com.gh/ghana-launches-national-healthcare-waste-management-and-guidelines-policy/

Ghana Statistical Service (2021) Ghana 2021 population and housing census, general report. Volume 3A. Ghana Statistical Service, Accra

Government of Ghana (GOG) (2011) renewable energy act, act 832. Government Printer, Accra

Greenpeace (2008) Poisoning the poor—electronic waste in Ghana. http://www.greenpeace.org/international/en/news/features/poisoning-the-poor-electroni/

Hemkhaus-Adelphi M (2020) From grave to cradle: e-waste management in Ghana (E-MAGIN Ghana. https://www.adelphi.de/en/project/grave-cradle-e-waste-management-ghana-e-magin-ghana. Morton Hemkhaus—Adelphi. (E-MAGIN Ghana)

ICLEI Africa (2021) Unravelling the stories of our cities: using urban metabolism to shape thriving African cities. https://africa.iclei.org/unravelling-the-stories-of-our-cities-using-urban-metabolism-to-shape-thriving-african-cities/

Inspire Clean Energy (2020) https://www.inspirecleanenergy.com/blog/clean-energy-101/advantages-disadvantages-fossil-fuels

Kapoor V, Tripathi S, Devi RS, Srivastava P, Bhadouria R (2020) Ecological economics of an urban settlement: an overview. In: Verma P, Singh P, Singh R, Raghubanshi AS (eds) Urban Ecology: Emerging patterns and socio-ecological systems. Elsevier, pp 91–110. https://doi.org/10.1016/B978-0-12-820730-7.00006-9

Keelson KB (2014) Evaluation of groundwater pollution with heavy metals at the Oblogo No.1 Dumpsite in Accra, Ghana. Int J Eng Technol Innov 4(3):170–181
Kennedy C, Cuddihy J, Engel-Yan J (2007) The changing metabolism of cities. J Ind Ecol 11(2):43e59. https://doi.org/10.1162/jie.2007.1107
Kissinger M, Stossel Z (2019) Towards an interspatial urban metabolism analysis in an interconnected world. Ecol Indic 101:1077–1085
Kumi E, Hemkhaus M, Hibler S, Boateng P, von Falkenhausen I (2019) Trust be key: exploring pathways to formalisation in Ghana's E-waste sector. Adelphi research GmbH, Berlin
Lundstedt S (2011) Recycling and disposal of electronic waste: health hazards and environmental impacts. Swedish Environmental Protection Agency, report 6417. http://www.naturvardsverkets e/Documents/publikationer6400/978-91-620-6417-4.pdf
Ministry of Health (2020) Health care waste management policy for Ghana. MOH, Accra
Ministry of Local Government and Rural Development (MLGRD) (2012) National urban policy. Government Printer, Accra
Ministry of Water Resources, Works and Housing (2015) National housing policy. Government Printer, Accra
Musango, JK, Currie P, Robinson B (2017) Urban metabolism for resource efficient cities: from theory to implementation; UN Environment, Economy Division: Paris, France
Newell JP, Cousins JJ, Baka JE (2017) Political-industrial ecology: an introduction. Geoforum 85:319e323. https://doi.org/10.1016/j.geoforum.2017.07.024
Orisakwe OE, Frazzoli C (2010) Electronic revolution and electronic wasteland: the West/waste Africa experience. J Nat Environ Sci 1(1):43–47
Oteng-Ababio M (2014) Rethinking waste as a resource: insights from a low-income community in Accra, Ghana City. Territory Archit 1(1):10
Otsuka M, Itai T, Asante KA, Muto M, Tanabe S (2012) Trace element contamination around the e-waste recycling site at Agbogbloshie, Accra City, Ghana. Interdisc Stud Environ Chem—Environ Pollute Ecotoxicol 6:161–167
Palczynski RJ, Scotia WN (2020). Study on solid waste management options for Africa. Project report. Final draft version. Prepared for African Development Bank Sustainable Development and Poverty Reduction Unit, Abidjan
Pearce BJ, Chertow M (2017) Scenarios for achieving absolute reductions in phosphorus consumption in Singapore. J Clean Prod 140:1587–1601
Prakash S, Manhart A, Amoyaw-Osei Y, Agyekum O (2010) Socio-economic assessment and feasibility study on sustainable e-waste management in Ghana. http://www.oeko.de/oekodoc/1057/2010-105-en.pdf. Accessed July 2012
Schluep M, Hagelueken C, Kuehr R, Magalini F, Maurer C, Meskers C, Mueller E, Wang F (2009) Recycling—from e-waste to resources, final report. http://www.unep.org/PDF/PressReleases/E_waste_publication_screen_FINALVERSION-sml-pdf
Stren R (1994) Urban research in the developing world. The World Bank, Washington, DC
Tan LM, Arbabi H, Brockway PE, Densley Tingley D, Mayfield M (2019) An ecological thermodynamic approach to urban metabolism: measuring resource utilization with open system network effectiveness analysis. Appl Energy 254:113618
Thomson G, Newman P (2018) Urban fabrics and urban metabolism—from sustainable to regenerative cities. Resour Conserv Recycl 132:218–229
Tysdenova O, Bengtsson M (2011) Chemical hazards associated with treatment of waste electrical and electronic equipment. Waste Manage 31(1):45–58
UNDP and UNEP (2020) Guidance note on mainstreaming environment into national development planning. http://www.nbsapcbw-seasi-01-undp-unep-guide-en.pdf
Voskamp IM, Spiller M, Streaked S, Bregt AK, Vreugdenhil C, Rijnaarts HHM (2018) Space time information analysis for resource-conscious urban planning and design: a stakeholder-based identification of urban metabolism data gaps. Resour Conserv Recycl 128:516–525

Wang HC (2022) Case studies of urban metabolism: what should be addressed next? Urban Aff Rev 1–20

Zhang Y (2019) Urban metabolism, 2nd edn. Elsevier, Oxford, UK

Zhang Y, Li Y, Liu G, Hao Y (2018) CO2 metabolic flow analysis in global trade based on ecological network analysis. J Clean Prod 170:34–41. https://doi.org/10.1016/j.jclepro

The manufacturer's authorised representative in the EU is Springer Nature Customer Service Centre GmbH, Europaplatz 3, 69115 Heidelberg, Germany. If you have any concerns regarding our products, please contact ProductSafety@springernature.com

Printed and bound by CPI Group (UK) Ltd, Croydon, CR0 4YY
01/10/2026
02225902-0011